GRAPHS &
DIGRAPHS

FOURTH EDITION

GRAPHS & DIGRAPHS

FOURTH EDITION

G. CHARTRAND
Western Michigan University

and

L. LESNIAK
Drew University

CHAPMAN & HALL/CRC

A CRC Press Company
Boca Raton   London   New York   Washington, D.C.

Library of Congress Cataloging-in-Publication Data

Chartrand, Gary.
  Graphs & digraphs.—4th ed. / Gary Chartrand and Linda Lesniak.
  p. cm.
  Includes bibliographical references and index.
  ISBN 1-58488-390-1 (alk. paper)
  1. Graph theory. 2. Directed graphs. I. Lesniak, Linda. II. Title.

QA166.C4525 2004
511'.5--dc22   2004052664

Visit the CRC Press Web site at www.crcpress.com

© 2005 by Chapman & Hall/CRC

No claim to original U.S. Government works
International Standard Book Number 1-58488-390-1
Library of Congress Card Number 2004052664
Printed in the United States of America 1 2 3 4 5 6 7 8 9 0
Printed on acid-free paper

Contents

Preface to the fourth edition

Not only is graph theory one of the major areas of combinatorics, it is developing into one of the major areas of mathematics. In addition to its growing interest and increased importance as a mathematical subject, graph theory has applications to many fields outside of mathematics, including computer science, chemistry, and communication and electrical networks.

As in the first three editions of *Graphs & Digraphs*, our major objective is to introduce and treat graph theory in the way we have always found it, namely, as the beautiful area of mathematics it is. We have striven to produce a reader-friendly, carefully written book that emphasizes the mathematical theory of graphs and digraphs.

One goal of the fourth edition is to make this book even friendlier. To accomplish this, we have endeavored to streamline the material. While we have added new material on list colorings of planar graphs, rainbow Ramsey numbers, and extremal graph theory and updated material on cages, we have also eliminated unenlightening proofs of theorems and deleted material in areas which have proved to be less active in recent years. We have also updated the extensive list of graph theory books so that the avid graph theory readers have many avenues to pursue their interests. In addition, with the significant assistance of Ping Zhang of Western Michigan University, to whom we are most grateful, we have compiled for instructors a manual containing solutions and hints for a large number of selected exercises in the text.

The text is intended for an introductory sequence in graph theory at the beginning graduate level or advanced undergraduate level, although a one-semester course can easily be designed by selecting topics of major importance and interest to the instructor and students. Only mathematical maturity, including a sound understanding of proof, is required as a prerequisite to understand and appreciate the material presented.

We greatly appreciate the suggestions given to us by a number of mathematicians: John Ganci, Texas Instruments; Christina (Kieka) Mynhardt, University of Victoria; Carsten Thomassen, the Technical University of Denmark; Ann Trenk, Wellesley College. Our sincere thanks to all of you. A special acknowledgement is due to Christine Spassione for her impeccable typing. Our thanks also go to Allen Schwenk, Western Michigan University, for his drawing of the graph that appears on the cover. Finally, we thank our editor Bob Stern and the staff of CRC Press for their interest in and assistance with the fourth edition.

G.C. & L.L.

1

Introduction to graphs and digraphs

We begin our study of graphs and digraphs by introducing many of the basic concepts that we shall encounter throughout our investigations.

1.1 GRAPHS

A *graph* G is a finite nonempty set of objects called *vertices* (the singular is *vertex*) together with a (possibly empty) set of unordered pairs of distinct vertices of G called *edges*. The *vertex set* of G is denoted by $V(G)$, while the *edge set* is denoted by $E(G)$.

The edge $e = \{u, v\}$ is said to *join* the vertices u and v. If $e = \{u, v\}$ is an edge of a graph G, then u and v are *adjacent vertices*, while u and e are *incident*, as are v and e. Furthermore, if e_1 and e_2 are distinct edges of G incident with a common vertex, then e_1 and e_2 are *adjacent edges*. It is convenient to henceforth denote an edge by uv or vu rather than by $\{u, v\}$.

The cardinality of the vertex set of a graph G is called the *order* of G and is commonly denoted by $n(G)$, or more simply by n when the graph under consideration is clear; while the cardinality of its edge set is the *size* of G and is often denoted by $m(G)$ or m.

It is customary to define or describe a graph G by means of a diagram in which each vertex of G is represented by a point (which we draw as a small circle) and each edge $e = uv$ of G is represented by a line segment or curve joining the points corresponding to u and v. We then refer to this diagram as the graph G itself.

A graph G with vertex set $V(G) = \{v_1, v_2, \ldots, v_n\}$ and edge set $E(G) = \{e_1, e_2, \ldots, e_m\}$ can also be described by means of matrices. One such matrix is the $n \times n$ *adjacency matrix* $A(G) = [a_{ij}]$, where

$$a_{ij} = \begin{cases} 1 & \text{if } v_i v_j \in E(G) \\ 0 & \text{if } v_i v_j \notin E(G). \end{cases}$$

Thus, the adjacency matrix of a graph G is a symmetric $(0, 1)$ matrix having zero entries along the main diagonal. Another matrix is the $n \times m$ *incidence matrix* $B(G) = [b_{ij}]$, where

$$b_{ij} = \begin{cases} 1 & \text{if } v_i \text{ and } e_j \text{ are incident} \\ 0 & \text{otherwise.} \end{cases}$$

For example, consider a graph G defined by the sets

$$V(G) = \{v_1, v_2, v_3, v_4\} \quad \text{and} \quad E(G) = \{e_1, e_2, e_3, e_4, e_5\},$$

where $e_1 = v_1v_2$, $e_2 = v_1v_3$, $e_3 = v_2v_3$, $e_4 = v_2v_4$, and $e_5 = v_3v_4$. The graph G and its adjacency and incidence matrices are shown in Figure 1.1.

With the exception of the order and the size, the parameter that we will encounter most frequently in the study of graphs is the degree of a vertex. The *degree of a vertex* v in a graph G is the number of edges of G incident with v, which is denoted by $\deg_G v$ or simply by $\deg v$ if G is clear from the context. The degree of v is also the number of vertices in G that are adjacent to v. A vertex is called *even* or *odd* according to whether its degree is even or odd. A vertex of degree 0 in G is also called an *isolated vertex*; while a vertex of degree 1 is also referred to as an *end-vertex* of G. The *minimum degree* of G is the minimum degree among the vertices of G and is denoted by $\delta(G)$. The *maximum degree* is defined similarly and is denoted by $\Delta(G)$. In Figure 1.2, a graph G is shown together with the degrees of its vertices. In this case, $\delta(G) = 1$ and $\Delta(G) = 5$.

For the graph G of Figure 1.2, $n = 9$ and $m = 11$, while the sum of the degrees of its nine vertices is 22. That this last number equals $2m$ illustrates a basic

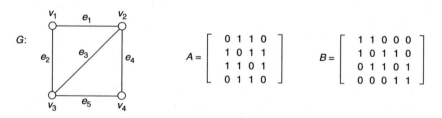

Figure 1.1 A graph and its adjacency and incidence matrices.

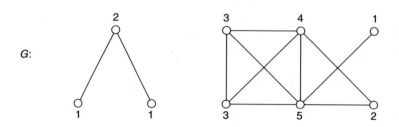

Figure 1.2 The degree of the vertices of a graph.

relationship involving the size of a graph and the degrees of its vertices. Every edge is incident with two vertices; hence, when the degrees of the vertices are summed, each edge is counted twice. We state this as our first theorem, which, not so coincidentally, is sometimes called *The First Theorem of Graph Theory.*

Theorem 1.1

Let G be a graph of order n and size m where $V(G) = \{v_1, v_2, \ldots, v_n\}$. Then

$$\sum_{i=1}^{n} \deg v_i = 2m.$$

This result has an interesting consequence.

Corollary 1.2

In any graph, there is an even number of odd vertices.

Proof

Let G be a graph of size m. Also, let W be the set of odd vertices of G and let U be the set of even vertices of G. By Theorem 1.1,

$$\sum_{v \in V(G)} \deg v = \sum_{v \in W} \deg v + \sum_{v \in U} \deg v = 2m.$$

Certainly, $\sum_{v \in U} \deg v$ is even; hence $\sum_{v \in W} \deg v$ is even, implying that $|W|$ is even and thereby proving the corollary. $\square$

Two graphs often have the same structure, differing only in the way their vertices and edges are labeled or in the way they are drawn. To make this idea more precise, we introduce the concept of isomorphism. A graph G_1 is *isomorphic* to a graph G_2 if there exists a one-to-one mapping ϕ, called an *isomorphism*, from $V(G_1)$ onto $V(G_2)$ such that ϕ preserves adjacency and nonadjacency; that is, $uv \in E(G_1)$ if and only if $\phi u \, \phi v \in E(G_2)$. It is easy to see that 'is isomorphic to' is an equivalence relation on graphs; hence, this relation divides the collection of all graphs into equivalence classes, two graphs being *nonisomorphic* if they belong to different equivalence classes.

If G_1 is isomorphic to G_2, then we say G_1 and G_2 are *isomorphic* and we denote this by writing $G_1 = G_2$. If G_1 is not isomorphic to G_2, then we write $G_1 \neq G_2$. If $G_1 = G_2$, then, by definition, there exists an isomorphism

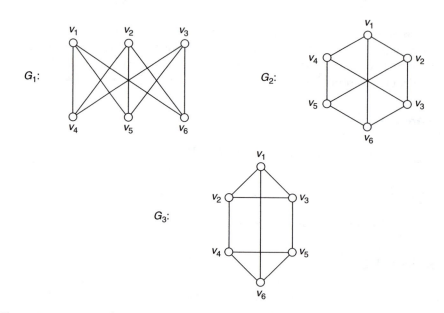

Figure 1.3 Isomorphic and nonisomorphic graphs.

$\phi: V(G_1) \rightarrow V(G_2)$. Since ϕ is a one-to-one and onto mapping, G_1 and G_2 have the same order. Since adjacent vertices in G_1 are mapped into adjacent vertices in G_2 and nonadjacent vertices of G_1 are mapped into nonadjacent vertices in G_2, the graphs G_1 and G_2 have the same size. Necessarily, every vertex v in G_1 and its image vertex ϕv in G_2 must have the same degree in their respective graphs. Therefore, the degrees of the vertices of G_1 are exactly the degrees of the vertices of G_2 (counting multiplicities). Although these conditions are necessary for G_1 and G_2 to be isomorphic, they are not sufficient. To illustrate this, consider the graphs G_i, $i = 1, 2, 3$, of Figure 1.3. Each G_i is a graph of order 6 and size 9 and the degree of every vertex of each graph is 3. Here, $G_1 = G_2$. For example, the mapping $\phi: V(G_1) \rightarrow V(G_2)$ defined by

$$\phi v_1 = v_1, \quad \phi v_2 = v_3, \quad \phi v_3 = v_5, \quad \phi v_4 = v_2, \quad \phi v_5 = v_4, \quad \phi v_6 = v_6$$

is an isomorphism, although there are many other isomorphisms. On the other hand, G_3 contains three pairwise adjacent vertices whereas G_1 does not; so there is no isomorphism from G_1 to G_3 and therefore $G_1 \neq G_3$. Of course, $G_2 \neq G_3$.

If G is a graph of order n and size m, then $n \geq 1$ and $0 \leq m \leq \binom{n}{2} = n(n-1)/2$. Although every graph must have at least one vertex, this is not the case with the number of edges. A graph with no edges is called an *empty graph*. There is only one graph of order 1 (up to isomorphism), and this is referred to as the *trivial graph*. A *nontrivial graph* then has $n \geq 2$. Thus far, whenever we have considered a graph

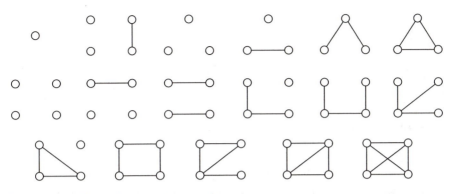

Figure 1.4 All graphs of order 4 or less.

G that is defined by a diagram, each point of this diagram (that is, each vertex of G) has been labeled. Therefore, the set of all vertex labels is $V(G)$. Here we are dealing with a *labeled graph*. There are occasions, however, when we are only interested in the structure of a graph defined by a diagram and the vertex set of the graph is irrelevant. In this case, we have an *unlabeled graph*. The distinct (nonisomorphic) graphs of order 4 or less are shown in Figure 1.4.

Frequently, a graph under study is contained within some larger graph also being investigated. We consider several instances of this now. A labeled graph H is a *subgraph* of a labeled graph G if $V(H) \subseteq V(G)$ and $E(H) \subseteq E(G)$; in such a case, we also say that G is a *supergraph* of H. For unlabeled graphs H and G, we say that H is a subgraph of G if the vertices of H and G can be labeled so that as labeled graphs, H is a subgraph of G. In Figure 1.5, H is then a subgraph of G but H is not a subgraph of F. If H is a subgraph of G, then we write $H \subseteq G$.

The simplest type of subgraph of a graph G is that obtained by deleting a vertex or edge. If $v \in V(G)$ and $|V(G)| \geq 2$, then $G - v$ denotes the subgraph with vertex set $V(G) - \{v\}$ and whose edges are all those of G not incident with v; if $e \in E(G)$, then $G - e$ is the subgraph having vertex set $V(G)$ and edge set $E(G) - \{e\}$. The deletion of a set of vertices or set of edges is defined analogously. These concepts are illustrated in Figure 1.6.

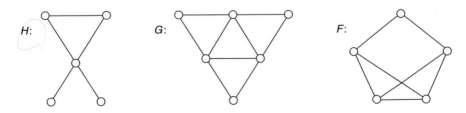

Figure 1.5 Subgraphs.

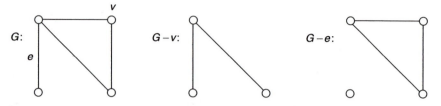

Figure 1.6 The deletion of a vertex or edge of a graph.

If u and v are nonadjacent vertices of a graph G, then $G+f$, where $f=uv$, denotes the graph with vertex set $V(G)$ and edge set $E(G)\cup\{f\}$. Clearly, $G\subseteq G+f$.

We have seen that $G-e$ has the same vertex set as G and that G has the same vertex set as $G+f$. Whenever a subgraph H of a graph G has the same order as G, then H is called a *spanning subgraph* of G.

Among the most important subgraphs we shall encounter are the 'induced subgraphs'. If U is a nonempty subset of the vertex set $V(G)$ of a graph G, then the subgraph $\langle U \rangle$ of G *induced* by U is the graph having vertex set U and whose edge set consists of those edges of G incident with two elements of U. A subgraph H of G is called *vertex-induced* or simply *induced* if $H = \langle U \rangle$ for some subset U of $V(G)$. Similarly, if X is a nonempty subset of $E(G)$, then the subgraph $\langle X \rangle$ *induced* by X is the graph whose vertex set consists of those vertices of G incident with at least one edge of X and whose edge set is X. A subgraph H of G is *edge-induced* if $H = \langle X \rangle$ for some subset X of $E(G)$. It is a simple consequence of the definitions that every induced subgraph of a graph G can be obtained by removing vertices from G while every subgraph of G can be obtained by deleting vertices and edges. These concepts are illustrated in Figure 1.7 for the graph G, where

$$V(G) = \{v_1, v_2, v_3, v_4, v_5, v_6\}, \quad U = \{v_1, v_2, v_5\} \quad \text{and} \quad X = \{v_1v_4, v_2v_5\},$$

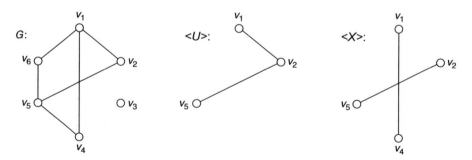

Figure 1.7 Induced and edge-induced subgraphs.

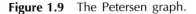

Figure 1.8 The regular graphs of order 4.

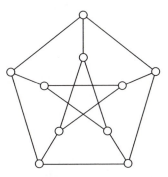

Figure 1.9 The Petersen graph.

There are certain classes of graphs that occur so often that they deserve special mention and in some cases, special notation. We describe the most prominent of these now.

A graph G is *regular of degree r* if deg $v = r$ for each vertex v of G. Such graphs are called *r-regular*. A graph is *complete* if every two of its vertices are adjacent. A complete graph of order n and size m is therefore a regular graph of degree $n - 1$ having $m = n(n - 1)/2$; we denote this graph by K_n. In Figure 1.8 are shown all (nonisomorphic) regular graphs with $n = 4$, including the complete graph $G_3 = K_4$.

A 3-regular graph is also called a *cubic graph*. The graphs of Figure 1.3 are cubic as is the complete graph K_4. One of the best known cubic graphs is the *Petersen graph*, shown in Figure 1.9. We will have many occasions to encounter this graph.

In 1936 Dénes König [K10] wrote the first book on graph theory. In it he proved that if G is a graph with $\Delta(G) = d$, there exists a d-regular graph H containing G as an induced subgraph. König's result actually first appeared in 1916 (see [K8]). His technique proves a somewhat stronger result.

Theorem 1.3

For every graph G and every integer $r \geq \Delta(G)$, there exists an r-regular graph containing G as an induced subgraph.

Proof

If G itself is r-regular, then there is nothing to prove. So we may assume that G is not r-regular. Let G' be another copy of G and join corresponding vertices whose degrees are less than r, calling the resulting graph G_1. If G_1 is r-regular, then G_1 has the desired properties. If not, we continue this procedure until arriving at an r-regular graph G_k where $k = r - \delta(G)$. (The proof of Theorem 1.3 is illustrated for the graph G of Figure 1.10 where $r = \Delta(G) = 3$.) □

Of course, we are not claiming that the r-regular graph constructed in the proof of Theorem 1.3 is one of smallest order with the desired property. Indeed, for the graph G of Figure 1.10, the graph G_2 has order 16, while the minimum order of a 3-regular graph containing G as an induced subgraph is actually 6. In fact, in 1963 Erdös and Kelly [EK1] produced a method for determining the minimum order of an r-regular graph H containing a given graph G as an induced subgraph.

The *complement* $\overline{G}$ of a graph G is that graph with vertex set $V(G)$ such that two vertices are adjacent in $\overline{G}$ if and only if these vertices are not adjacent in G. Hence, if G is a graph of order n and size m, then $\overline{G}$ is a graph of order n and size $\overline{m}$, where $m + \overline{m} = \binom{n}{2}$. In Figure 1.8, the graphs G_0 and G_3 are complementary, as are G_1 and G_2. Thus the complement $\overline{K}_n$ of the complete graph K_n is the empty graph of order n. A graph G is *self-complementary* if $G = \overline{G}$. Certainly, if G is a self-complementary graph of order n, then its size is $m = n(n-1)/4$. Since only one of n and $n-1$ is even, either $4 \mid n$ or $4 \mid n-1$; that

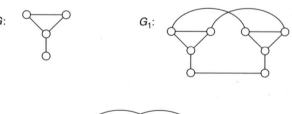

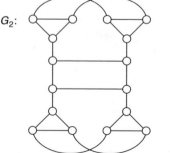

Figure 1.10 A 3-regular graph containing G as an induced subgraph.

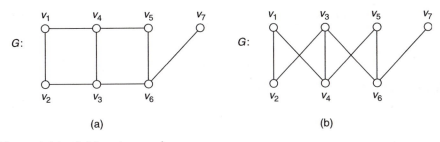

Figure 1.11 A bipartite graph.

is, if G is a self-complementary graph of order n, then either $n \equiv 0 \pmod{4}$ or $n \equiv 1 \pmod{4}$.

A graph G is *k-partite*, $k \geq 1$, if it is possible to partition $V(G)$ into k subsets $V_1, V_2, \ldots, V_k$ (called *partite sets*) such that every element of $E(G)$ joins a vertex of V_i to a vertex of V_j, $i \neq j$. Every graph is k-partite for some k; indeed if n is the order of G then G is n-partite. If G is a 1-partite graph of order n, then $G = \overline{K}_n$. For $k = 2$, such graphs are called *bipartite graphs*; this class of graphs is particularly important and will be encountered many times. In Figure 1.11(a), a bipartite graph G is given. Then G is redrawn in Figure 1.11(b) to emphasize that it is bipartite. If G is an *r-regular* bipartite graph, $r \geq 1$, with partite sets V_1 and V_2, then $|V_1| = |V_2|$. This follows since its size $m = r|V_1| = r|V_2|$.

A *complete k-partite graph* G is a k-partite graph with partite sets $V_1, V_2, \ldots, V_k$ having the added property that if $u \in V_i$ and $v \in V_j$, $i \neq j$, then $uv \in E(G)$. If $|V_i| = n_i$, then this graph is denoted by $K(n_1, n_2, \ldots, n_k)$ or $K_{n_1, n_2, \ldots, n_k}$. (The order in which the numbers $n_1, n_2, \ldots, n_k$ are written is not important.) Note that a complete k-partite graph is complete if and only if $n_i = 1$ for all i, in which case it is K_k. A *complete bipartite graph* with partite sets V_1 and V_2, where $|V_1| = r$ and $|V_2| = s$, is then denoted by $K(r, s)$ or more commonly $K_{r,s}$. The graph $K_{1,s}$ is called a *star*. A graph is a *complete multipartite graph* if it is a complete k-partite graph for some $k \geq 2$.

There are many ways of combining graphs to produce new graphs. We next describe some binary operations defined on graphs. This discussion introduces notation that will prove useful in giving examples. In the following definitions, we assume that G_1 and G_2 are two graphs with disjoint vertex sets.

The *union* $G = G_1 \cup G_2$ has $V(G) = V(G_1) \cup V(G_2)$ and $E(G) = E(G_1) \cup E(G_2)$. If a graph G consists of k (≥ 2) disjoint copies of a graph H, then we write $G = kH$. The graph $2K_1 \cup 3K_2 \cup K_{1,3}$ is shown in Figure 1.12.

The *join* $G = G_1 + G_2$ has $V(G) = V(G_1) \cup V(G_2)$ and

$$E(G) = E(G_1) \cup E(G_2) \cup \{uv \mid u \in V(G_1) \text{ and } v \in V(G_2)\}.$$

Using the join operation, we see that $K_{r,s} = \overline{K}_r + \overline{K}_s$. Another illustration is given in Figure 1.13.

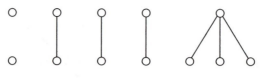

Figure 1.12 The union of graphs.

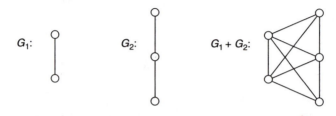

Figure 1.13 The join of two graphs.

The *Cartesian product* $G = G_1 \times G_2$ has $V(G) = V(G_1) \times V(G_2)$, and two vertices (u_1, u_2) and (v_1, v_2) of G are adjacent if and only if either

$$u_1 = v_1 \quad \text{and} \quad u_2 v_2 \in E(G_2)$$

or

$$u_2 = v_2 \quad \text{and} \quad u_1 v_1 \in E(G_1).$$

A convenient way of drawing $G_1 \times G_2$ is first to place a copy of G_2 at each vertex of G_1 (Figure 1.14[b]) and then to join corresponding vertices of G_2 in those copies of G_2 placed at adjacent vertices of G_1 (Figure 1.14[c]). Equivalently, $G_1 \times G_2$ can be constructed by placing a copy of G_1 at each vertex of G_2 and adding the appropriate edges. As expected, $G_1 \times G_2 = G_2 \times G_1$ for all graphs G_1 and G_2.

An important class of graphs is defined in terms of Cartesian products. The *n-cube* Q_n is the graph K_2 if $n = 1$, while for $n \geq 2$, Q_n is defined recursively as

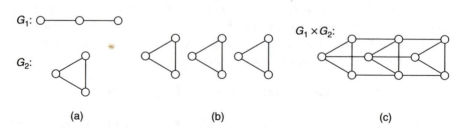

Figure 1.14 The Cartesian product of two graphs.

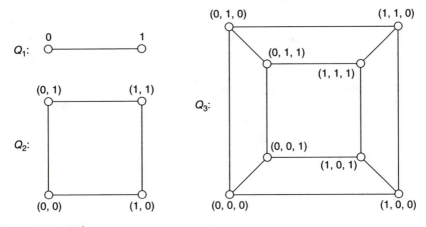

Figure 1.15 Cubes.

$Q_{n-1} \times K_2$. The n-cube Q_n can also be considered as that graph whose vertices are labeled by the binary n-tuples $(a_1, a_2, \ldots, a_n)$ (that is, a_i is 0 or 1 for $1 \leq i \leq n$) and such that two vertices are adjacent if and only if their corresponding n-tuples differ at precisely one coordinate. The graph Q_n is an n-regular graph of order 2^n. The n-cubes, $n = 1, 2$ and 3, are shown in Figure 1.15 with appropriate labelings. The graphs Q_n are often called *hypercubes*.

EXERCISES 1.1

1.1 Determine all nonisomorphic graphs of order 5.

1.2 Let n be a given positive integer, and let r and s be nonnegative integers such that $r + s = n$ and s is even. Show that there exists a graph G of order n having r even vertices and s odd vertices.

1.3 Figure 1.3 shows two regular nonisomorphic graphs of order 6 and size 9. Give another example of two nonisomorphic regular graphs of the same order and same size.

1.4 For each integer $k \geq 2$, give an example of k nonisomorphic regular graphs, all of the same order and same size.

1.5 A nontrivial graph G is called *irregular* if no two vertices of G have the same degrees. Prove that no graph is irregular.

1.6 Let G be a graph of order n all of whose vertices have degree r, where r is a positive integer, except for exactly one vertex of each of the degrees $r - 1, r - 2, \ldots, r - j$, where $1 < j < r$. By König's proof of Theorem 1.3, there is an r-regular graph of order $2jn$ containing G as an induced subgraph. Show, in fact, that there exists an r-regular graph of order $2n$ containing G as an induced subgraph.

1.7 Let G_1 and G_2 be self-complementary graphs, where G_2 has even order n. Now let G be the graph obtained from G_1 and G_2 by joining each vertex of G_2 whose degree is less than $n/2$ to every vertex G_1. Show that G is self-complementary.

1.8 Determine all self-complementary graphs of order 5 or less.

1.9 Prove that there exists a self-complementary graph of order n for every positive integer n with $n \equiv 0 \pmod 4$ or $n \equiv 1 \pmod 4$.

1.10 Let G be a self-complementary graph of order n, where $n \equiv 1 \pmod 4$. Prove that G contains at least one vertex of degree $(n-1)/2$. (*Hint:* Prove the stronger result that G contains an odd number of vertices of degree $(n-1)/2$.)

1.11 Let G be a nonempty graph with the property that whenever $uv \notin E(G)$ and $vw \notin E(G)$, then $uw \notin E(G)$. Prove that G has this property if and only if G is a complete k-partite graph for some $k \geq 2$. (*Hint:* Consider $\overline{G}$.)

1.2 DEGREE SEQUENCES

In this section, we investigate the concept of degree in more detail. A sequence d_1, $d_2, \ldots, d_n$ of nonnegative integers is called a *degree sequence* of a graph G if the vertices of G can be labeled $v_1, v_2, \ldots, v_n$ so that $\deg v_i = d_i$ for all i. For example, a degree sequence of the graph of Figure 1.16 is 4, 3, 2, 2, 1 (or 1, 2, 2, 3, 4, or 2, 1, 4, 2, 3, etc.).

Given a graph G, a degree sequence of G can be easily determined, of course. On the other hand, if a sequence s: $d_1, d_2, \ldots, d_n$ of nonnegative integers is given, then under what conditions is s a degree sequence of some graph? If such a graph exists, then s is called a *graphical sequence*. Certainly the conditions $d_i \leq n-1$ for all i and $\sum_{i=1}^{n} d_i$ is even are necessary for a sequence to be graphical and should be checked first, but these conditions are not sufficient. The sequence 3, 3, 3, 1 is not graphical, for example. A necessary and sufficient condition for a sequence to be graphical was found by Havel [H7] and later rediscovered by Hakimi [H3].

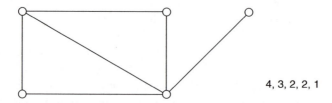

4, 3, 2, 2, 1

Figure 1.16 A degree sequence of a graph.

Theorem 1.4

A sequence s: $d_1, d_2, \ldots, d_n$ of nonnegative integers with $d_1 \geq d_2 \geq \cdots \geq d_n$, $n \geq 2$, $d_1 \geq 1$, is graphical if and only if the sequence s_1: $d_2 - 1$, $d_3 - 1, \ldots, d_{d_1+1} - 1$, $d_{d_1+2}, \ldots, d_n$ is graphical.

Proof

Assume first that s_1 is a graphical sequence. Then there exists a graph G_1 of order $n-1$ such that s_1 is a degree sequence of G_1. Thus, the vertices of G_1 can be labeled as $v_2, v_3, \ldots, v_n$ so that

$$\deg v_i = \begin{cases} d_i - 1 & 2 \leq i \leq d_1 + 1 \\ d_i & d_1 + 2 \leq i \leq n. \end{cases}$$

A new graph G can now be constructed by adding a new vertex v_1 and the d_1 edges $v_1 v_i, 2 \leq i \leq d_1 + 1$. Then in G, deg $v_i = d_i$ for $1 \leq i \leq n$, and so s: $d_1, d_2, \ldots, d_n$ is graphical.

Conversely, let s be a graphical sequence. Hence there exist graphs of order n with degree sequence s. Among all such graphs let G be one such that $V(G) = \{v_1, v_2, \ldots, v_n\}$, deg $v_i = d_i$ for $i = 1, 2, \ldots, n$, and the sum of the degrees of the vertices adjacent to v_1 is maximum. We show first that v_1 is adjacent to vertices having degrees $d_2, d_3, \ldots, d_{d_1+1}$.

Suppose, to the contrary, that v_1 is not adjacent to vertices having degrees $d_2, d_3, \ldots, d_{d_1+1}$. Then there exist vertices v_r and v_s with $d_r > d_s$ such that v_1 is adjacent to v_s but not to v_r. Since the degree of v_r exceeds that of v_s, there exists a vertex v_t such that v_t is adjacent to v_r but not to v_s. Removing the edges $v_1 v_s$ and $v_r v_t$ and adding the edges $v_1 v_r$ and $v_s v_t$ results in a graph G' having the same degree sequence as G. However, in G' the sum of the degrees of the vertices adjacent to v_1 is larger than that in G, contradicting the choice of G.

Thus, v_1 is adjacent to vertices having degrees $d_2, d_3, \ldots, d_{d_1+1}$, and the graph $G - v_1$ has degree sequence s_1; so s_1 is graphical. $\square$

Theorem 1.4 actually provides an algorithm for determining whether a given finite sequence of nonnegative integers is graphical. If, upon repeated application of Theorem 1.4, we arrive at a sequence every term of which is 0, then the original sequence is graphical. On the other hand, if we arrive at a sequence containing a negative integer, then the given sequence is not graphical.

We now illustrate Theorem 1.4 with the sequence

$$s: 5, 3, 3, 3, 3, 2, 2, 2, 1, 1, 1.$$

After one application of Theorem 1.4, we get

$$s_1': 2, 2, 2, 2, 1, 2, 2, 1, 1, 1.$$

Reordering this sequence, we obtain

s_1: 2, 2, 2, 2, 2, 2, 1, 1, 1, 1.

Continuing, we have

s_2': 1, 1, 2, 2, 2, 1, 1, 1, 1

s_2: 2, 2, 2, 1, 1, 1, 1, 1, 1

$s_3' = s_3$: 1, 1, 1, 1, 1, 1, 1, 1

s_4': 0, 1, 1, 1, 1, 1, 1

s_4: 1, 1, 1, 1, 1, 1, 0

s_5': 0, 1, 1, 1, 1, 0

s_5: 1, 1, 1, 1, 0, 0

s_6': 0, 1, 1, 0, 0

s_6: 1, 1, 0, 0, 0

$s_7' = s_7$: 0, 0, 0, 0.

Therefore, s is graphical. Of course, if we observe that some sequence prior to s_7 is graphical, then we can conclude by Theorem 1.4 that s is graphical. For example, the sequence s_3 is easily seen to be graphical since it is the degree sequence of the graph G_3 of Figure 1.17. By Theorem 1.4, each of the sequences s_2, s_1, and s is in turn graphical. To construct a graph with degree sequence s_2, we proceed in reverse from s_3' to s_2, observing that a vertex should be added to G_3 so that it is adjacent to two vertices of degree 1. We thus obtain a graph G_2 with degree sequence s_2 (or s_2'). Proceeding from s_2' to s_1, we again add a new vertex joining it to two vertices of degree 1 in G_2. This gives a graph G_1 with degree sequence s_1 (or s_1'). Finally, we obtain a graph G with degree sequence s by considering s_1'; that is, a new vertex is added to G_1, joining it to vertices of degrees 2, 2, 2, 2, 1.

It should be pointed out that the graph G in Figure 1.17 is not the only graph with degree sequence s. Indeed, there are graphs that cannot be produced by the method used to construct graph G of Figure 1.17. For example, the graph H of Figure 1.18 is such a graph.

Another result that determines which sequences are graphical is due to Erdös and Gallai [EG3].

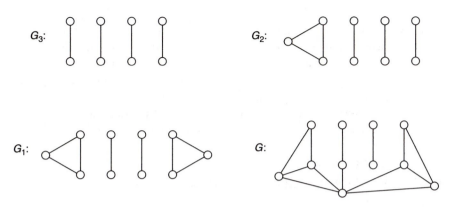

Figure 1.17 Construction of a graph G with given degree sequence.

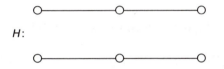

Figure 1.18 A graph that cannot be constructed by the method following Theorem 1.4.

Theorem 1.5

A sequence $d_1, d_2, \ldots, d_n$ $(n \geq 2)$ of nonnegative integers with $d_1 \geq d_2 \geq \cdots \geq d_n$ is graphical if and only if $\sum_{i=1}^{n} d_i$ is even and for each integer $k, 1 \leq k \leq n-1$,

$$\sum_{i=1}^{k} d_i \leq k(k-1) + \sum_{i=k+1}^{n} \min\{k, d_i\}.$$

EXERCISES 1.2

1.12 Determine whether the following sequences are graphical. If so, construct a graph with the appropriate degree sequence.

    (a)   4, 4, 3, 2, 1
    (b)   3, 3, 2, 2, 2, 2, 1, 1
    (c)   7, 7, 6, 5, 4, 4, 3, 2
    (d)   7, 6, 6, 5, 4, 3, 2, 1
    (e)   7, 4, 3, 3, 2, 2, 2, 1, 1, 1

1.13 Show that the sequence $d_1, d_2, \ldots, d_n$ is graphical if and only if the sequence $n - d_1 - 1, n - d_2 - 1, \ldots, n - d_n - 1$ is graphical.

1.14 (a) Using Theorem 1.4, show that $s: 7, 6, 5, 4, 4, 3, 2, 1$ is graphical.
(b) Prove that there exists exactly one graph with degree sequence s.

1.15 Show that for every finite set S of positive integers, there exists a positive integer k such that the sequence obtained by listing each element of S a total k times is graphical. Find the minimum such k for $S = \{2, 6, 7\}$.

1.3 CONNECTED GRAPHS AND DISTANCE

Let u and v be (not necessarily distinct) vertices of a graph G. A u–v *walk* W of G is a finite, alternating sequence

$$W: u = u_0, e_1, u_1, e_2, \ldots, u_{k-1}, e_k, u_k = v$$

of vertices and edges, beginning with vertex u and ending with vertex v, such that $e_i = u_{i-1} u_i$ for $i = 1, 2, \ldots, k$. The number k (the number of occurrences of edges) is called the *length* of W. A *trivial walk* contains no edges, that is, $k = 0$. We note that there may be repetition of vertices and edges in a walk. Often only the vertices of a walk are indicated since the edges present are then evident. Two u–v walks W_1: $u = u_0, u_1, \ldots, u_k = v$ and W_2: $u = v_0, v_1, \ldots, v_\ell = v$ are considered to be *equal* if and only if $k = \ell$ and $u_i = v_i$ for $0 \le i \le k$; otherwise, W_1 and W_2 are *different*. Observe that the edges of two different u–v walks of G may very well induce the same subgraph of G.

A u–v walk is *closed* or *open* depending on whether $u = v$ or $u \ne v$. A u–v *trail* is a u–v walk in which no edge is repeated, while a u–v *path* is a u–v walk in which no vertex is repeated. A vertex u forms the *trivial u–u path*. Every path is therefore a trail. In the graph G of Figure 1.19, W: $v_1, v_2, v_3, v_2, v_5, v_3, v_4$ is a v_1–v_4 walk that is not a trail, T: $v_1, v_2, v_5, v_1, v_3, v_4$ is a v_1–v_4 trail that is not a path and P: v_1, v_3, v_4 is a v_1–v_4 path.

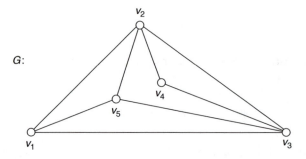

Figure 1.19 Walks, trails and paths.

By definition, every path is a walk. Although the converse of this statement is not true in general, we do have the following theorem. A walk W is said to *contain* a walk W' if W' is a subsequence of W.

Theorem 1.6

Every $u–v$ walk in a graph contains a $u–v$ path.

Proof

Let W be a $u–v$ walk in a graph G. If W is closed, the result is trivial. Let W: $u = u_0, u_1, u_2, \ldots, u_k = v$ be an open $u–v$ walk of a graph G. (A vertex may have received more than one label.) If no vertex of G occurs in W more than once, then W is a $u–v$ path. Otherwise, there are vertices of G that occur in W twice or more. Let i and j be distinct positive integers, with $i < j$ say, such that $u_i = u_j$. If the terms $u_i, u_{i+1}, \ldots, u_{j-1}$ are deleted from W, a $u–v$ walk W_1 is obtained having fewer terms than that of W. If there is no repetition of vertices in W_1, then W_1 is a $u–v$ path. If this is not the case, we continue the above procedure until finally arriving at a $u–v$ walk that is a $u–v$ path. □

As the next theorem indicates, the powers of the adjacency matrix of a graph can be used to compute the number of walks of various lengths in the graph.

Theorem 1.7

If A is the adjacency matrix of a graph G with $V(G) = \{v_1, v_2, \ldots, v_n\}$, then the (i, j) entry of A^k, $k \geq 1$, is the number of different $v_i–v_j$ walks of length k in G.

Proof

The proof is by induction on k. The result is obvious for $k = 1$ since there exists a $v_i–v_j$ walk of length 1 if and only if $v_i v_j \in E(G)$. Let $A^{k-1} = [a_{ij}^{(k-1)}]$ and assume that $a_{ij}^{(k-1)}$ is the number of different $v_i–v_j$ walks of length $k-1$ in G; furthermore, let $A^k = [a_{ij}^{(k)}]$. Since $A^k = A^{k-1} A$, we have

$$a_{ij}^{(k)} = \sum_{\ell=1}^{n} a_{i\ell}^{(k-1)} a_{\ell j}. \tag{1.1}$$

Every $v_i–v_j$ walk of length k in G consists of a $v_i–v_\ell$ walk of length $k-1$, where v_ℓ is adjacent to v_j, followed by the edge $v_\ell v_j$ and the vertex v_j. Thus by the inductive hypothesis and (1.1), we have the desired result. □

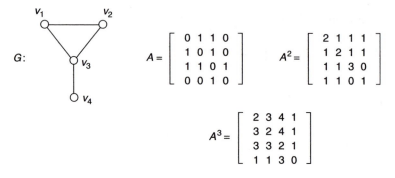

$$A = \begin{bmatrix} 0 & 1 & 1 & 0 \\ 1 & 0 & 1 & 0 \\ 1 & 1 & 0 & 1 \\ 0 & 0 & 1 & 0 \end{bmatrix} \qquad A^2 = \begin{bmatrix} 2 & 1 & 1 & 1 \\ 1 & 2 & 1 & 1 \\ 1 & 1 & 3 & 0 \\ 1 & 1 & 0 & 1 \end{bmatrix}$$

$$A^3 = \begin{bmatrix} 2 & 3 & 4 & 1 \\ 3 & 2 & 4 & 1 \\ 3 & 3 & 2 & 1 \\ 1 & 1 & 3 & 0 \end{bmatrix}$$

Figure 1.20 A graph and powers of its adjacency matrix.

As an illustration of Theorem 1.8, consider the graph G of Figure 1.20 having adjacency matrix A. We can determine A^2 without matrix multiplication by observing that the (i, i) entry of $A^2, 1 \le i \le 4$, is $\deg v_i$ and the (i, j) entry of $A^2, i \ne j$, is the number of different v_i–v_j paths of length 2. We now turn to A^3. Since the different v_1–v_3 walks of length 3 in G are

$$W_1: v_1, v_3, v_1, v_3, ; \quad W_2: v_1, v_2, v_1, v_3; \quad W_3: v_1, v_3, v_2, v_3; \quad W_4: v_1, v_3, v_4, v_3,$$

the $(1, 3)$ entry of A^3 is 4. The entire matrix A^3 can be computed in this manner.

A nontrivial closed trail of a graph G is referred to as a *circuit* of G, and a circuit $v_1, v_2, \ldots, v_n, v_1$ $(n \ge 3)$ whose n vertices v_i are distinct is called a *cycle*. An *acyclic graph* has no cycles. The subgraph of a graph G induced by the edges of a trail, path, circuit or cycle is also referred to as a *trail, path, circuit* or *cycle* of G. A cycle is *even* if its length is even; otherwise it is *odd*. A cycle of length n is an *n-cycle*; a 3-cycle is also called a *triangle*. A graph of order n that is a path or a cycle is denoted by P_n or C_n, respectively.

We now consider a very basic concept in graph theory, namely connected and disconnected graphs. A vertex u is said to be *connected* to a vertex v in a graph G if there exists a u–v path in G. A graph G is *connected* if every two of its vertices are connected. A graph that is not connected is *disconnected*. The relation 'is connected to' is an equivalence relation on the vertex set of every graph G. Each subgraph induced by the vertices in a resulting equivalence class is called a *connected component* or simply a *component* of G. Equivalently, a component of a graph G is a connected subgraph of G not properly contained in any other connected subgraph of G; that is, a component of G is a subgraph that is maximal with respect to the property of being connected. Hence, a connected subgraph F of a graph G is a component of G if for each connected graph H with $F \subseteq H \subseteq G$ where $V(F) \subseteq V(H)$ and $E(F) \subseteq E(H)$, it follows that $F = H$. The number of components of G is denoted by $k(G)$; of course, $k(G) = 1$ if and only if G is connected. For the graph G of Figure 1.21, $k(G) = 6$.

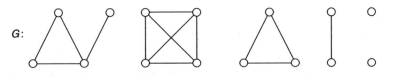

Figure 1.21 A graph with six components.

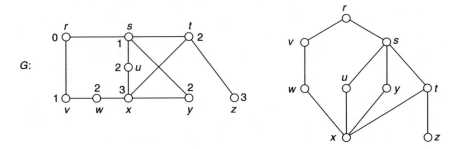

Figure 1.22 The distance levels from the vertex r.

For a connected graph G, we define the *distance* $d(u, v)$ between two vertices u and v as the minimum of the lengths of the u–v paths of G. A u–v path of length $d(u, v)$ is called a u–v *geodesic*. Under this distance function, the set $V(G)$ is a metric space, that is, the following properties hold:

1. $d(u, v) \geq 0$ for all pairs, u, v of vertices of G, and $d(u, v) = 0$ if and only if $u = v$;
2. (symmetric property) $d(u, v) = d(v, u)$ for all pairs u, v of vertices of G;
3. (triangle inequality) $d(u, v) + d(v, w) \geq d(u, w)$ for all triples u, v, w of vertices of G.

Each vertex of the graph G of Figure 1.22 is labeled with its distance from r. The graph G is then redrawn to illustrate these distances better. The vertices of G are thus partitioned into levels, according to their distance from r. There are a number of instances when it is useful to draw a graph in this manner.

If G is a disconnected graph, then we can define distance as above between vertices in the same component of G. If u and v are vertices in distinct components of G, then $d(u, v)$ is undefined (or we could define $d(u, v) = \infty$).

We are now prepared to present a useful characterization of bipartite graphs.

Theorem 1.8

A nontrivial graph is bipartite if and only if it contains no odd cycles.

Proof

Let G be a bipartite graph with partite sets V_1 and V_2. Suppose that $C: v_1$, $v_2, \ldots, v_k, v_1$ is a cycle of G. Without loss of generality, we may assume that $v_1 \in V_1$. However, then $v_2 \in V_2$, $v_3 \in V_1$, $v_4 \in V_2$, and so on. This implies $k = 2s$ for some positive integer s; hence, C has even length. *the length*

For the converse, it suffices to prove that every nontrivial connected graph G without odd cycles is bipartite, since a nontrivial graph is bipartite if and only if each of its nontrivial components is bipartite. Let $v \in V(G)$ and denote by V_1 the subset of $V(G)$ consisting of all vertices u of G such that $d(v, u)$ is even. Let $V_2 = V(G) - V_1$. We claim that V_1 and V_2 are partite sets of G.

Assume, to the contrary, that this is not the case. Then either V_1 or V_2 contain two adjacent vertices, say V_1 contains adjacent vertices u and w. Neither u nor w can be v. Suppose that $d(v, u) = 2s$ and $d(v, w) = 2t$, where s, $t \geq 1$. Let P_1: $v = u_0, u_1, \ldots, u_{2s} = u$, and P_2: $v = w_0, w_1, \ldots, w_{2t} = w$ be a v–u geodesic and v–w geodesic, respectively. Let w' be the last vertex that P_1 and P_2 have in common. Then $w' = u_i = w_i$ for some integer i and C': $u_i, u_{i+1}, \ldots,$ $u_{2s}, w_{2t}, w_{2t-1}, \ldots, w_i = u_i$ is a cycle of length $(2s - i) + (2t - i) + 1 = 2(s + t - i) + 1$. Thus C' is an odd cycle, which is a contradiction. □

The *eccentricity* $e(v)$ of a vertex v of a connected graph G is the number $\max_{u \in V(G)} d(u, v)$. That is, $e(v)$ is the distance between v and a vertex farthest from v. The *radius* rad G of G is the minimum eccentricity among the vertices of G, while the *diameter* diam G of G is the maximum eccentricity. Consequently, diam G is the greatest distance between any two vertices of G. Also, a graph G has radius 1 if and only if G contains a vertex adjacent to all other vertices of G. A vertex v is a *central vertex* if $e(v) = $ rad G and the *center* Cen(G) is the subgraph of G induced by its central vertices.

For the graph G of Figure 1.23, rad $G = 3$ and diam $G = 5$. Here, Cen$(G) = K_3$. A vertex v is a *peripheral vertex* if $e(v) = $ diam G, while the *periphery*

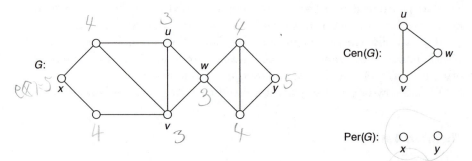

Figure 1.23 A graph with radius 3 and diameter 5.

Per(G) is the subgraph of G induced by its peripheral vertices. For the graph G of Figure 1.23, the vertices x and y are peripheral vertices and Per(G) $= 2K_1$.

The radius and diameter are related by the following inequalities.

Theorem 1.9

For every connected graph G,

$$\text{rad } G \leq \text{diam } G \leq 2 \text{ rad } G.$$

Proof

The inequality rad $G \leq$ diam G is a direct consequence of the definitions. In order to verify the second inequality, select vertices u and v in G such that $d(u, v) =$ diam G. Furthermore, let w be a central vertex of G. Since d is a metric on $V(G)$,

$$d(u, v) \leq d(u, w) + d(w, v) \leq 2e(w) = 2 \text{ rad } G. \qquad \square$$

Theorem 1.9 gives a lower bound (namely, rad G) for the diameter of a connected graph G as well as an upper bound (namely, 2 rad G). This is the first of many results we shall encounter for which a question of 'sharpness' is involved. In other words, just how good is this result? Ordinarily, there are many interpretations of such a question. We shall consider some possible interpretations in the case of the upper bound.

Certainly, the upper bound in Theorem 1.9 would not be considered sharp if diam $G < 2$ rad G for every graph G; however, it would be considered sharp indeed if diam $G = 2$ rad G for every graph G. In the latter case, we would have a formula, not just a bound. Actually, there are graphs G for which diam $G < 2$ rad G and graphs H for which diam $H = 2$ rad H. This alone may be a satisfactory definition of 'sharpness'. A more likely interpretation is the existence of an infinite class H of graphs H such that diam $H = 2$ rad H for each $H \in$ H. Such a class exists; for example, let H consist of the graphs of the type $K_t + \overline{K}_2$. One disadvantage of this example is that for each $H \in$ H, diam $H = 2$ and rad $H = 1$. Perhaps a more satisfactory class (which fills a more satisfactory requirement for sharpness) is the class of paths P_{2k+1}, $k \geq 1$. In this case, diam $P_{2k+1} = 2k$ and rad $P_{2k+1} = k$; that is, for each positive integer k, there exists a graph G such that diam $G = 2$ rad $G = 2k$ (Exercise 1.29).

In the graph G of Figure 1.23, we saw that Cen(G) $= K_3$. It is not difficult to see that Cen(P_{2k+1}) $= K_1$ and Cen(P_{2k}) $= K_2$ for all $k \geq 1$. Also, Cen(C_n) $= C_n$ for all $n \geq 3$. Hence there are many graphs that are centers of graphs. Hedetniemi (see Buckley, Miller and Slater [BMS1]) showed that there is no restriction on which graphs are centers.

Theorem 1.10

Every graph is the center of some connected graph.

Proof

Let G be a given graph. We construct a graph H from G by adding four new vertices u_1, v_1, u_2, v_2 and for $i=1$, 2, every vertex of G is joined to v_i, and u_i is joined to v_i. (This construction is illustrated in Figure 1.24.) Since $e(u_i)=4$ and $e(v_i)=3$ for $i=1$, 2, while $e_H(x)=2$ for every vertex x of G, it follows that $\mathrm{Cen}(H)=G$. □

The center of a connected graph G was introduced as one means of describing the 'middle' of a graph. Since this concept is vague and is open to interpretation, other attempts have been made to identify the middle of a graph. We describe another of these. The *total distance td*(u) of a vertex u in a connected graph G is defined by

$$td(u) = \sum_{v \in V(G)} d(u, v).$$

A vertex v in G is called a *median vertex* if v has the minimum total distance among the vertices of G. Equivalently, v is a median vertex if v has the minimum average distance to all vertices of G. The *median* $\mathrm{Med}(G)$ of G is then the subgraph of G induced by its median vertices. In the graph G of Figure 1.25, each vertex is labeled with its total distance. The median of G is also shown.

There is also no restriction on which graphs can be medians. Slater [S5] showed that every graph is the median of some connected graph. Indeed, Hendry [H10] showed that for every two graphs G_1 and G_2, there exists a connected graph H such that $\mathrm{Cen}(H)=G_1$ and $\mathrm{Med}(H)=G_2$.

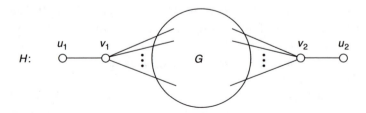

Figure 1.24 A graph with given center.

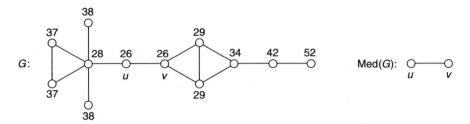

Figure 1.25 The median of a graph.

Not every graph is the periphery of some graph, as Bielak and Syslo [BS1] verified.

Theorem 1.11

A graph G is the periphery of some connected graph if and only if every vertex of G has eccentricity 1 or no vertex of G has eccentricity 1.

Proof

Suppose first that every vertex of G has eccentricity 1. Then $\text{Per}(G) = G$ and G is the periphery of itself. Next, suppose that no vertex of G has eccentricity 1. Define H to be the graph obtained from G by adding a new vertex v to G and joining v to every vertex of G. Then $e(v) = 1$. Let $x \in V(G)$. Since $e_G(x) \neq 1$, there is some vertex y in G such that $xy \notin E(G)$. However, $d_H(x, y) = 2$ since x, v, y is a path of length 2 in H. Thus $e_H(x) = 2$. Indeed, then, every vertex of G has eccentricity 2 in H. Therefore $\text{Per}(H) = G$.

For the converse, assume that G is a graph for which some but not all vertices have eccentricity 1, and suppose, to the contrary, that G is the periphery of a connected graph H. Certainly G is a proper subgraph of H. Therefore, for each vertex x of G, it follows that $e_H(x) = \text{diam } H \geq 2$. Let u be a vertex of G having eccentricity 1 in G. Thus, u is adjacent to all other vertices of G. Let v be a vertex of H such that $d_H(u, v) = e_H(u) = \text{diam } H \geq 2$. Therefore $e_H(v) = \text{diam } H$ and v is a peripheral vertex of H. On the other hand, since $d_H(u, v) \geq 2$, the vertex v is not adjacent to u and so v is not in G, which produces a contradiction. □

EXERCISES 1.3

1.16 Let u and v be arbitrary vertices of a connected graph G. Show that there exists a u–v walk containing all vertices of G.

1.17 Prove that 'is connected to' is an equivalence relation on the vertex set of a graph.

1.18 (a) Let G be a graph of order n such that $\deg v \geq (n-1)/2$ for every $v \in V(G)$. Prove that G is connected.

(b) Examine the sharpness of the bound in (a).

1.19 Let $n \geq 2$ be an integer. Determine the minimum positive integer m such that *every* graph of order n and size m is connected.

1.20 Prove that a graph G is connected if and only if for every partition $V(G) = V_1 \cup V_2$, there exists an edge of G joining a vertex of V_1 and a vertex of V_2.

1.21 Prove that if G is a graph with $\delta(G) \geq 2$, then G contains a cycle.

1.22 Prove that every graph G has a path of length $\delta(G)$.

1.23 Show that if G is a graph of order n and size $n^2/4$, then either G contains an odd cycle or $G = K_{n/2,n/2}$.

1.24 (a) Show that there are exactly two 4-regular graphs G of order 7. (Hint: Consider $\overline{G}$.)

(b) How many 6-regular graphs of order 9 are there?

1.25 Characterize those graphs G having the property that every induced subgraph of G is a connected subgraph of G.

1.26 (a) Show that if G is a connected graph such that the degree of every vertex is one of three distinct numbers, then there is a path P in G containing three vertices whose degrees are distinct.

(b) Is the statement in (a) true if 'three' is replaced by 'four'?

1.27 Let G be a nontrivial connected graph that is not bipartite. Show that G contains adjacent vertices u and v such that $\deg u + \deg v$ is even.

1.28 Prove that if G is a disconnected graph, then $\overline{G}$ is connected and, in fact, diam $\overline{G} \leq 2$.

1.29 Let a and b be positive integers with $a \leq b \leq 2a$. Show that there exists a graph G with rad $G = a$ and diam $G = b$.

1.30 Define the *central appendage number* of a graph G to be the minimum number of vertices that must be added to G to produce a connected graph H with $\text{Cen}(H) = G$. Show that the central appendage number of a graph (a) can be 0, (b) can never be 1, and (c) is at most 4.

1.31 Let G be a connected graph.

(a) If u and v are adjacent vertices of G, then show that $|e(u) - e(v)| \leq 1$.

(b) If k is an integer such that rad $G \leq k \leq$ diam G, then show that there is a vertex w such that $e(w) = k$.

(c) If k is an integer such that rad $G < k \leq$ diam G, then show that there are at least two vertices of G with eccentricity k. (*Hint*: Let w be a vertex with $e(w) = k$, and let u be a vertex with $d(w, u) = e(w) = k$. For a central vertex v of G, let P be a v–u path of length $d(v, u)$. Show that $e(v) < k \leq e(u)$. Then show that there is a vertex x (distinct from w) on P such that $e(x) = k$.)

1.32 Show that for every pair r, s of positive integers, there exists a positive integer n such that for every connected graph G of order n, either $\Delta(G) \geq r$ or diam $G \geq s$.

1.33 Let F and H be two subgraphs in a connected graph G. Define the distance $d(F, H)$ between F and H as

$$d(F, H) = \min\{d(u, v) | u \in V(F), v \in V(H)\}.$$

Show that for every positive integer k, there exists a connected graph G such that $d(\mathrm{Cen}(G), \mathrm{Med}(G)) = k$.

1.34 Every complete graph is the periphery of itself. Can a complete graph be the periphery of a connected graph G with diam $G \geq 2$?

1.4 DIGRAPHS AND MULTIGRAPHS

There are occasions when the standard definition of graph does not serve our purposes. This remark leads us to our next topics. When the symmetric nature of graphs does not satisfy our requirements, we are led to directed graphs. A *directed graph* or *digraph D* is a finite nonempty set of objects called *vertices* together with a (possibly empty) set of ordered pairs of distinct vertices of D called *arcs* or *directed edges*. As with graphs, the vertex set of D is denoted by $V(D)$ and the arc set is denoted by $E(D)$. A digraph D with $V(D) = \{u, v, w\}$ and $E(D) = \{(u, w), (w, u), (u, v)\}$ is illustrated in Figure 1.26. Observe that when a digraph is described by means of a diagram, the 'direction' of each arc is indicated by an arrowhead.

The terminology used in discussing digraphs is quite similar to that used for graphs. The cardinality of the vertex set of a digraph D is called the *order* of D and is denoted by $n(D)$, or simply n. The *size* $m(D)$ (or m) of D is the cardinality of its arc set.

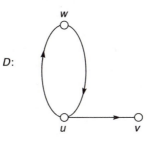

Figure 1.26 A digraph.

If $a = (u, v)$ is an arc of a digraph D, then a is said to *join* u and v. We further say that a is *incident from* u and *incident to* v, while u is *incident to* a and v is *incident from* a. Moreover, u is said to be *adjacent to* v and v is *adjacent from* u. In the digraph D of Figure 1.26, vertex u is adjacent to vertex v, but v is *not* adjacent to u. Two vertices u and v of a digraph D are *nonadjacent* if u is neither adjacent to nor adjacent from v in D.

The *outdegree* od v of a vertex v of a digraph D is the number of vertices of D that are adjacent from v. The *indegree* id v of v is the number of vertices of D adjacent to v. The *degree* deg v of a vertex v of D is defined by

$$\deg v = \operatorname{od} v + \operatorname{id} v.$$

In the digraph D of Figure 1.26, od $u = 2$, id $u =$ id $v =$ id $w =$ od $w = 1$, while od $v = 0$. For the same digraph, deg $u = 3$, deg $w = 2$ and deg $v = 1$.

We now present *The First Theorem of Digraph Theory.*

Theorem 1.12

If D is a digraph of order n and size m with $V(D) = \{v_1, v_2, \ldots, v_n\}$, then

$$\sum_{i=1}^{n} \operatorname{od} v_i = \sum_{i=1}^{n} \operatorname{id} v_i = m.$$

Proof

When the outdegrees of the vertices are summed, each arc is counted once, since every arc is incident *from* exactly one vertex. Similarly, when the indegrees are summed, an arc is counted just once since every arc is incident *to* a single vertex. ☐

A digraph D_1 is *isomorphic* to a digraph D_2 if there exists a one-to-one mapping ϕ, called an *isomorphism*, from $V(D_1)$ onto $V(D_2)$ such that $(u, v) \in E(D_1)$ if and only if $(\phi u, \phi v) \in E(D_2)$. The relation 'is isomorphic to' is an equivalence relation on digraphs. Thus, this relation partitions the set of all digraphs into equivalence classes; two digraphs are *nonisomorphic* if they belong to different equivalence classes. If D_1 is isomorphic to D_2, then we say D_1 and D_2 are *isomorphic* and write $D_1 = D_2$.

There is only one digraph of order 1 (up to isomorphism); this is the *trivial digraph*. Also, there is only one digraph of order 2 and size m for each m with $0 \leq m \leq 2$. There are four digraphs of order 3 and size 3, and these are shown in Figure 1.27.

A labeled digraph D_1 is a *subdigraph* of a labeled digraph D if $V(D_1) \subseteq V(D)$ and $E(D_1) \subseteq E(D)$. For unlabeled digraphs D_1 and D_3 we say that D_1 is a *subdigraph* of D if D_1 and D can be labeled so that, as labeled digraphs, D_1 is

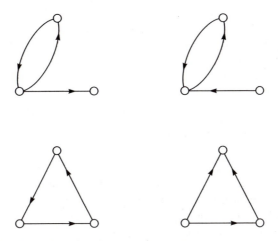

Figure 1.27 The digraphs of order 3 and size 3.

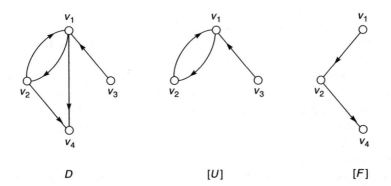

Figure 1.28 Induced and arc-induced subdigraphs.

a subdigraph of D. We write $D_1 \subseteq D$ to indicate that D_1 is a subdigraph of D. A subdigraph D_1 of D is a *spanning subdigraph* if D_1 has the same order as D. Vertex-deleted, arc-deleted, induced and arc-induced subdigraphs are defined in the expected manner. These last two concepts are illustrated for the digraph D of Figure 1.28, where

$$V(D) = \{v_1, v_2, v_3, v_4\}, \quad U = \{v_1, v_2, v_3\}, \quad \text{and} \quad F = \{(v_1, v_2), (v_2, v_4)\}.$$

We now consider certain types of digraphs that occur regularly in our discussions. A digraph D is called *symmetric* if, whenever (u, v) is an arc of D, then (v, u) is also. There is a natural one-to-one correspondence between the set of symmetric digraphs and the set of graphs. A digraph D is called an *asymmetric digraph* or an *oriented graph* if whenever (u, v) is an arc of D, then (v, u) is *not* an arc of D. Thus, an oriented graph D can be obtained from a graph G by assigning a

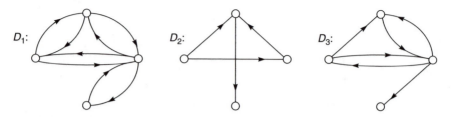

Figure 1.29 Symmetric and asymmetric digraphs.

direction to (or by 'orienting') each edge of G, thereby transforming each edge of G into an arc and transforming G itself into an asymmetric digraph; D is also called an *orientation* of G. The digraph D_1 of Figure 1.29 is symmetric while D_2 is asymmetric; the digraph D_3 has neither property.

A digraph D is called *complete* if for every two distinct vertices u and v of D, at least one of the arcs (u, v) and (v, u) is present in D. The *complete symmetric digraph* of order n has both arcs (u, v) and (v, u) for every two distinct vertices u and v and is denoted by K_n^*. Indeed, if G is a graph, then G^* denotes the symmetric digraph obtained by replacing each edge of G by a symmetric pair of arcs. The digraph K_n^* has size $n(n-1)$ and od $v = $ id $v = n-1$ for every vertex v of K_n^*. The digraphs K_1^*, K_2^*, K_3^* and K_4^* are shown in Figure 1.30. The *underlying graph* of a digraph D is that graph obtained by replacing each arc (u, v) or symmetric pairs (u, v), (v, u) of arcs by the edge of uv. Certainly the underlying graph of G^* is G.

A complete asymmetric digraph is called a *tournament* and will be studied in some detail in Chapter 5.

A digraph D is called *regular of degree r* or *r-regular* if od $v = $ id $v = r$ for every vertex v of D. The digraph K_n^* is $(n-1)$-regular. A 1-regular digraph D_1 and 2-regular digraph D_2 are shown in Figure 1.31. The digraph D_2 is a tournament.

The terms walk, open and closed walk, trail, path, circuit and cycle for graphs have natural counterparts in digraph theory, the important difference being that the directions of the arcs must be followed in each of these walks. In particular,

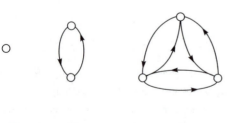
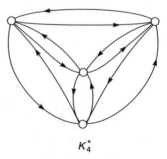

K_1^* $\qquad$ K_2^* $\qquad\qquad$ K_3^* $\qquad\qquad\qquad$ K_4^*

Figure 1.30 Complete symmetric digraphs.

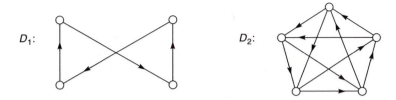

Figure 1.31 Regular digraphs.

when referring to digraphs, the terms *directed path* and *directed cycle* are synonymous with path and cycle. More formally, for vertices u and v in a digraph D, a *u–v walk* in D is a finite, alternating sequence

$$u = u_0, a_1, u_1, a_2, \ldots, u_{k-1}, a_k, u_k = v$$

of vertices and arcs, beginning with u and ending with v, such that $a_i = (u_{i-1}, u_i)$ for $i = 1, 2, \ldots, k$. The number k is the *length* of the walk.

A digraph D is *connected* (or *weakly connected*) if the underlying graph of D is connected. A digraph D is *strong* (or *strongly connected*) if for every pair u, v of vertices, D contains both a *u–v* path and a *v–u* path. While all digraphs of Figure 1.32 are connected only D_1 is strong.

Distance can be defined in digraphs as well. For vertices u and v in a digraph D containing a *u–v* path, the *(directed) distance* $d(u, v)$ from u to v is the length of a shortest *u–v* path in D. Thus the distances $d(u, v)$ and $d(v, u)$ are defined for all pairs u, v of vertices in a strong digraph. This distance is not a metric, in general. Although the distance satisfies the triangle inequality, it is not symmetric—unless D is symmetric, in which case D is, in actuality, a graph. Eccentricity can be defined as before, as well as radius and diameter. The *eccentricity* $e(u)$ of a vertex u in D is the distance from u to a vertex farthest from u. The minimum eccentricity of the vertices of D is the *radius* rad D of D, while the *diameter* diam D is the greatest eccentricity.

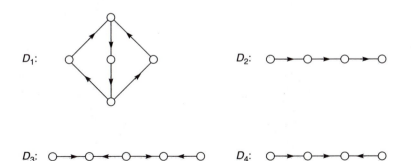

Figure 1.32 Connectedness properties of digraphs.

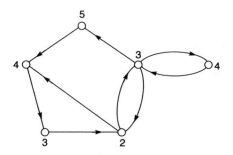

Figure 1.33 Eccentricities in a strong digraph.

The vertices of the strong digraph D of Figure 1.33 are labeled with their eccentricities. Observe that rad $D = 2$ and diam $D = 5$, so it is not true, in general, that diam $D \leq 2$ rad D, as is the case with graphs.

In the definition of a graph G, either one edge or no edge joins a pair of distinct vertices of G. For a digraph D, two (directed) edges can join distinct vertices of D—if they are directed oppositely. There are occasions when we will want to permit more than one edge to join distinct vertices (and in the same direction in the case of digraphs).

If one allows more than one edge (but yet a finite number) between the same pair of vertices in a graph, the resulting structure is a *multigraph*. Such edges are called *parallel edges*. If more than one arc in the same direction is permitted to join two vertices in a digraph, a *multidigraph* results. A *loop* is an edge (or arc) that joins a vertex to itself. Although loops are permitted to occur in multigraphs and multidigraphs, these will occur rarely. Figures 1.34(a)–(d) show a multigraph and multidigraph without loops and a multigraph and multidigraph with loops.

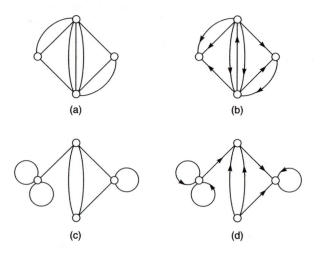

Figure 1.34 Multigraphs, multidigraphs, pseudographs and pseudodigraphs.

EXERCISES 1.4

1.35 Determine all (pairwise nonisomorphic) digraphs of order 4 and size 4.

1.36 Prove or disprove: For every integer $n \geq 2$, there exists a digraph D of order n such that for every two distinct vertices u and v of D, od $u \neq$ od v and id $u \neq$ id v.

1.37 Prove or disprove: No digraph contains an odd number of vertices of odd outdegree or an odd number of vertices of odd indegree.

1.38 Prove or disprove: If D_1 and D_2 are two digraphs with $V(D_1) = \{u_1, u_2, \ldots, u_n\}$ and $V(D_2) = \{v_1, v_2, \ldots, v_n\}$ such that $\mathrm{id}_{D_1} u_i = \mathrm{id}_{D_2} v_i$ and $\mathrm{od}_{D_1} u_i = \mathrm{od}_{D_2} v_i$ for $i = 1, 2, \ldots, n$, then $D_1 = D_2$.

1.39 Prove that if every proper induced subdigraph of a digraph D of order $n \geq 4$ is regular, then $E(D) = \emptyset$ or $D = K_n^*$.

1.40 Prove that there exist regular tournaments of every odd order but there are no regular tournaments of even order.

1.41 The *adjacency matrix* $A(D)$ of a digraph D with $V(D) = \{v_1, v_2, \ldots, v_n\}$ is the $n \times n$ matrix $[a_{ij}]$ defined by $a_{ij} = 1$ if $(v_i, v_j) \in E(D)$ and $a_{ij} = 0$ otherwise.

   (a) What information do the row sums and column sums of the adjacency matrix of a digraph provide?

   (b) Characterize matrices that are adjacency matrices of digraphs.

1.42 Prove that if D is a digraph with od $v \geq 1$ and id $v \geq 1$ for every vertex v of D, then D contains a cycle.

1.43 Prove that for every two positive integers a and b with $a \leq b$, there exists a strong digraph D with rad $D = a$ and diam $D = b$.

1.44 The *center* $\mathrm{Cen}(D)$ of a strong digraph D is the subdigraph induced by those vertices v with $e(v) = \mathrm{rad}\, D$. Prove that for every asymmetric digraph D_1, there exists a strong asymmetric digraph D such that $\mathrm{Cen}(D) = D_1$.

2

The structure of graphs

Although being connected is the most basic structural property that a graph may possess, more information about its structure is provided by special vertices, edges and subgraphs it contains and as well as their location within the graph.

2.1 CUT-VERTICES, BRIDGES AND BLOCKS

Some graphs are connected so slightly that they can be disconnected by the removal of a single vertex or a single edge. Such vertices and edges play a special role in graph theory, and we discuss these next.

A vertex v of a graph G is called a *cut-vertex* of G if $k(G - v) > k(G)$. Thus, a vertex of a connected graph is a cut-vertex if its removal produces a disconnected graph. In general, a vertex v of a graph G is a cut-vertex of G if its removal disconnects a component of G. The following theorem characterizes cut-vertices.

Theorem 2.1

A vertex v of a graph G is a cut-vertex of G if and only if there exist vertices u and w $(u, w \neq v)$ such that v is on every u–w path of G.

Proof

It suffices to prove the theorem for connected graphs. Let v be a cut-vertex of G; so the graph $G - v$ is disconnected. If u and w are vertices in different components of $G - v$, then there are no u–w paths in $G - v$. However, since G is connected, there are u–w paths in G. Therefore, every u–w path of G contains v.

Conversely, assume that there exist vertices u and w in G such that the vertex v lies on every u–w path of G. Then there are no u–w paths in $G - v$, implying that $G - v$ is disconnected and that v is a cut-vertex of G. $\qquad\square$

The complete graphs have no cut-vertices while, at the other extreme, each nontrivial path contains only two vertices that are not cut-vertices. In order to see that this is the other extreme, we prove the following theorem.

Theorem 2.2

Every nontrivial connected graph contains at least two vertices that are not cut-vertices.

Proof

Assume that the theorem is false. Then there exists a nontrivial connected graph G containing at most one vertex that is not a cut-vertex; that is, every vertex of G, with at most one exception, is a cut-vertex. Let u and v be vertices of G such that $d(u, v) = \text{diam } G$.

At least one of u and v is a cut-vertex, say v. Let w be a vertex belonging to a component of $G - v$ not containing u. Since every u–w path in G contains v, we conclude that

$$d(u, w) > d(u, v) = \text{diam } G,$$

which is impossible. The desired result now follows. $\qquad\square$

Analogous to the cut-vertex is the concept of a bridge. A *bridge* of a graph G is an edge e such that $k(G - e) > k(G)$. If e is a bridge of G, then it is immediately evident that $k(G - e) = k(G) + 1$. Furthermore, if $e = uv$, then u is a cut-vertex of G if and only if $\deg u > 1$. Indeed, the complete graph K_2 is the only connected graph containing a bridge but no cut-vertices. Bridges are characterized in a manner similar to that of cut-vertices.

Theorem 2.3

An edge of a graph G is a bridge of G if and only if there exist vertices u and w such that e is on every u–w path of G.

For bridges, there is another useful characterization.

Theorem 2.4

An edge e of a graph G is a bridge of G if and only if e lies on no cycle of G.

Proof

Assume, without loss of generality, that G is connected. Let $e = uv$ be an edge of G, and suppose that e lies on a cycle C of G. Furthermore, let w_1 and w_2 be arbitrary distinct vertices of G. If e does not lie on a w_1–w_2 path P of G, then P is also a w_1–w_2 path of $G - e$. If, however, e lies on a w_1–w_2 path Q of G, then replacing e by the u–v path (or v–u path) on C not containing e produces a w_1–w_2 walk in $G - e$. By Theorem 1.6, there is a w_1–w_2 path in $G - e$. Thus, w_1 and w_2 are connected in $G - e$ and so e is not a bridge.

Conversely, suppose that e is not a bridge of G. Thus $G - e$ is connected. Hence there exists a u–v path P in $G - e$; however, P together with e produce a cycle in G containing e. □

A *cycle edge* is an edge that lies on a cycle. From Theorem 2.4, a cycle edge of a graph G is an edge that is not a bridge of G. A bridge incident with an end-vertex is called a *pendant edge*.

Many of the graphs we encounter do not contain cut-vertices; we discuss these next. A nontrivial connected graph with no cut-vertices is called a *nonseparable graph*. Nontrivial graphs with cut-vertices contain special subgraphs in which we are also interested. A *block* of a graph G is a maximal nonseparable subgraph of G. A block is necessarily an induced subgraph, and, moreover, the blocks of a graph partition its edge set. If a connected graph G contains a single block, then G is nonseparable. For this reason, a nonseparable graph is also referred to as a block itself. Every two blocks have at most one vertex in common, namely a cut-vertex. The graph of Figure 2.1 has five blocks B_i, $1 \le i \le 5$, as indicated. The vertices v_3, v_5 and v_8 are cut-vertices, while $v_3 \, v_5$ and $v_4 \, v_5$ are bridges; moreover, $v_4 \, v_5$ is a pendant edge.

Two useful criteria for a graph to be nonseparable are now presented.

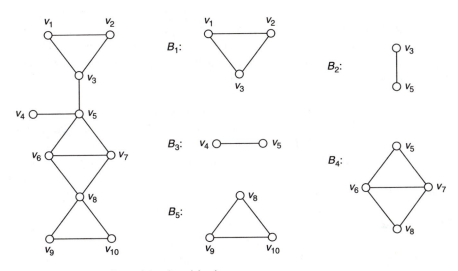

Figure 2.1 A graph and its five blocks.

Theorem 2.5

A graph G of order at least 3 is nonseparable if and only if every two vertices of G lie on a common cycle of G.

Proof

Let G be a graph such that each two of its vertices lie on a cycle. Thus G is connected. Suppose that G is not nonseparable. Hence G contains a cut-vertex v. By Theorem 2.1, there exist vertices u and w such that v is on every u–w path in G. Let C be a cycle of G containing u and w. The cycle C determines two distinct u–w paths, one of which does not contain v, contradicting the fact that every u–w path contains v. Therefore, G is nonseparable.

Conversely, let G be a nonseparable graph with at least three vertices. We show that every two vertices of G lie on a common cycle of G. Let u be an arbitrary vertex of G, and denote by U the set of all vertices that lie on a cycle containing u. We now show that $U = V = V(G)$. Assume that $U \neq V$; so there exists a vertex $v \in V - U$. Since G is nonseparable, it contains no cut-vertices, and furthermore, since the order of G is at least 3, the graph G contains no bridge. By Theorem 2.4, every edge of G lies on a cycle of G; hence, every vertex adjacent to u is an element of U.

Since G is connected, there exists a u–v path $u = u_0, u_1, u_2, \ldots, u_k = v$ in G. Let i be the smallest integer, $2 \leq i \leq k$, such that $u_i \notin U$; thus $u_{i-1} \in U$. Let C be a cycle containing u and u_{i-1}. Because u_{i-1} is not a cut-vertex of G, there exists a u_i–u path P: $u_i = v_0, v_1, v_2, \ldots, v_\ell = u$ not containing u_{i-1}. If the only vertex common to P and C is u, then a cycle containing u and u_i exists, which produces a contradiction. Hence P and C have a vertex in common different from u. Let j be the smallest integer, $1 \leq j \leq \ell$, such that v_j belongs to both P and C. A cycle containing u and u_i can now be constructed by beginning with the u_i–v_j subpath of P, proceeding along C from v_j to u and then to u_{i-1}, and finally taking the edge $u_{i-1}u_i$ back to u_i. Thus, a contradiction arises again, implying that the vertex v does not exist and that every two vertices lie on a cycle. $\qquad\square$

An *internal vertex* of a u–v path P is any vertex of P different from u or v. A collection $\{P_1, P_2, \ldots, P_k\}$ of paths is called *internally disjoint* if each internal vertex of P_i ($i = 1, 2, \ldots, k$) lies on no path P_j ($j \neq i$). In particular, two u–v paths are internally disjoint if they have no vertices in common, other than u and v. *Edge-disjoint* u–v paths have no edges in common. A second characterization of nonseparable graphs is now apparent.

Corollary 2.6

A graph G of order at least 3 is nonseparable if and only if there exist two internally disjoint u–v paths for every two distinct vertices u and v of G.

Theorem 2.5 suggests the following definitions: A block of order at least 3 is called a *cyclic block*, while the block K_2 is called the *acyclic block*.

We now state a theorem of which Theorem 2.2 is a corollary.

Theorem 2.7

Let G be a graph with one or more cut-vertices. Then among the blocks of G, there are at least two which contain exactly one cut-vertex of G.

In view of Theorem 2.7, we define an *end-block* of a graph G as a block containing exactly one cut-vertex of G. Hence every connected graph with at least one cut-vertex contains at least two end-blocks. In this context, another result that is often useful is presented. Its proof will become evident in the next chapter.

Theorem 2.8

Let G be a graph with at least one cut-vertex. Then G contains a cut-vertex v with the property that, with at most one exception, all blocks of G containing v are end-blocks.

Another interesting property of blocks of graphs was pointed out by Harary and Norman [HN2].

Theorem 2.9

The center of every connected graph G lies in a single block of G.

Proof

Suppose that G is a connected graph whose center $\mathrm{Cen}(G)$ does not lie within a single block of G. Then G has a cut-vertex v such that $G-v$ contains components G_1 and G_2, each of which contains vertices of $\mathrm{Cen}(G)$. Let u be a vertex such that $d(u, v) = e(v)$, and let P_1 be a v–u geodesic. At least one of G_1 and G_2, say G_2, contains no vertices of P_1. Let w be a vertex of $\mathrm{Cen}(G)$ belonging to G_2, and let P_2 be a w–v geodesic. The paths P_1 and P_2 together form a u–w path P_3, which is necessarily a u–w path of length $d(u, w)$. However, then $e(w) > e(v)$, which contradicts the fact that w is a central vertex. Thus $\mathrm{Cen}(G)$ lies in a single block of G. $\square$

EXERCISES 2.1

2.1 Prove that if v is a cut-vertex of a connected graph G, then v is *not* a cut-vertex of $\overline{G}$.

2.2 Prove Theorem 2.3.

2.3 Prove that every graph containing only even vertices is bridgeless.

2.4 Prove Corollary 2.6.

2.5 Let u and v be distinct vertices of a nonseparable graph G of order $n \geq 3$. If P is a given u–v path of G, does there always exist a u–v path Q such that P and Q are internally disjoint u–v paths?

2.6 Let G and H be graphs with $V(G) = \{v_1, v_2, \ldots, v_n\}$ and $V(H) = \{u_1, u_2, \ldots, u_n\}$, $n \geq 3$.

   (a) Vertices u_i and u_j are adjacent in H if and only if v_i and v_j belong to a common cycle in G. Characterize those graphs G for which H is complete.

   (b) Vertices u_i and u_j are adjacent in H if and only if $\deg_G v_i + \deg_G v_j$ is odd. Prove that H is bipartite.

2.7 An *element* of a graph G is a vertex or an edge of G. Prove that a graph G of order at least 3 is nonseparable if and only if every pair of elements of G lie on a common cycle of G.

2.8 Let G be a graph of order $n \geq 3$ with the property that $\deg u + \deg v \geq n$ for every pair u, v of nonadjacent vertices of G. Show that G is nonseparable.

2.2 THE AUTOMORPHISM GROUP OF A GRAPH

We have already described one way of studying the structure of graphs, namely, by determining the number and location of special vertices, edges and subgraphs. Another natural way of studying the structure of graphs is by investigating their symmetries. A common method of doing this is by means of groups.

An *automorphism* of a graph G is an isomorphism between G and itself. Thus an automorphism of G is a permutation of $V(G)$ that preserves adjacency (and nonadjacency). Of course, the identity function on $V(G)$ is an automorphism of G. The inverse of an automorphism of G is also an automorphism of G, as is the composition of two automorphisms of G. These observations lead us to the fact that the set of all automorphisms of a graph G form a group (under the operation of composition), called the *automorphism group* or simply the *group* of G and denoted by $\text{Aut}(G)$.

The automorphism group of the graph G_1 of Figure 2.2 is cyclic of order 2, which we write as $\text{Aut}(G_1) \cong Z_2$. In addition to the identity permutation, on $V(G_1)$, the group $\text{Aut}(G_1)$ contains the 'reflection' $\alpha = (uy)(vx)$, where α is expressed as 'permutation cycles'. The graph G_2 of Figure 2.2 has only the identity automorphism, so $\text{Aut}(G_2) \cong Z_1$.

Every permutation of the vertex set of K_n is an automorphism and so $\text{Aut}(K_n)$ is the symmetric group S_n of order $n!$ On the other hand, the automorphism group of

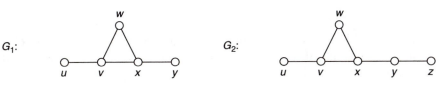

G_1: G_2:

Figure 2.2   Graphs with cyclic automorphism groups of orders 2 and 1.

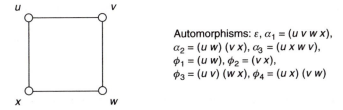

Automorphisms: ε, $\alpha_1 = (u\, v\, w\, x)$,
$\alpha_2 = (u\, w)(v\, x)$, $\alpha_3 = (u\, x\, w\, v)$,
$\phi_1 = (u\, w)$, $\phi_2 = (v\, x)$,
$\phi_3 = (u\, v)(w\, x)$, $\phi_4 = (u\, x)(v\, w)$

Figure 2.3   The 4-cycle and its automorphism group.

C_n, $n \geq 3$, is the dihedral group D_n of order $2n$, consisting of n rotations and n reflections. The 4-cycle C_4 and its automorphism group are illustrated in Figure 2.3.

Next we present a few basic facts concerning automorphism groups of graphs. We have already noted that every automorphism of a graph preserves both adjacency and nonadjacency. This observation leads to the following result.

Theorem 2.10

For every graph G, $\text{Aut}(G) \cong \text{Aut}(\overline{G})$.

We mentioned previously that $\text{Aut}(K_n) \cong S_n$ for every positive integer n. Certainly, if G is a graph of order n containing adjacent vertices as well as nonadjacent vertices, then $\text{Aut}(G)$ is isomorphic to a proper subgroup of the symmetric group S_n. Combining this observation with Theorem 2.10 and Lagrange's Theorem on the order of a subgroup of a finite group, we arrive at the following.

Theorem 2.11

The order $|\text{Aut}(G)|$ of the automorphism group of a graph G of order n is a divisor of $n!$ and equals $n!$ if and only if $G = K_n$ or $G = \overline{K}_n$.

Two *labelings* of a graph G of order n from the same set of n labels are considered *distinct* if they do not produce the same edge set. With the aid of the automorphism group of a graph G of order n, it is possible to determine the number of distinct labelings of G.

Theorem 2.12

The number of distinct labelings of a graph G of order n from a set of n labels is $n!/|\operatorname{Aut}(G)|$.

Proof

Let S be a set of n labels. Certainly, there exist $n!$ labelings of G using the elements of S without regard to which labelings are distinct. For a given labeling of G, each automorphism of G gives rise to an identical labeling of G; that is, each labeling of G from S determines $|\operatorname{Aut}(G)|$ identical labelings of G. Hence there are $n!/|\operatorname{Aut}(G)|$ distinct labelings of G. □

As an illustration of Theorem 2.12, consider the graph $G = P_3$ of Figure 2.4 and the set $S = \{1, 2, 3\}$. Since $\operatorname{Aut}(G) \cong Z_2$, the number of distinct labelings of G is $3!/2 = 3$. The three distinct labelings of G are shown in Figure 2.4.

If a relation R is defined on the vertex set of a graph G by uRv if $\phi u = v$ for some automorphism ϕ of G, then R is an equivalence relation. The equivalence classes of R are referred to as *orbits*, and two vertices belonging to the same orbit are called *similar vertices*. Of course, two similar vertices have the same degree. The automorphism group of the graph G of Figure 2.5 is cyclic of order 3 and G has four orbits.

In Chapter 1 we defined the total distance $td(u)$ of a vertex u in a connected graph G as the sum of the distances between u and all vertices of G. For the graph G of Figure 1.25 we computed the total distance of each vertex. This graph is shown again in Figure 2.6. The orbits of this graph are $\{r_1, r_2\}$, $\{s\}$, $\{t_1, t_2\}$, $\{u\}$, $\{v\}$, $\{w_1, w_2\}$, $\{x\}$, $\{y\}$ and $\{z\}$. Since r_1 and r_2 are similar vertices, they necessarily have

G: ⚬――⚬――⚬ $S = \{1, 2, 3\}$

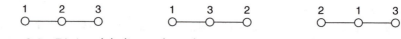

Figure 2.4 Distinct labelings of graphs.

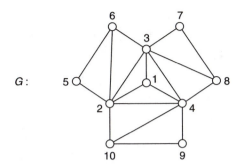

Figure 2.5 The orbits of a graph.

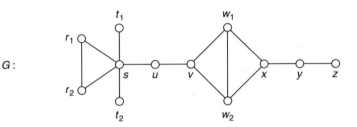

Figure 2.6 Similar vertices in a graph.

the same total distances. The same can be said of t_1 and t_2, as well as of w_1 and w_2. Of course, it is possible for vertices that are not similar to have the same total distances, as is the case with u and v. For the graph G of Figure 2.6, $\text{Aut}(G) \cong Z_2 \times Z_2 \times Z_2$.

A graph that contains a single orbit is called *vertex-transitive*. Thus a graph G is vertex-transitive if and only if for every two vertices u and v of G, there exists an automorphism ϕ of G such that $\phi u = v$. Necessarily, every vertex-transitive graph is regular. The graphs K_n ($n \geq 1$), C_n ($n \geq 3$) and $K_{r,r}$ ($r \geq 1$) are vertex-transitive. The Petersen graph (Figure 1.9) is vertex-transitive. Also the regular graphs $G_1 = C_5 \times K_2$ and $G_2 = K_{2,2,2}$ of Figure 2.7 are vertex-transitive, while the regular graphs G_3 and G_4 are not vertex-transitive.

Every digraph also has an automorphism group. An *automorphism* of a digraph D is an isomorphism of D with itself, that is, an automorphism of D is a permutation α on $V(D)$ such that (u, v) is an arc of D if and only if $(\alpha u, \alpha v)$ is an arc of D. The set of all automorphisms under composition forms a group, called the *automorphism group* of D and denoted by $\text{Aut}(D)$. While we have seen (the graph G of Figure 2.5) that it is not necessarily easy to find a graph G with $\text{Aut}(G) \cong Z_3$, this is actually quite easy for digraphs. For digraphs D_1 and D_2 of Figure 2.8, $\text{Aut}(D_1) \cong Z_3$ and $\text{Aut}(D_2) \cong Z_5$.

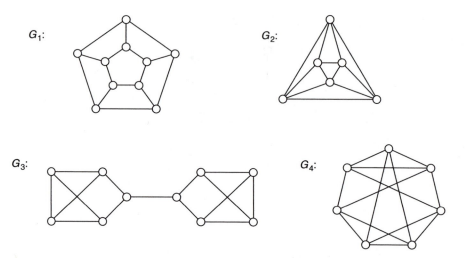

Figure 2.7 Vertex-transitive graphs and regular graphs that are not vertex-transitive.

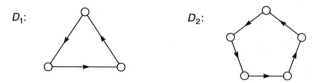

Figure 2.8 Digraphs with cyclic automorphism groups.

EXERCISES 2.2

2.9 For the graphs G_1 and G_2 below, describe the automorphisms of G_1 and of G_2 in terms of permutation cycles.

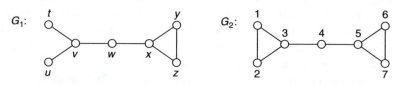

2.10 Describe the elements of Aut(C_5).
2.11 Determine the number of distinct labelings of $K_{r,r}$.
2.12 For which pairs k, n of positive integers with $k \le n$ does there exist a graph G of order n having k orbits?
2.13 For which pairs k, n of positive integers does there exist a graph G of order n and a vertex v of G such that there are exactly k vertices similar to v?

2.14 Describe the automorphism groups of the digraphs below.

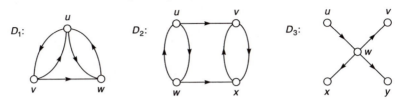

2.3 CAYLEY COLOR GRAPHS

We have seen that we can associate a group with every graph or digraph. We now consider the reverse question of associating a graph and a digraph with a given group. We consider only finite groups in this context. A nontrivial group Γ is said to be *generated* by nonidentity elements $h_1, h_2, \ldots, h_k$ (and these elements are called *generators*) of Γ if every element of Γ can be expressed as a (finite) product of generators. Every nontrivial finite group has a finite generating set (often several such sets) since the set of all nonidentity elements of the group is always a generating set for Γ.

Let Γ be a given nontrivial finite group with $\Delta = \{h_1, h_2, \ldots, h_k\}$ a generating set for Γ. We associate a digraph with Γ and Δ called the *Cayley color graph of* Γ *with respect to* Δ and denoted by $D_\Delta(\Gamma)$. The vertex set of $D_\Delta(\Gamma)$ is the set of group elements of Γ; therefore, $D_\Delta(\Gamma)$ has order $|\Gamma|$. Each generator h_i is now regarded as a color. For $g_1, g_2 \in \Gamma$, there exists an arc (g_1, g_2) colored h_i in $D_\Delta(\Gamma)$ if and only if $g_2 = g_1 h_i$. If h_i is a group element of order 2 (and is therefore self-inverse) and $g_2 = g_1 h_i$, then necessarily $g_1 = g_2 h_i$. When a Cayley color graph $D_\Delta(\Gamma)$ contains each of the arcs (g_1, g_2) and (g_2, g_1), both colored h_i, then it is customary to represent this symmetric pair of arcs by the single edge $g_1 g_2$.

We now illustrate the concepts just introduced. Let Γ denote the symmetric group S_3 of all permutations on the set $\{1, 2, 3\}$, and let $\Delta = \{a, b\}$, where $a = (123)$ and $b = (12)$. The Cayley color graph $D_\Delta(\Gamma)$ in this case is shown in Figure 2.9.

If the generating set Δ of a given nontrivial finite group Γ with n elements is chosen to be the set of all nonidentity group elements, then for every two vertices g_1, g_2 of $D_\Delta(\Gamma)$, both (g_1, g_2) and (g_2, g_1) are arcs (although not necessarily of the same color) and $D_\Delta(\Gamma)$ is the complete symmetric digraph K_n^* in this case.

Let Γ be a nontrivial finite group with generating set Δ. An element $\alpha \in \text{Aut}(D_\Delta(\Gamma))$ is said to be *color-preserving* if for every arc (g_1, g_2) of $D_\Delta(\Gamma)$, the arcs (g_1, g_2) and $(\alpha g_1, \alpha g_2)$ have the same color. For a given nontrivial finite group Γ with generating set Δ, it is a routine exercise to prove that the set of all color-preserving

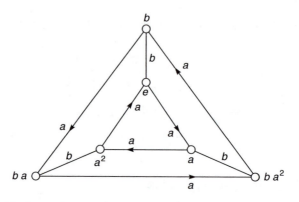

Figure 2.9 A Cayley color graph.

automorphisms of $D_\Delta(\Gamma)$ forms a subgroup of $\mathrm{Aut}(D_\Delta(\Gamma))$. A useful characterization of color-preserving automorphisms is given in the next result.

Theorem 2.13

Let Γ be a nontrivial finite group with generating set Δ and let α be a permutation of $V(D_\Delta(\Gamma))$. Then α is a color-preserving automorphism of $D_\Delta(\Gamma)$ if and only if

$$\alpha(gh) = (\alpha g)h$$

for every $g \in \Gamma$ and $h \in \Delta$.

The major significance of the group of color-preserving automorphisms of a Cayley color graph is contained in the following theorem.

Theorem 2.14

Let Γ be a nontrivial finite group with generating set Δ. Then the group of color-preserving automorphisms of $D_\Delta(\Gamma)$ is isomorphic to Γ.

Proof

Let $\Gamma = \{g_1, g_2, \ldots, g_n\}$. For $i = 1, 2, \ldots, n$, define $\alpha_i \colon V(D_\Delta(\Gamma)) \to V(D_\Delta(\Gamma))$ by $\alpha_i g_s = g_i g_s$ for $1 \leq s \leq n$. Since Γ is a group, the mapping α_i is one-to-one and onto. Let $h \in \Delta$. Then for each i, $1 \leq i \leq n$, and for each s, $1 \leq s \leq n$,

$$\alpha_i(g_s h) = g_i(g_s h) = (g_i g_s)h = (\alpha_i g_s)h.$$

Hence, by Theorem 2.13, α_i is a color-preserving automorphism of $D_\Delta(\Gamma)$.

Let α be an arbitrary color-preserving automorphism of $D_\Delta(\Gamma)$. Let g_1 be the identity of Γ. Suppose that $\alpha(g_1) = g_r$. Let $g_s \in \Gamma$. Sinse g_s can be expressed as a product of generators, $g_s = h_1 h_2 \ldots h_t$ where $h_j \in \Delta$, $1 \leq j \leq t$. Therefore,

$$\alpha_1(g_s) = \alpha(g_1 h_1 h_2 \ldots h_t) = \alpha(g_1 h_1 h_2 \ldots h_{t-1})h_t$$
$$= \alpha(g_1 h_1, \ldots, h_{t-2})h_{t-1}h_t = \ldots = \alpha(g_1)h_1 h_2 \ldots h_t = g_r g_s.$$

Thus $\alpha = \alpha_r$.

We now show that the mapping ϕ defined by $\phi g_i = \alpha_i$ is an isomorphism from Γ to the group of color-preserving automorphisms of $D_\Delta(\Gamma)$. The mapping ϕ is already one-to-one and onto. It remains to show that ϕ is operation-preserving, namely that $\phi(g_i g_j) = \phi(g_i)\phi(g_j)$ for g_i, $g_j \in \Gamma$. Let $g_i g_j = g_k$. Then $\phi(g_i g_j) = \phi(g_k) = \alpha_k$ and $\phi(g_i)\,\phi(g_j) = \alpha_i \alpha_j$. Now

$$\alpha_k(g_s) = g_k g_s = (g_i g_j)g_s = g_i(g_j g_s) = \alpha_i(g_j g_s) = \alpha_i(\alpha_j g_s) = (\alpha_i \alpha_j)g_s$$

and so $\alpha_k = \alpha_i \alpha_j$. $\qquad\qquad\square$

In 1936 the first book on graph theory was published. In this book the author König [K10, p. 5] proposed the problem of determining all finite groups Γ for which there exists a graph G such that $\mathrm{Aut}(G) \cong \Gamma$. The problem was solved in 1938 by Frucht [F4] who proved that *every* finite group has this property. We are now in a position to present a proof of this result.

If Γ is the trivial group, then $\mathrm{Aut}(G) \cong \Gamma$ for $G = K_1$. Therefore, let $\Gamma = \{g_1, g_2, \ldots, g_n\}$, $n \geq 2$, be a given finite group, and let $\Delta = \{h_1, h_2, \ldots, h_t\}$, $1 \leq t \leq n$, be a generating set for Γ. We first construct the Cayley color graph $D_\Delta(\Gamma)$ of Γ with respect to Δ; the Cayley color graph is actually a digraph, of course. By Theorem 2.14, the group of color-preserving automorphisms of $D_\Delta(\Gamma)$ is isomorphic to Γ. We now transform the digraph $D_\Delta(\Gamma)$ into a graph G by the following technique. Let (g_i, g_j) be an arc of $D_\Delta(\Gamma)$ colored h_k. Delete this arc and replace it by the 'graphical' path $g_i, u_{ij}, u'_{ij}, g_j$. At the vertex u_{ij} we construct a new path P_{ij} of length $2k - 1$ and at the vertex u'_{ij} a path P'_{ij} of length $2k$. This construction is now performed with every arc of $D_\Delta(\Gamma)$, and is illustrated in Figure 2.10 for $k = 1, 2$ and 3.

The addition of the paths P_{ij} and P'_{ij} in the formation of G is, in a sense, equivalent to the direction and the color of the arcs in the construction of $D_\Delta(\Gamma)$.

It now remains to observe that every color-preserving automorphism of $D_\Delta(\Gamma)$ induces an automorphism of G, and conversely. Thus we have a proof of Frucht's theorem.

Theorem 2.15

For every finite group Γ, there exists a graph G such that $\mathrm{Aut}(G) \cong \Gamma$.

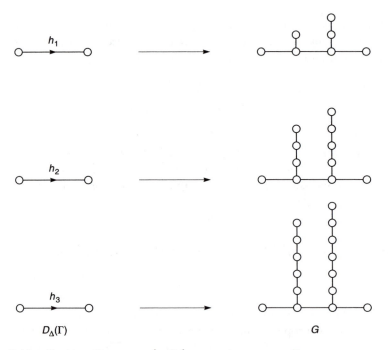

Figure 2.10 Constructing a graph G from a given group Γ.

The condition of having a given group prescribed is not a particularly stringent one for graphs. For example, Izbicki [I2] showed that for every finite group Γ and integer $r \geq 3$, there exists an r-regular graph G with $\mathrm{Aut}(G) \cong \Gamma$.

We have now seen that for every finite group Γ and generating set Δ there is an associated digraph, namely the Cayley color graph $D_\Delta(\Gamma)$. The underlying graph of a Cayley color graph $D_\Delta(\Gamma)$ is called a *Cayley graph* and is denoted by $G_\Delta(\Gamma)$. Thus a graph G is a Cayley graph if and only if there exists a finite group Γ and a generating set Δ for Γ such that, $G = G_\Delta(\Gamma)$, that is, the vertices of G are the elements of Γ and two vertices g_1 and g_2 of G are adjacent if and only if either $g_1 = g_2 h$ or $g_2 = g_1 h$ for some $h \in \Delta$.

As observed earlier, K_n^* is a Cayley color graph; consequently, every complete graph is a Cayley graph. Since $K_2 \times K_3$ is the underlying graph of the Cayley color graph of Figure 2.8, $K_2 \times K_3$ is a Cayley graph. Every Cayley graph is regular. Indeed every Cayley graph is vertex-transitive. The converse is not true, however. For example, the Petersen graph (Figure 1.9) is vertex-transitive but it is not a Cayley graph.

EXERCISES 2.3

2.15 Construct the Cayley color graph of the cyclic group of order 4 when the generating set Δ has (a) one element and (b) three elements.

2.16 Prove Theorem 2.13.

2.17 Determine the group of color-preserving automorphisms for the Cayley color graph $D_\Delta(\Gamma)$ below.

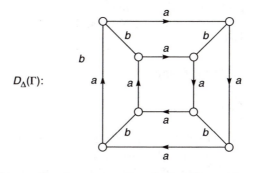

2.18 Determine the smallest integer $n > 1$ such that there exists a connected graph G of order n such that $|\text{Aut}(G)| = 1$.

2.19 Find a nonseparable graph G whose automorphism group is isomorphic to the cyclic group of order 4.

2.20 For a given finite group Γ, determine an infinite number of mutually nonisomorphic graphs whose groups are isomorphic to Γ.

2.21 Show that every n-cycle is a Cayley graph.

2.22 Show that the cube Q_3 is a Cayley graph.

2.4 THE RECONSTRUCTION PROBLEM

If ϕ is an automorphism of a nontrivial graph G and u is a vertex of G, then $G - u = G - \phi u$, that is, if u and v are similar vertices, then $G - u = G - v$. The converse of this statement is not true, however. Indeed, for the vertices u and v of the graph G of Figure 2.11, $G - u = G - v$, but u and v are *not* similar vertices of G.

These observations suggest the problem of determining how much structure of a graph G is discernible from its vertex-deleted subgraphs. This, in fact, brings us to a famous problem in graph theory.

Probably the foremost unsolved problem in graph theory is the *Reconstruction Problem*. This problem is due to P. J. Kelly and S. M. Ulam and its origin dates back to 1941. We discuss it briefly in this section.

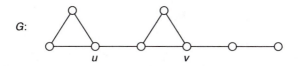

Figure 2.11 A graph with nonsimilar vertices whose vertex-deleted subgraphs are isomorphic.

A graph G with $V(G) = \{v_1, v_2, \ldots, v_n\}$, $n \geq 2$, is said to be *reconstructible* if for every graph H having $V(H) = \{u_1, u_2, \ldots, u_n\}$, $G - v_i = H - u_i$ for $i = 1, 2, \ldots, n$ implies that $G = H$. Hence, if G is a reconstructible graph, then the subgraphs $G - v$, $v \in V(G)$, determine G uniquely.

We now state a conjecture of Kelly and Ulam, the following formulation of which is due to F. Harary.

The Reconstruction Conjecture

Every graph of order at least 3 is reconstructible.

The Reconstruction Problem is thus to determine the truth or falsity of the Reconstruction Conjecture. The condition on the order in the Reconstruction Conjecture is needed since if $G_1 = K_2$, then G_1 is *not* reconstructible. If $G_2 = 2K_1$, then the subgraphs $G_1 - v$, where $v \in V(G_1)$, and the subgraphs $G_2 - v$, for $v \in V(G_2)$, are precisely the same. Thus G_1 is not uniquely determined by its subgraphs $G_1 - v$, $v \in V(G_1)$. By the same reasoning, $G_2 = 2K_1$ is also non-reconstructible. The Reconstruction Conjecture claims that K_2 and $2K_1$ are the only nonreconstructible graphs.

Before proceeding further, we note that there is a related problem that we shall not consider. Given graphs $G_1, G_2, \ldots, G_n$, does there exist a graph G with $V(G) = \{v_1, v_2, \ldots, v_n\}$ such that $G_i = G - v_i$ for $i = 1, 2, \ldots, n$? The answer to this question is not known in general. Although there is a similarity between this question and the Reconstruction Problem, the question is quite distinct from the problem in which we are interested.

If there is a counterexample to the Reconstruction Conjecture, then it must have order at least 12, for, with the aid of computers, McKay [M5] and Nijenhuis [N2] have shown that all graphs of order less than 12 (and greater than 2) are reconstructible. The graph G of Figure 2.12 is therefore reconstructible since its

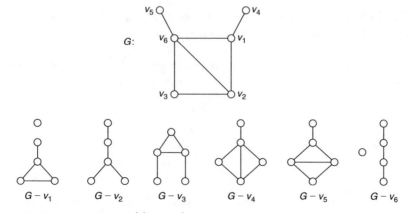

Figure 2.12 A reconstructible graph.

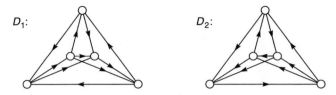

D_1: D_2:

Figure 2.13   Two nonreconstructible digraphs.

order is less than 12. Hence the graphs $G - v_i$ $(1 \le i \le 6)$ uniquely determine G. However, there exists a graph H with $V(H) = \{v_1, v_2, \ldots, v_6\}$ such that $G - v_i = H - v_i$ for $1 \le i \le 5$, but $G - v_6 \ne H - v_6$. Therefore, the graphs $G - v_i$ $(1 \le i \le 5)$ do *not* uniquely determine G. On the other hand, the graphs $G - v_i$ $(4 \le i \le 6)$ do uniquely determine G.

Digraphs are not reconstructible, however. The vertex-deleted subdigraphs of the digraphs D_1 and D_2 of Figure 2.13 are the same; yet $D_1 \ne D_2$. Indeed, for digraphs, there are infinitely many pairs of counterexamples (Stockmeyer [S7]).

There are several properties of a graph G that can be found by considering the subgraphs $G - v$, $v \in V(G)$. We begin with the most elementary of these.

Theorem 2.16

If G is a graph of order $n \ge 3$ and size m, then n and m as well as the degrees of the vertices of G are determined from the n subgraphs $G - v$, $v \in V(G)$.

Proof

It is trivial to determine the number n, which is necessarily one greater than the order of any subgraph $G - v$. Also, n is equal to the number of subgraphs $G - v$.

To determine m, label these subgraphs by G_i, $i = 1, 2, \ldots, n$. Let $V(G) = \{v_1, v_2, \ldots, v_n\}$ and suppose that $G_i = G - v_i$, where $v_i \in V(G)$. Let m_i denote the size of G_i. Consider an arbitrary edge e of G, say $e = v_j v_k$. Then e belongs to $n - 2$ of the subgraphs G_i, namely all except G_j and G_k. Hence, $\sum_{i=1}^{n} m_i$ counts each edge $n - 2$ times; that is, $\sum_{i=1}^{n} m_i = (n - 2)m$. Therefore,

$$m = \frac{\sum_{i=1}^{n} m_i}{n - 2}.$$

The degrees of the vertices of G can be determined by simply noting that $\deg v_i = m - m_i$, $i = 1, 2, \ldots, n$. $\square$

We illustrate Theorem 2.16 with the six subgraphs $G - v$ shown in Figure 2.14 of some unspecified graph G. From these subgraphs we determine n, m and $\deg v_i$

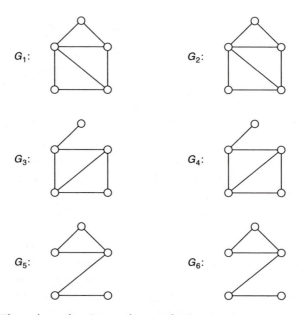

Figure 2.14 The subgraphs $G - v$ of a graph G.

for $i = 1, 2, \ldots, 6$. Clearly, $n = 6$. By calculating the integers m_i, $i = 1, 2, \ldots, 6$, we find that $m = 9$. Thus, deg $v_1 =$ deg $v_2 = 2$, deg $v_3 =$ deg $v_4 = 3$, and deg $v_5 =$ deg $v_6 = 4$.

We say that a graphical parameter or graphical property is *recognizable* if, for each graph G of order at least 3, it is possible to determine the value of the parameter for G or whether G has the property from the subgraphs $G - v$, $v \in V(G)$. Theorem 2.16 thus states that for a graph of order at least 3, the order, the size, and the degrees of its vertices are recognizable parameters. From Theorem 2.16, it also follows that the property of graph regularity is recognizable; indeed, the degree of regularity is a recognizable parameter. For regular graphs, much more can be said. $\quad\square$

Theorem 2.17

Every regular graph of order at least 3 is reconstructible.

Proof

As we have already mentioned, regularity and the degree of regularity are recognizable. Thus, without loss of generality, we may assume that G is an r-regular graph with $V(G) = \{v_1, v_2, \ldots, v_n\}$, $n \geq 3$. It remains to show that G is uniquely determined by its subgraphs $G - v_i$, $i = 1, 2, \ldots, n$. Consider $G - v_1$, say.

Add vertex v_1 to $G - v_1$ together with all those edges $v_1 v$ where $\deg_{G - v_1} v = r - 1$. This produces G.   □

If G has order $n \geq 3$, then it is discernible whether G is connected from the n subgraphs $G - v$, $v \in V(G)$.

Theorem 2.18

For graphs of order at least 3, connectedness is a recognizable property. In particular, if G is a graph with $V(G) = \{v_1, v_2, \ldots, v_n\}$, $n \geq 3$, then G is connected if and only if at least two of the subgraphs $G - v_i$ are connected.

Proof

Let G be a connected graph. By Theorem 2.2, G contains at least two vertices that are not cut-vertices, implying the result.

Conversely, assume that there exist vertices $v_1, v_2, \in V(G)$ such that both $G - v_1$ and $G - v_2$ are connected. Thus, in $G - v_1$ and also in G, vertex v_2 is connected to v_i, $i \geq 3$. Moreover, in $G - v_2$ (and thus in G), v_1 is connected to each v_i, $i \geq 3$. Hence every pair of vertices of G are connected and so G is connected.   □

Since connectedness is a recognizable property, it is possible to determine from the subgraphs $G - v$, $v \in V(G)$, whether a graph G of order at least 3 is disconnected. We now show that disconnected graphs are reconstructible. There have been several proofs of this fact. The proof given here is due to Manvel [M2].

Theorem 2.19

Disconnected graphs of order at least 3 are reconstructible.

Proof

We have already noted that disconnectedness in graphs of order at least 3 is a recognizable property. Thus, we assume without loss of generality that G is a disconnected graph with $V(G) = \{v_1, v_2, \ldots, v_n\}$, $n \geq 3$. Further, let $G_i = G - v_i$ for $i = 1, 2, \ldots, n$. From Theorem 2.16, the degrees of the vertices v_i, $i = 1, 2, , \ldots, n$, can be determined from the graphs $G - v_i$. Hence, if G contains an isolated vertex, then G is reconstructible. Assume then that G has no isolated vertices.

Since every component of G is nontrivial, it follows that $k(G_i) \geq k(G)$ for $i = 1, 2, \ldots, n$ and that $k(G_j) = k(G)$ for some integer j satisfying $1 \leq j \leq n$. Hence the number of components of G is $\min\{k(G_i) \mid i = 1, 2, \ldots, n\}$. Suppose that F is a component of G of maximum order. Necessarily, F is a component of maximum order among the components of the graphs G_i, that is, F is recognizable. Delete a vertex that is not a cut-vertex from F, obtaining F'.

Assume that there are r (≥ 1) components of G isomorphic to F. The number r is recognizable, as we shall see. Let

$$S = \{G_i \mid k(G_i) = k(G)\},$$

and let S' be the subset of S consisting of all those graphs G_i having a minimum number ℓ of components isomorphic to F. (Observe that if $r = 1$, then there exist graphs G_i in S containing no components isomorphic to F; that is, $\ell = 0$.) In general, then, $r = \ell + 1$. Next let S'' denote the set of those graphs G_i in S' having a maximum number of components isomorphic to F'.

Assume that $G_1, G_2, \ldots, G_t$ ($t \geq 1$) are the elements of S''. Each graph G_i in S'' has $k(G)$ components. Since each graph G_i ($1 \leq i \leq t$) has a minimum number of components isomorphic to F, each vertex v_i ($1 \leq i \leq t$) belongs to a component F_i of G isomorphic to F, where the components F_i of G ($1 \leq i \leq t$) are not necessarily distinct. Further, since each graph G_i ($1 \leq i \leq t$) has a maximum number of components isomorphic to F', it follows that $F_i - v_i = F'$ for each $i = 1, 2, \ldots, t$. Hence, every two of the graphs $G_1, G_2, \ldots, G_t$ are isomorphic, and G can be produced from G_1, say, by replacing a component of G_1 isomorphic to F' by a component isomorphic to F. □

It can be shown that (connected) graphs of order at least 3 whose complements are disconnected are reconstructible (Exercise 2.27). However, it remains to be shown that *all* connected graphs of order at least 3 are reconstructible.

EXERCISES 2.4

2.23 Reconstruct the graph G whose subgraphs $G - v$, $v \in V(G)$, are given in Figure 2.14.

2.24 Reconstruct the graph G whose subgraphs $G - v$, $v \in V(G)$, are given in the accompanying figure.

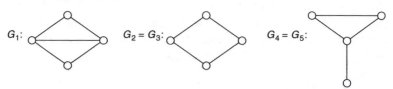

2.25 Let G be a graph with $V(G) = \{v_1, v_2, \ldots, v_7\}$ such that $G - v_i = K_{2,4}$ for $i = 1, 2, 3$ and $G - v_i = K_{3,3}$ for $i = 4, 5, 6, 7$. Show that G is reconstructible.

2.26 Show that the digraphs of Figure 2.13 are not isomorphic.

2.27 (a) Prove that if G is reconstructible, then $\overline{G}$ is reconstructible.

     (b) Prove that every graph of order n (≥ 3) whose complement is disconnected is reconstructible.

2.28 Prove that bipartiteness is a recognizable property.

2.29 Reconstruct the graph G whose subgraphs $G - v$, $v \in V(G)$, are given in the accompanying figure.

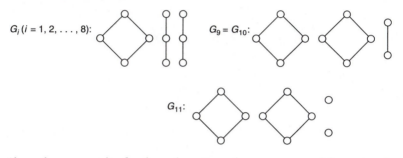

2.30 Show that no graph of order at least 3 can be reconstructed from exactly two of the subgraphs $G - v$, $v \in V(G)$.

3
Trees and connectivity

Connectedness of graphs is explored more fully in this chapter. Among the connected graphs, the simplest yet most important are the trees. Several results and concepts involving trees are presented. Connectivity and edge-connectivity are described and related theorems by Menger are stated.

3.1 ELEMENTARY PROPERTIES OF TREES

A *tree* is an acyclic connected graph; while a *forest* is an acyclic graph. Thus every component of a forest is a tree. There are several observations that can be made regarding trees. First, by Theorem 2.4, it follows that every edge of a tree T is a bridge; that is, every block of T is acyclic. Conversely, if every edge of a connected graph G is a bridge, then G is a tree.

There is one tree of each of the orders 1, 2 and 3; while there are two trees of order 4, three trees of order 5, and six trees of order 6. Figure 3.1 shows all trees of order 6.

If u and v are any two nonadjacent vertices of a tree T, then $T+uv$ contains precisely one cycle C. If, in turn, e is any edge of C in $T+uv$, then the graph $T+uv-e$ is once again a tree.

In a nontrivial tree T, it is immediate that the number of blocks to which a vertex v of T belongs equals $\deg v$. So $T-v$ is a forest with $\deg v$ components. If $\deg v = 1$, then $T-v$ is a tree. Thus, every vertex of T that is not an end-vertex belongs to at least two blocks and is necessarily a cut-vertex. The next result is a basic property of trees that is often useful when using mathematical induction to prove theorems dealing with trees.

Theorem 3.1

Every nontrivial tree has at least two end-vertices.

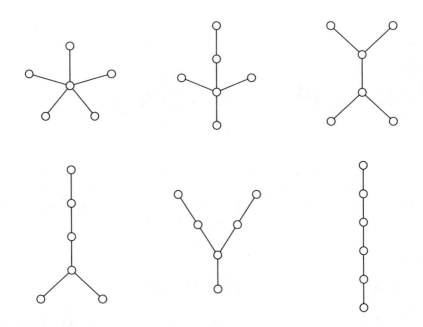

Figure 3.1   The trees of order 6.

Proof

Let T be a nontrivial tree and let P be a longest path of T. Suppose that P is a u–v path. Since P is a longest path, neither u nor v is adjacent to any vertex not on P. Certainly, u is adjacent to the vertex immediately following it on P, and v is adjacent to the vertex immediately preceding it on P; however, neither u nor v is adjacent to any other vertex of P since T contains no cycles. Therefore, $\deg u = \deg v = 1$.   □

There are a number of ways to characterize trees (for example, see Berge [B6, p. 152] and Harary [H6, p. 32]). Three of these are particularly useful.

Theorem 3.2

A graph G of order n and size m is a tree if and only if G is acyclic and $n = m + 1$.

Proof

Assume that G is a tree. Then G is acyclic by definition. We show that $m = n - 1$ by induction on n. For $n = 1$, the result (and graph) is trivial. Assume, then, that the equality $m = n - 1$ holds for all trees with n (≥ 1) vertices and m edges, and let T be a tree with $n+1$ vertices. Let v be an end-vertex of T. The graph $T' = T - v$ is a tree of order n, and so T' has $m = n - 1$ edges by the inductive hypothesis. Since T has one more edge than T', it follows that T has $m + 1 = n$ edges. Since $n + 1 = (m + 1) + 1$, the desired result follows.

Conversely, let G be an acyclic graph of order n and size $m = n - 1$. To show that G is a tree, we need only verify that G is connected. Denote by $G_1, G_2, \ldots, G_k$ the components of G, where G_i $(1 \leq i \leq k)$ has order n_i and size m_i. Since each graph G_i is a tree, $m_i = n_i - 1$. Hence,

$$n - 1 = m = \sum_{i=1}^{k} m_i = \sum_{i=1}^{k} (n_i - 1) = n - k$$

so that $k = 1$ and G is connected. ☐

The proof of Theorem 3.2 provides us with another result.

Corollary 3.3

A forest F of order n has n − k(F) edges.

A *spanning tree* of a graph G is a spanning subgraph of G that is a tree. *Every connected graph* G contains a spanning tree. If G is itself a tree, then this observation is trivial. If G is not a tree, then a spanning tree T of G can be obtained by removing cycle edges from G one at a time until only bridges remain. If G has order n and size m, then since T has size $n - 1$, it is necessary to delete a total of $m - (n - 1) = m - n + 1$ edges to produce T. This, of course, implies that $m \geq n - 1$, that is, every connected graph of order n has at least $n - 1$ edges.

Another characterization of trees is presented next.

Theorem 3.4

A graph G of order n and size m is a tree if and only if G is connected and $m = n - 1$.

Proof

Let T be a tree of order n and size m. By definition, T is connected and by Theorem 3.2, $m = n - 1$. For the converse, we assume that G is a connected graph of order n and size m with $m = n - 1$. It suffices to show that G is acyclic. If G contains a cycle C and e is an edge of C, then $G - e$ is a connected graph of order n having $n - 2$ edges, which is impossible as we have observed. Therefore, G is acyclic and is a tree. ☐

Hence, if G is a graph of order n and size m, then any two of the properties (i) G is connected, (ii) G is acyclic and (iii) $m = n - 1$ characterize G as a tree. There is yet another interesting characterization of trees that deserves mention.

Theorem 3.5

A graph G is a tree if and only if every two distinct vertices of G are connected by a unique path of G.

Proof

If G is a tree, then certainly every two vertices u and v are connected by at least one path. If u and v are connected by two different paths, then G contains a cycle, producing a contradiction.

On the other hand, suppose that G is a graph for which every two distinct vertices are connected by a unique path. This implies that G is connected. If G has a cycle C containing vertices u and v, then u and v are connected by at least two paths. This contradicts our hypothesis. Thus, G is acyclic and so G is a tree. $\square$

We now discuss some other properties of trees, particularly related to the degrees of its vertices. It is easy to determine whether a sequence of positive integers is the degree sequence of a tree.

Theorem 3.6

A sequence $d_1, d_2, \ldots, d_n$ of $n \geq 2$ positive integers is the degree sequence of a tree of order n if and only if $\sum_{i=1}^{n} d_i = 2n - 2$.

Proof

First, let T be a tree of order n and size m with degree sequence $d_1, d_2, \ldots, d_n$. Then $\sum_{i=1}^{n} d_i = 2m = 2(n - 1) = 2n - 2$. We verify the converse by induction on n. For $n = 2$, the only sequence of two positive integers with sum equal to 2 is 1,1, and this is the degree sequence of the tree K_2. Assume now that whenever a sequence of $n - 1 \geq 2$ positive integers has the sum $2(n - 2) = 2n - 4$, then it is the degree sequence of a tree of order $n - 1$.

Let $d_1, d_2, \ldots, d_n$ be a sequence of n positive integers with $\sum_{i=1}^{n} d_i = 2n - 2$. We show that this is a degree sequence of a tree. Suppose that $d_1 \geq d_2 \geq \cdots \geq d_n$. Since each term d_i is a positive integer and $\sum_{i=1}^{n} d_i = 2n - 2$, it follows that $2 \leq d_1 \leq n - 1$ and $d_{n-1} = d_n = 1$. Hence $d_1 - 1, d_2, d_3, \ldots, d_{n-1}$ is a sequence of $n - 1$ positive integers whose sum is $2n - 4$. By the inductive hypothesis, then, there exists a tree T' of order $n-1$ with $V(T') = \{v_1, v_2, \ldots, v_{n-1}\}$ such that $\deg v_1 = d_1 - 1$ and $\deg v_i = d_i$ for $2 \leq i \leq n-1$. Let T be the tree obtained from T' by adding a new vertex v_n and joining it to v_1. Thus $d_1, d_2, \ldots, d_n$ is a degree sequence of T. $\qquad\square$

There is a simple connection between the number of end-vertices in a tree and the number of vertices of the various degrees exceeding 2.

Theorem 3.7

Let T be a nontrivial tree with $\Delta(T) = k$, and let n_i be the number of vertices of degree i for $i = 1, 2, \ldots, k$. Then

$$n_1 = n_3 + 2n_4 + 3n_5 + \cdots + (k-2)n_k + 2.$$

Proof

Suppose that T has order n and size m. Then $\sum_{i=1}^{k} n_i = n$ and

$$\sum_{i=1}^{k} i n_i = 2m = 2n - 2 = 2 \sum_{i=1}^{k} n_i - 2$$

or

$$\sum_{i=1}^{k} (i-2)n_i + 2 = 0. \qquad (3.1)$$

Solving (3.1) for n_1 gives the desired result. $\qquad\square$

There are some classes of trees with which we should be familiar. Of course, the paths P_n and stars $K_{1,s}$ are trees. A tree T is a *double star* if it contains exactly two vertices that are not end-vertices; necessarily, these vertices are adjacent. A *caterpillar* is a tree T with the property that the removal of the end-vertices of T results in a path. This path is referred to as the *spine* of the caterpillar. If the spine is trivial, the caterpillar is a star; if the spine is K_2, then the caterpillar is a double star. Examples of a path, star, double star and caterpillar are shown in Figure 3.2.

Knowledge of the properties of trees is often useful when attempting to prove certain results about graphs in general. Because of the simplicity of the structure of trees, every graph ordinarily contains a number of trees as subgraphs. Of course, every tree of order n or less is a subgraph of K_n. A more general result is given next.

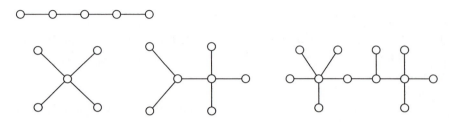

Figure 3.2 Paths, stars, double stars and caterpillars.

Theorem 3.8

Let T be a tree of order k, and let G be a graph with $\delta(G) \geq k - 1$. Then G contains a subgraph isomorphic to T.

Proof

We proceed by induction on k. The result is obvious for $k = 1$ since K_1 is a subgraph of every graph and for $k = 2$ since K_2 is a subgraph of every nonempty graph.

Assume for each tree T' of order $k - 1$, $k \geq 3$, and every graph H with $\delta(H) \geq k - 2$ that H contains a subgraph isomorphic to T'. Let T be a tree of order k and let G be a graph with $\delta(G) \geq k - 1$. We show that G contains a subgraph isomorphic to T.

Let v be an end-vertex of T and let u be the vertex of T adjacent to v. The graph $T - v$ is necessarily a tree of order $k - 1$. The graph G has $\delta(G) \geq k - 1 > k - 2$; thus by the inductive hypothesis, G contains a subgraph F isomorphic to $T - v$. Let u' denote the vertex of F that corresponds to u. Since $\deg_G u' \geq k - 1$ and $T - v$ has order $k - 1$, the vertex u' is adjacent to a vertex of G that does not belong to F. Therefore, G contains a subgraph isomorphic to T. ☐

Although no convenient closed formula is known for the number of nonisomorphic trees of order n, a formula does exist for the number of distinct labeled trees (whose vertices are labeled from a fixed set of cardinality n). For $n = 3$ and $n = 4$, the answer is sufficiently simple that we can actually draw all three distinct trees of order 3 whose vertices are labeled with elements of the set $\{1, 2, 3\}$ and all 16 distinct trees of order 4 whose vertices are labeled with elements of the set $\{1, 2, 3, 4\}$. These are shown in Figure 3.3.

In general, the number of distinct trees of order n whose vertices are labeled with the same set of n labels is n^{n-2}. This result is due to Cayley [C3]. There have been a number of proofs of Cayley's theorem. The one that we describe here is due to Prüfer [P5]. The proof consists of showing the existence of a one-to-one correspondence between the trees of order n whose vertices are labeled with

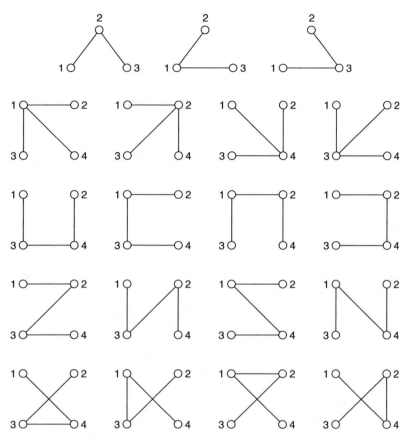

Figure 3.3 The labeled trees of orders 3 and 4.

elements of the set $\{1, 2, \ldots, n\}$ and the sequences (called *Prüfer sequences*) of length $n - 2$ whose entries are from the set $\{1, 2, \ldots, n\}$. Since the number of such sequences is n^{n-2}, once the one-to-one correspondence has been established, the proof is complete. □

Before stating Cayley's theorem formally, we illustrate the technique with an example. Consider the tree T of Figure 3.4 of order $n = 8$ whose vertices are labeled with elements of $\{1, 2, \ldots, 8\}$. The end-vertex of $T_0 = T$ having the smallest label is found, its neighbor is the first term of the Prüfer sequence for T, and this end-vertex is deleted, producing a new tree T_1. The neighbor of the end-vertex of T_1 having the smallest label is the second term of the Prüfer sequence for T; this end-vertex is deleted, producing the tree T_2. We continue this until we arrive at $T_{n-2} = K_2$. The resulting sequence of length $n - 2$ is the Prüfer sequence for T.

In the example just described, observe that every vertex v of T appears in its Prüfer sequence $\deg v - 1$ times. This is true in general. Therefore, no end-vertex

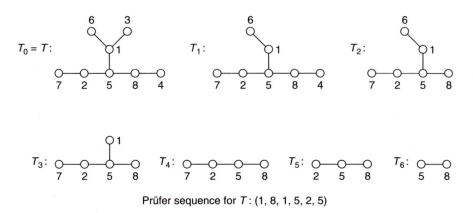

Prüfer sequence for T: (1, 8, 1, 5, 2, 5)

Figure 3.4 Determining the Prüfer sequence of a tree.

of T appears in the Prüfer sequence for T. So, if T is a tree of order n and size m, then the number of terms in its Prüfer sequence is

$$\sum_{v \in V(T)} (\deg v - 1) = 2m - n = 2(n-1) - n = n - 2.$$

We now consider the converse question, that is, if $(a_1, a_2, \ldots, a_{n-2})$ is a sequence of length $n-2$ such that each $a_i \in \{1, 2, \ldots, n\}$, then we construct a labeled tree T of order n such that the given sequence is the Prüfer sequence for T. Suppose that we are given the sequence (1, 8, 1, 5, 2, 5). We determine the smallest element of the set $\{1, 2, \ldots, 8\}$ not appearing in this sequence. This element is 3. In T, we join 3 to 1 (the first element of the sequence). The first term is deleted and the reduced sequence (8, 1, 5, 2, 5) is now considered. Also, the element 3 is deleted from the set $\{1, 2, \ldots, 8\}$, and the smallest element of this set not appearing in (8, 1, 5, 2, 5) is found, which is 4, and is joined to 8. This procedure is continued until two elements of the set remain. These two vertices are joined and T is constructed. This is illustrated in Figure 3.5.

Since a step in the second procedure is simply the reverse of a step in the first procedure, we have the desired one-to-one correspondence. We have now described a technique for proving of Cayley's theorem.

Theorem 3.9

There are n^{n-2} distinct labeled trees of order n.

Theorem 3.9 might be considered as a formula for determining the number of distinct spanning trees in the labeled graph K_n. We now consider the same question for graphs in general.

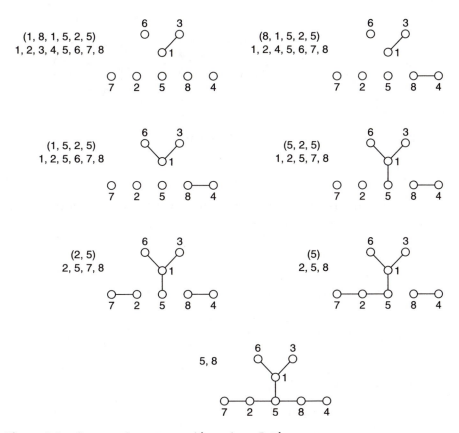

Figure 3.5 Constructing a tree with a given Prüfer sequence.

The next result, namely Theorem 3.10, is due to Kirchhoff [K4] and is often referred to as the *Matrix-Tree Theorem*. Let G be a graph with $V(G) = \{v_1, v_2, \ldots, v_n\}$. The *degree matrix* $D(G) = [d_{ij}]$ in the $n \times n$ matrix with $d_{ii} = \deg v_i$ and $d_{ij} = 0$ for $i \neq j$. We now state the Matrix-Tree Theorem.

Theorem 3.10

If G is a nontrivial labeled graph with adjacency matrix A and degree matrix D, then the number of distinct spanning trees of G is the value of any cofactor of the matrix $D - A$.

We illustrate the Matrix-Tree Theorem for the graph G of Figure 3.6, where the matrixes D and $D - A$ are also shown.

To calculate a cofactor of $D - A$, we delete the entries in row i and column j for some i and j with $1 \leq i, j \leq 4$ and multiply $(-1)^{i+j}$ and the determinant of the

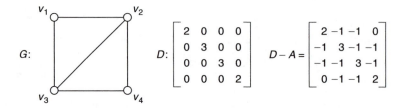

Figure 3.6 Illustrating the Matrix-Tree Theorem.

resulting submatrix. For example, the cofactor of the $(2, 3)$ entry in the matrix $D - A$ in Figure 3.6 is

$$(-1)^{2+3} \begin{vmatrix} 2 & -1 & 0 \\ -1 & -1 & -1 \\ 0 & -1 & 2 \end{vmatrix}$$

Expanding by the first row, we obtain

$$-\left(2 \begin{vmatrix} -1 & -1 \\ -1 & 2 \end{vmatrix} - (-1) \begin{vmatrix} -1 & -1 \\ 0 & 2 \end{vmatrix} + 0 \begin{vmatrix} -1 & -1 \\ 0 & -1 \end{vmatrix}\right)$$
$$= -(2(-3) + 1(-2) + 0) = 8$$

Consequently, there are eight distinct spanning trees of the graph G of Figure 3.6, all of which are shown in Figure 3.7.

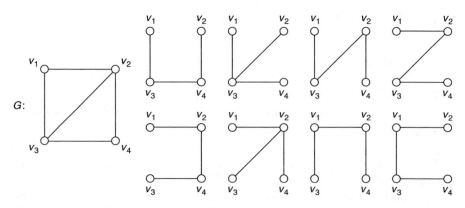

Figure 3.7 The distinct spanning trees of a graph.

EXERCISES 3.1

3.1 Draw all forests of order 6.

3.2 Prove that a graph G is a forest if and only if every induced subgraph of G contains a vertex of degree at most 1.

3.3 Characterize those graphs with the property that every connected subgraph is an induced subgraph.

3.4 A tree is called *central* if its center is K_1 and *bicentral* if its center is K_2. Show that every tree is central or bicentral.

3.5 Determine the Prüfer sequences of the trees in Figure 3.3.

3.6 (a) Which trees have constant Prüfer sequences?
   (b) Which trees have Prüfer sequences each term of which is one of two numbers?
   (c) Which trees have Prüfer sequences with distinct terms?

3.7 Determine the labeled tree having Prüfer sequence (4, 5, 7, 2, 1, 1, 6, 6, 7).

3.8 Let G be the labeled graph below.

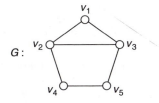

   (a) Use the Matrix-Tree Theorem to compute the number of distinct labeled spanning trees of G.
   (b) Draw all the distinct labeled spanning trees of G.

3.9 (a) Let $G = K_4$ with $V(G) = \{v_1, v_2, v_3, v_4\}$. Draw all spanning trees of G in which v_4 is an end-vertex.
   (b) Let v be a fixed vertex of $G = K_n$. Determine the number of spanning trees of G in which v is an end-vertex.

3.10 Prove Theorem 3.9 as a corollary to Theorem 3.10.

3.2 ARBORICITY AND VERTEX-ARBORICITY

One of the most common problems in graph theory deals with decomposition of a graph into various subgraphs possessing some prescribed property. There are ordinarily two problems of this type, one dealing with a decomposition of the vertex set and the other with a decomposition of the edge set. One such property that has been the subject of investigation is that of being acyclic, which we now consider.

For a graph G, it is always possible to partition $V(G)$ into subsets V_i, $1 \le i \le k$, such that each induced subgraph $\langle V_i \rangle$ is acyclic, that is, a forest. One way to accomplish this is by selecting each subset V_i so that $|V_i| \le 2$; however, the major problem is to partition $V(G)$ so that as few subsets as possible are involved. This suggests our next concept. The *vertex-arboricity* $a(G)$ of a graph G is the minimum number of subsets into which $V(G)$ can be partitioned so that each subset induces an acyclic subgraph. It is obvious that $a(G) = 1$ if and only if G is acyclic. For a few classes of graphs, the vertex-arboricity is easily determined. For example, $a(C_n) = 2$. If n is even, $a(K_n) = n/2$; while if n is odd, $a(K_n) = (n+1)/2$. So, in general, $a(K_n) = \lceil n/2 \rceil$. Also, $a(K_{r,s}) = 1$ if $r = 1$ or $s = 1$, and $a(K_{r,s}) = 2$ otherwise. No formula is known in general, however, for the vertex-arboricity of a graph although some bounds for this number exist. First, it is clear that for every graph G of order n,

$$a(G) \le \left\lceil \frac{n}{2} \right\rceil. \tag{3.2}$$

The bound (3.2) is not a particularly good one in general. In order to present a better bound, a new concept is introduced at this point.

A graph G is called *critical with respect to vertex-arboricity* if $a(G - v) < a(G)$ for all vertices v of G. This is the first of several occasions when a graph will be defined as critical with respect to a certain parameter. In order to avoid cumbersome phrases, we will simply use the term 'critical' when the parameter involved is clear by context. In particular, a graph G that is critical with respect to vertex-arboricity will be referred to in this section as a critical graph and, further, as a *k-critical* graph if $a(G) = k$. A k-critical graph necessarily has $k \ge 2$. The complete graph K_{2k-1} is k-critical while each cycle is 2-critical. It is not difficult to give examples of critical graphs; indeed, every graph G with $a(G) = k \ge 2$ contains an induced k-critical subgraph. In fact, every induced subgraph G' of G of minimum order with $a(G') = k$ is k-critical.

Before presenting the aforementioned bound for $a(G)$, we give another result.

Theorem 3.11

If G is a graph having $a(G) = k \ge 2$ that is critical with respect to vertex-arboricity, then $\delta(G) \ge 2k - 2$.

Proof

Let G be a k-critical graph, $k \ge 2$, and suppose that G contains a vertex v of degree $2k - 3$ or less. Since G is k-critical, $a(G - v) = k - 1$ and there is a partition

$V_1, V_2, \ldots, V_{k-1}$ of the vertex set of $G - v$ such that each subgraph $\langle V_i \rangle$ is acyclic. Because $\deg v \leq 2k - 3$, at least one of these subsets, say V_j, contains at most one vertex adjacent with v in G. The subgraph $\langle V_j \cup \{v\} \rangle$ is necessarily acyclic. Hence $V_1, V_2, \ldots, V_j \cup \{v\}, \ldots, V_{k-1}$ is a partition of the vertex set of G into $k - 1$ subsets, each of which induces an acyclic subgraph. This contradicts the fact that $a(G) = k$. □

We are now in a position to present the desired upper bound [CK1].

Theorem 3.12

For each graph G,

$$a(G) \leq 1 + \left\lfloor \frac{\max \delta(G')}{2} \right\rfloor,$$

where the maximum is taken over all induced subgraphs G' of G.

Proof

The result is obvious for acyclic graphs; thus, let G be a graph with $a(G) = k \geq 2$. Now let H be an induced k-critical subgraph of G. Since H itself is an induced subgraph of G,

$$\delta(H) \leq \max \delta(G'), \tag{3.3}$$

where the maximum is taken over all induced subgraphs G' of G. By Theorem 3.11, $\delta(H) \geq 2k - 2$, so by (3.3),

$$\max \delta(G') \geq 2k - 2 = 2a(G) - 2.$$

This inequality now produces the desired result. □

Since $\delta(G') \leq \Delta(G)$ for each induced subgraph G' of G, we note the following consequence of the preceding result.

Corollary 3.13

For every graph G,

$$a(G) \geq 1 + \left\lfloor \frac{\Delta(G)}{2} \right\rfloor.$$

We now turn to the second decomposition problem. The *edge-arboricity*, or simply the *arboricity* $a_1(G)$, of a nonempty graph G is the minimum number of subsets into which $E(G)$ can be partitioned so that each subset induces an acyclic subgraph. As with vertex-arboricity, a nonempty graph has arboricity 1 if and only if it is a forest. The following lower bound for the arboricity of a graph was established by Burr [B12].

Theorem 3.14

For each graph G,

$$a_1(G) \geq \left\lceil \frac{1 + \max \delta(G')}{2} \right\rceil,$$

where the maximum is taken over all induced subgraphs G' of G.

Proof

Let G_1 be an induced subgraph of G having order n_1 and size m_1. Thus G_1 can be decomposed into $a_1(G)$ or fewer acyclic subgraphs, each of which has size at most $n_1 - 1$. Then

$$\delta(G_1) \leq \frac{2m_1}{n_1} \leq \frac{2(n_1 - 1)a_1(G)}{n_1} < 2a_1(G).$$

Hence $\max \delta(G') < 2a_1(G)$, where the maximum is taken over all induced subgraphs G' of G. Therefore, $2a_1(G) \geq 1 + \max \delta(G')$ which yields the desired result. □

We now have a result that relates $a(G)$ and $a_1(G)$, also due to Burr [B12].

Corollary 3.15

For every graph G, $a(G) \leq a_1(G)$.

Proof

By Theorems 3.12 and 3.14, we have

$$2a(G) - 2 \leq \max \delta(G') \leq 2a_1(G) - 1,$$

where the maximum is taken over all induced subgraphs G' of G. So $a(G) \leq a_1(G) + \frac{1}{2}$. Since $a(G)$ and $a_1(G)$ are integers, the result follows. □

Unlike vertex-arboricity there is a formula for the arboricity of a graph, which was discovered by Nash-Williams [N1].

Theorem 3.16

For every nonempty graph G,

$$a_1(G) = \max \left\lceil \frac{|E(H)|}{|V(H)| - 1} \right\rceil,$$

where the maximum is taken over all nontrivial induced subgraphs H of G.

As a consequence of Theorem 3.16, it follows that

$$a_1(K_n) = \left\lceil \frac{n}{2} \right\rceil \quad \text{and} \quad a_1(K_{r,s}) = \left\lceil \frac{rs}{r+s-1} \right\rceil.$$

EXERCISES 3.2

3.11 Are there graphs G of order n other than K_n with $a(G) = \lceil n/2 \rceil$?

3.12 For all pairs k, n of positive integers with $k \leq \lceil n/2 \rceil$, give an example of a graph G of order n with $a(G) = k$.

3.13 What upper bounds for $a(K_{1,s})$ are given by Theorem 3.12 and Corollary 3.13?

3.14 Show that the formula given for $a_1(G)$ for a nonempty graph G in Theorem 3.17 is, in fact, equivalent to the statement: The arboricity of G is at most k if and only in $|E(H)| \leq k(|V(H)| - 1)$ for all induced subgraphs H of G.

3.15 Give an example of a graph G that has a nonempty induced subgraph H such that

$$\left\lceil \frac{|E(G)|}{|V(G)| - 1} \right\rceil < \left\lceil \frac{|E(H)|}{|V(H)| - 1} \right\rceil,$$

thereby proving that, in general, $a_1(G) \neq \lceil |E(G)|/(|V(G)|) - 1 \rceil$. Determine $a_1(G)$ for this graph.

3.3 CONNECTIVITY AND EDGE-CONNECTIVITY

A *vertex-cut* in a graph G is a set U of vertices of G such that $G - U$ is disconnected. Every graph that is not complete has a vertex-cut. Indeed, the set of all vertices distinct from two nonadjacent vertices is a vertex-cut. Of course, the removal of any proper subset of vertices from a complete graph leaves another complete graph. The *vertex-connectivity* or simply the *connectivity* $\kappa(G)$ of a graph G is the minimum cardinality of a vertex-cut of G if G is not complete, and $\kappa(G) = n - 1$ if $G = K_n$ for some positive integer n. Hence $\kappa(G)$ is the minimum number of vertices whose removal results in a disconnected or trivial graph. It is an immediate consequence of the definition that a nontrivial graph has connectivity 0 if and only if G is disconnected. Furthermore, a graph G has connectivity 1 if and only if $G = K_2$ or G is a connected graph with cut-vertices; $\kappa(G) \geq 2$ if and only if G is nonseparable of order 3 or more.

Connectivity has an edge analogue. An *edge-cut* in a graph G is a set X of edges of G such that $G - X$ is disconnected. An edge-cut X is minimal if no proper subset of X is also an edge-cut. If X is a minimal edge-cut of a connected graph G, then, necessarily, $G - X$ contains exactly two components. Every nontrivial graph has an edge-cut. The *edge-connectivity* $\kappa_1(G)$ of a graph G is the minimum cardinality of an edge-cut of G if G is nontrivial, and $\kappa_1(K_1) = 0$. So $\kappa_1(G)$ is the minimum number of edges whose removal from G results in a disconnected or trivial graph. Thus $\kappa_1(G) = 0$ if and only if G is disconnected or trivial; while $\kappa_1(G) = 1$ if and only if G is connected and contains a bridge.

We now describe a basic relationship between vertex-cuts and edge-cuts in graphs. The following result is due to Brualdi and Csima [BC4].

Theorem 3.17

Let G be a connected graph of order $n \geq 3$ that is not complete. For each edge-cut X of G, there is a vertex-cut U of G such that $|U| \leq |X|$.

Proof

Assume, without loss of generality, that X is a minimal edge-cut of G. Then $G - X$ is a disconnected graph containing exactly two components G_1 and G_2, where G_i has order n_i $(i = 1, 2)$. Thus $n_1 + n_2 = n$. We consider two cases.

Case 1. *Every vertex of G_1 is adjacent to every vertex of G_2.* Then $|X| = n_1 n_2$. Since $(n_1 - 1)(n_2 - 1) \geq 0$, it follows that $n_1 n_2 \geq n_1 + n_2 - 1 = n - 1$ and so $|X| \geq n - 1$. Since G is not complete, G contains two nonadjacent vertices u and v. Then $U = V(G) - \{u, v\}$ is a vertex-cut of cardinality $n - 2$ and $|U| < |X|$.

Case 2. *There are vertices u in G_1 and v in G_2 that are not adjacent in G.* For each edge e in X, we select a vertex for U in the following way. If u is incident with e, then choose the other vertex (in G_2) incident with e for U; otherwise, select for U the vertex that is incident with e and belongs to G_1. Now $|U| \leq |X|$.

Furthermore, $u, v \in V(G - U)$, but $G - U$ contains no $u - v$ path, so U is a vertex-cut. $\qquad \square$

We are now in a position to present a result due to Whitney [W5].

Theorem 3.18

For every graph G,

$$\kappa(G) \leq \kappa_1(G) \leq \delta(G).$$

Proof

If G is trivial or disconnected, then $\kappa(G) = \kappa_1(G) = 0$; so we can assume that G is a nontrivial connected graph. Let v be a vertex of G such that $\deg v = \delta(G)$. The removal of the $\delta(G)$ edges of G incident with v results in a graph G' in which v is isolated, so G' is either disconnected or trivial. Therefore, $\kappa_1(G) \leq \delta(G)$.

We now verify the other inequality. If $G = K_n$ for some positive integer n, then $\kappa(K_n) = \kappa_1(K_n) = n - 1$. Suppose next that G is not complete, and let X be an edge-cut such that $|X| = \kappa_1(G)$. By Theorem 3.17, there exists a vertex-cut U such that $|U| \leq |X|$. Thus

$$\kappa(G) \leq |U| \leq |X| = \kappa_1(G). \qquad \square$$

Figure 3.8 shows a graph G with $\kappa(G) = 2$, $\kappa_1(G) = 3$, and $\delta(G) = 4$. It can be shown (Exercise 3.31) that if a, b and c are positive integers with $a \leq b \leq c$, then there is a graph G with $\kappa(G) = a$, $\kappa_1(G) = b$ and $\delta(G) = c$.

A graph G is said to be *k-connected*, $k \geq 1$, if $\kappa(G) \geq k$. Thus G is 1-connected if and only if G is nontrivial and connected; while G is 2-connected if and only if G is nonseparable and has order at least 3. In general, G is k-connected if and only if the removal of fewer than k vertices results in neither a disconnected nor trivial graph.

It is often the case that knowing that a graph is k-connected for some specified positive integer k is as valuable as knowing the actual connectivity of the graph. As would be expected, the higher the degrees of the vertices of a graph, the more

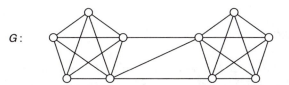

Figure 3.8 A graph G with $\kappa(G) = 2$, $\kappa_1(G) = 3$ and $\delta(G) = 4$.

likely it is that the graph has large connectivity. There are several sufficient conditions of this type. We present one of the simplest of these, originally presented in [CH1].

Theorem 3.19

Let G be a graph of order $n \geq 2$, and let k be an integer such that $1 \leq k \leq n-1$. If

$$\deg v \geq \left\lceil \frac{n+k-2}{2} \right\rceil$$

for every vertex v of G, then G is k-connected.

Proof

Suppose that the theorem is false. Then there is a graph G satisfying the hypothesis of the theorem such that G is not k-connected. Certainly, then, G is not a complete graph. Hence there exists a vertex-cut U of G such that $|U| = \ell \leq k-1$. The graph $G - U$ is therefore disconnected of order $n - \ell$.

Let G_1 be a component of $G - U$ of smallest order, say n_1. Thus $n_1 \leq \lfloor (n-\ell)/2 \rfloor$. Let v be a vertex of G_1. Necessarily, v is adjacent in G only to vertices of U or to other vertices of G_1. Hence

$$\deg v \leq \ell + (n_1 - 1) \leq \ell + \lfloor (n-\ell)/2 \rfloor - 1$$
$$= \lfloor (n+\ell-2)/2 \rfloor \leq \lfloor (n+k-3)/2 \rfloor$$

contrary to the hypothesis. □

A graph G is *k-edge-connected*, $k \geq 1$, if $\kappa_1(G) \geq k$. Equivalently, G is k-edge-connected if the removal of fewer than k edges from G results in neither a disconnected graph nor a trivial graph. The class of k-edge-connected graphs is characterized in the following simple but useful theorem.

Theorem 3.20

A nontrivial graph G is k-edge-connected if and only if there exists no nonempty proper subset W of V(G) such that the number of edges joining W and V(G) − W is less than k.

Proof

First, assume that there exists no nonempty proper subset W of $V(G)$ for which the number of edges joining W and $V(G) - W$ is less than k but that G is not k-edge-connected. Since G is nontrivial, this implies that there exist ℓ edges, $0 < \ell < k$, such that their deletion from G results in a disconnected graph H. Let H_1 be a component of H. Since the number of edges joining $V(H_1)$ and $V(G) - V(H_1)$ is at most ℓ, where $\ell < k$, this is a contradiction.

Conversely, suppose that G is a k-edge-connected graph. If there should exist a subset W of $V(G)$ such that j edges, $j < k$, join W and $V(G) - W$, then the deletion of these j edges produces a disconnectd graph–again a contradiction. The characterization now follows. □

According to Theorem 3.18, $\kappa_1(G) \le \delta(G)$ for every graph G. The following theorem of Plesník [P4] gives a sufficient condition for equality to hold.

Theorem 3.21

If G is a graph of diameter 2, then $\kappa_1(G) = \delta(G)$.

Proof

Let S be an edge-cut with $|S| - \kappa_1(G)$, and let H_1 and H_2 be the two components of $G - S$, where n_i is the order of H_i ($i = 1, 2$) with $n_1 \le n_2$.

Suppose first that H_1 contains a vertex u that is adjacent no vertex of H_2. Since diam $G = 2$, every vertex of H_2 is adjacent to some vertex of H_1. Hence we can conclude that either each vertex of H_1 is adjacent to some vertex of H_2 or each vertex of H_2 is adjacent to some vertex of H_1. Therefore,

$$\kappa_1(G) = |S| \ge \min\{n_1, n_2\} = n_1.$$

For $u \in V(H_1)$, let $d_i(u)$ denote the number of vertices of H_i ($i = 1, 2$) adjacent to u in G. Thus $\deg_G u = d_1(u) + d_2(u)$. In particular,

$$\delta(G) \le \deg u = d_1(u) + d_2(u) \le n_1 - 1 + d_2(u)$$
$$\le \kappa_1(G) - 1 + d_2(u) \le \delta(G) - 1 + d_2(u).$$

Hence $d_2(u) \ge 1$ for each $u \in V(G_1)$; that is, each vertex of H_1 is adjacent to some vertex of H_2. Let $V(H_1) = \{u_1, u_2, \ldots, u_k\}$, where then $k = n_1$. Now

$$\kappa_1(G) = |S| = \sum_{i=1}^{k} d_2(u_i) = \sum_{i=1}^{k-1} d_2(u_i) + d_2(u_k) \ge (k-1) + d_2(u_k)$$
$$= n_1 - 1 + d_2(u_k).$$

So, $n_1 - 1 + d_2(u_k) \le \kappa_1(G) \le \delta(G) \le n_1 - 1 + d_2(u_k)$. Thus, $\kappa_1(G) = \delta(G)$. □

Corollary 3.22

If G is a graph of order $n \geq 2$ such that

$$\deg u + \deg v \geq n - 1,$$

for each pair u, v of nonadjacent vertices G, then $\kappa_1(G) = \delta(G)$.

EXERCISES 3.3

3.16 Determine the connectivity and edge-connectivity of each complete k-partite graph.

3.17 Let $v_1, v_2, \ldots, v_k$ be k distinct vertices of a k-connected graph G. Let H be the graph formed from G by adding a new vertex of degree k that is adjacent to each of $v_1, v_2, \ldots, v_k$. Show $\kappa(H) = k$.

3.18 Let $H = G + K_1$, where G is k-connected. Prove that H is $(k+1)$-connected.

3.19 Let G be a graph with degree sequence $d_1, d_2, \ldots, d_n$, where $d_1 \geq d_2 \geq \ldots \geq d_n$. Define $H = G + K_1$. Determine $\kappa_1(H)$.

3.20 Show that every k-connected graph contains every tree of order $k + 1$ as a subgraph.

3.21 Let G be a noncomplete graph of order n and connectivity k such that $\deg v \geq (n + 2k - 2)/3$ for every vertex v of G. Show that if S is a vertex-cut of cardinality $\kappa(G)$, then $G - S$ has exactly two components.

3.22 For a graph G of order $n \geq 2$, define the *k-connectivity* $\kappa_k(G)$ of G, $2 \leq k \leq n$, as the minimum number of vertices whose removal from G results in a graph with at least k components or a graph of order less than k. (Note that $\kappa_2(G) = \kappa(G)$) A graph G is defined to be (ℓ, k)-connected if $\kappa_k(G) \geq \ell$. Let G be a graph of order n containing a set of at least k pairwise nonadjacent vertices. Show that if

$$\deg_G v \geq \left\lceil \frac{n + (k-1)(\ell - 2)}{k} \right\rceil$$

for every $v \in V(G)$, then G is (ℓ, k)-connected.

3.23 Prove Corollary 3.22.

3.24 Let G be a graph of diameter 2. Show that if S is a set of $\kappa_1(G)$ edges whose removal disconnects G, then at least one of the components of $G - S$ is isomorphic to K_1 or $K_{\delta(G)}$.

3.25 Let a, b, and c be positive integers with $a \leq b \leq c$. Prove that there exists a graph G with $\kappa(G) = a$, $\kappa_1(G) = b$ and $\delta(G) = c$.

3.26 Verify that Theorem 3.19 is best possible by showing that for each positive integer k, there exists a graph G of order n ($\geq k+1$) such that $\delta(G) = \lceil (n + k - 3)/2 \rceil$ and $\kappa(G) < k$.

3.27 Verify that Theorem 3.21 is best possible by finding an infinite class of graphs G of diameter 3 for which $\kappa_1(G) \neq \delta(G)$.

3.28 The *connection number* $c(G)$ of a connected graph G of order $n \geq 2$ is the smallest integer k with $2 \leq k \leq n$ such that *every* induced subgraph of order k in G is connected. State and prove a theorem that gives a relationship between $\kappa(G)$ and $c(G)$ for a graph G of order n.

3.4 MENGER'S THEOREM

A nontrivial graph G is connected (or, equivalently, 1-connected) if between every two distinct vertices of G there exists at least one path. This fact can be generalized in many ways, most of which involve, either directly or indirectly, a theorem due to Menger [M7]. In this section, we discuss the major ones of these, beginning with Dirac's proof [D8] of Menger's theorem itself.

A set S of vertices (or edges) of a graph G is said to *separate* two vertices u and v of G if the removal of the elements of S from G produces a disconnected graph in which u and v lie in different components. Certainly then, S is a vertex-cut (edge-cut) of G.

In the graph G of Figure 3.9, there is a set $S = \{w_1, w_2, w_3\}$ of vertices of G that separate the vertices u and v. No set with fewer than three vertices separates u and v. As is guaranteed by Menger's theorem [M7], stated next, there are three internally disjoint u–v paths in G.

Theorem 3.23

Let u and v be nonadjacent vertices in a graph G. Then the minimum number of vertices that separate u and v is equal to the maximum number of internally disjoint $u - v$ paths in G.

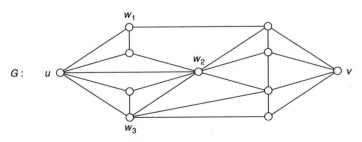

Figure 3.9 A graph illustrating Menger's theorem.

Proof

First, the result is true if u and v lie in different components of G or if u and v lie in different blocks of G; so we may assume that the graphs under consideration are connected and that u and v lie in the same block. If the minimum number of vertices that separate u and v is k (≥ 1), then the maximum number of internally disjoint u–v paths in G is at most k. Thus, if $k = 1$, the result is true (since we are assuming that G is connected). Denote by $S_k(u, v)$ the statement that the minimum number of vertices that separate u and v is k.

Suppose that the theorem is false. Then there exists a smallest positive integer t (≥ 2) such that $S_t(u, v)$ is true in some graph G but the maximum number of internally disjoint u–v paths is less than t. Among all such graphs G of smallest order, let H be one of minimum size.

We now establish three properties of the graph H.

1. *For every two adjacent vertices v_1 and v_2 of H, where neither v_1 nor v_2 is u or v, there exists a set U of $t-1$ vertices of H such that $U \cup \{v_i\}$, $i = 1, 2$, separates u and v.*

To see this, let $e = v_1 v_2$ and observe that $S_t(u, v)$ is false for $H - e$. However, we claim that $S_{t-1}(u,v)$ is true for $H - e$. If not, there exists a set W of vertices that separates u and v in $H - e$, where $|W| \leq t - 2$. Then W separates u and v in both $H - v_1$ and $H - v_2$; so $W \cup \{v_i\}$, $i = 1, 2$, separates u and v in H, which is impossible. Thus, as claimed, $S_{t-1}(u, v)$ is true in $H - e$. So there exists a set U that separates u and v in $H - e$, where $|U| = t - 1$. However, then $U \cup \{v_i\}$, $i = 1, 2$, separates u and v in H.

2. *For each vertex w ($\neq u,v$) in H, not both uw and vw are edges of H.*

Suppose that this is not true. Then $S_{t-1}(u, v)$ is true for $H - w$. However, then, $H - w$ contains $t - 1$ internally disjoint u–v paths. So H contains t internally disjoint u–v paths, which is a contradiction.

3. *If $W = \{w_1, w_2, \ldots, w_t\}$ is a set of vertices that separates u and v in H, then either $u w_i \in E(H)$ for all i ($1 \leq i \leq t$) or $v w_i \in E(H)$ for all i ($1 \leq i \leq t$).*

Suppose that statement 3 is false. Define H_u as the subgraph induced by the edges on all u–w_i paths in H that contain only one vertex of W. Define H_v similarly. Observe that $V(H_u) \cap V(H_v) = W$. Suppose that the above statement is not true. Then both H_u and H_v have order at least $t + 2$. Define H_u^* to consist of H_u, a new vertex v^* together with all edges $v^* w_i$. Also, define H_v^* to consist of H_v, a new vertex u^* together with all edges $u^* w_i$. Observe that H_u^* and H_v^* have smaller order than H. So $S_t(u, v^*)$ is true in H_u^* and $S_t(u^*, v)$ is true in H_v^*. Therefore, there exist t internally disjoint u–v^* paths in H_u^* and t internally disjoint u^*–v paths in H_v^*. These $2t$ paths produce t internally disjoint u–v paths in H, a contradiction.

Let P be a u–v path in H of length $d(u, v)$. By statement 2, $d(u, v) \geq 3$. Thus we may write P: $u, u_1, u_2, \ldots, v$ where $u_1, u_2 \neq v$. By statement 1, there exists a set U of $t-1$ vertices such that both $U \cup \{u_1\}$ and $U \cup \{u_2\}$ separate u and v. Since v is not adjacent to u_1, every vertex of $U \cup \{u_1\}$ is adjacent to u. Since u is not adjacent to u_2, every vertex of $U \cup \{u_2\}$ is adjacent to v. Thus $d(u, v) = 2$, which is impossible. $\qquad\square$

With the aid of Menger's theorem, it is now possible to present Whitney's characterization [W5] of k-connected graphs.

Theorem 3.24

A graph G of order $n \geq 2$ is k-connected $(1 \leq k \leq n-1)$ if and only if for each pair u, v of distinct vertices there are at least k internally disjoint u–v paths in G.

Proof

Let G be a k-connected graph and assume, to the contrary, that there are two vertices u and v such that the maximum number of internally disjoint u–v paths in G is ℓ, where $\ell < k$. If $uv \notin E(G)$ then, by Theorem 3.24, $\kappa(G) \leq \ell < k$, which is contrary to hypothesis. If $uv \in E(G)$, then the maximum number of internally disjoint u–v paths in $G - uv$ is $\ell - 1 < k - 1$; hence $\kappa(G - uv) < k - 1$. Therefore, there exists a set U of fewer than $k-1$ vertices such that $G - uv - U$ is a disconnected graph. Therefore, at least one of $G - (U \cup \{u\})$ and $G - (U \cup \{v\})$ is disconnected, implying that $\kappa(G) < k$. This also produces a contradiction.

Conversely, suppose that G is a nontrivial graph that is not k-connected but in which every pair of distinct vertices are connected by at least k internally disjoint paths. Certainly, G is not complete.

Since G is not k-connected, $\kappa(G) < k$. Let W be a set of $\kappa(G)$ vertices of G such that $G - W$ is disconnected, and let u and v be in different components of $G - W$. The vertices u and v are necessarily nonadjacent; however, by hypothesis, there are at least k internally disjoint u–v paths. By Theorem 3.23, u and v cannot be separated by fewer than k vertices, so a contradiction arises. $\qquad\square$

With the aid of Theorem 3.24, the following result can now be established rather easily.

Theorem 3.25

If G is a k-connected graph and $v, v_1, v_2, \ldots, v_k$ are $k+1$ distinct vertices of G, then there exist internally disjoint $v-v_i$ paths $(1 \le i \le k)$.

Proof

Construct a new graph H from G by adding a new vertex u to G together with the edges $uv_i, i = 1, 2, \ldots, k$. Since G is k-connected, H is k-connected (Exercise 3.23). By Theorem 3.24, there exist k internally disjoint $u-v$ paths in H. The restriction of these paths to G yields the desired internally disjoint $v-v_i$ paths. □

One of the interesting properties of 2-connected graphs is that every two vertices of such graphs lie on a common cycle. (This is a direct consequence of Theorem 2.5.) There is a generalization of this fact to k-connected graphs by Dirac [D5].

Theorem 3.26

Let G be a k-connected graph, $k \ge 2$. Then every k vertices of G lie on a common cycle of G.

Proof

For $k = 2$, the result follows from Theorem 2.5; hence, we assume that $k \ge 3$. Let W be a set of k vertices of G. Among all cycles of G, let C be a cycle containing a maximum number, say ℓ, of vertices of W. We observe that $\ell \ge 2$. We wish to show that $\ell = k$. Assume, to the contrary, that $\ell < k$. Let w be a vertex of W such that w does not lie on C.

Necessarily, C contains at least $\ell + 1$ vertices; for if this were not the case, then the vertices of C could be labeled so that $C: w_1, w_2, \ldots, w_\ell, w_1$, where $w_i \in W$ for $1 \le i \le \ell$. By Theorem 3.25, there exist internally disjoint $w-w_i$ paths Q_i, $1 \le i \le \ell$. Replacing the edge $w_1 w_2$ on C by the w_1-w_2 path determined by Q_1 and Q_2, we obtain a cycle containing at least $\ell + 1$ vertices of W, which is a contradiction. Therefore, C contains at least $\ell + 1$ vertices.

Thus we may assume that C contains vertices $w_1, w_2, \ldots, w_\ell, w_{\ell+1}$, such that $w_i \in W$ for $1 \le i \le \ell$ and $w_{\ell+1} \notin W$. Since $k \ge \ell + 1$, we may apply Theorem 3.25 again to conclude that there exist $\ell + 1$ internally disjoint $w-w_i$ paths P_i $(1 \le i \le \ell + 1)$. For each $i = 1, 2, \ldots, \ell + 1$, let v_i be the first vertex on P_i that belongs to C (possibly $v_i = w_i$) and let P_i' denote the $w-v_i$ subpath of P_i. Since C

contains exactly ℓ vertices of W, there are distinct integers s and t, $1 \leq s, t \leq \ell + 1$, such that one of the two v_s–v_t paths, say P, determined by C contains no interior vertex belonging to W. Replacing P by the v_s–v_t path determined by P'_s and P'_t, we obtain a cycle of G containing at least $\ell + 1$ vertices of W. This contradiction gives the desired result that $\ell = k$. □

Both Theorems 3.23 and 3.24 have 'edge' analogues; the analogue to Theorem 3.23 was proved in Elias, Feinstein and Shannon [EFS1] and Ford and Fulkerson [FF1]. It is not surprising that the edge analogue of Theorem 3.24 can be proved in a manner that bears a striking similarity to the proof of Theorem 3.23.

Theorem 3.27

The minimum number of edges that separate two vertices u and v in a graph G equals the maximum number of edge-disjoint u–v paths in G.

Proof

We actually prove a stronger result here by allowing G to be a multigraph. If u and v are vertices in different components of a multigraph G, then the theorem is true. Thus, without loss of generality, we may assume that the multigraphs under consideration are connected. If the minimum number of edges that separate u and v is k, where $k \geq 1$, then the maximum number of edge-disjoint u–v paths is at most k. Thus, the result is true if $k = 1$.

For vertices u and v of a multigraph G, let $S_k(u,v)$ denote the statement that the minimum number of edges that separate u and v is k. If the theorem is not true, then there exists a positive integer $\ell(\geq 2)$ for which there are multigraphs G containing vertices u and v such that $S_\ell(u, v)$ is true, but there is no set of ℓ edge-disjoint u–v paths. Among all such multigraphs G, let F denote one of minimum size.

If every u–v path of F has length 1 or 2, then since the minimum number of edges of F that separate u and v is ℓ, it follows that there are ℓ edge-disjoint u–v paths in F, producing a contradiction. Thus F contains at least one u–v path P of length 3 or more. Let e_1 be an edge of P incident with neither u nor v. Then for $F - e_1$, the statement $S_\ell(u, v)$ is false but $S_{\ell-1}(u, v)$ is true. This implies that e_1 belongs to a set of ℓ edges of F that separate u and v, say $\{e_1, e_2, \ldots, e_\ell\}$. We now subdivide each of the edges e_i, $1 \leq i \leq \ell$, that is, let $e_i = u_i v_i$, replace each e_i by a new vertex w_i, and add the 2ℓ edges $u_i w_i$ and $w_i v_i$. The vertices w_i are now identified, producing a new vertex w and a new multigraph H. The vertex w in H is a cut-vertex, and every u–v path of H contains w.

Denote by H_u the submultigraph of H determined by all u–w paths of H; the submultigraph H_v is defined similarly. Each of the multigraphs H_u and H_v has fewer edges than does F (since e_1 was chosen to be an edge of a u–v path in F incident with neither u nor v). Also, the minimum number of edges separating u and w in H_u is ℓ, and the minimum number of edges separating u and w in H_v is ℓ. Thus, the multigraph H_u satisfies $S_\ell(u, w)$ and the multigraph H_v satisfies $S_\ell(w, v)$. This implies that H_u contains a set of ℓ edge-disjoint u–w paths and H_v contains a set of ℓ edge-disjoint w–v paths. For each $i = 1, 2, \ldots, \ell$, a u–w path and w–v path can be paired off to produce a u–v path in H containing the two edges $u_i w$ and $w v_i$. These ℓ u–v paths of H are edge-disjoint. The process of subdividing the edges $e_i = u_i v_i$ of F and identifying the vertices w_i to obtain w can now be reversed to produce ℓ edge-disjoint u–v paths in F. This, however, produces a contradiction.

Since the theorem has been proved for multigraphs G, its validity follows in the case where G is a graph. □

With the aid of Theorem 3.27, it is now possible to present an edge analogue of Theorem 3.24.

Theorem 3.28

A graph G is k-edge-connected if and only if for every two distinct vertices u and v of G, there exist at least k edge-disjoint u–v paths in G.

EXERCISES 3.4

3.29 Prove that a graph G of order $n \geq k + 1 \geq 3$ is k-connected if and only if for each set S of k distinct vertices of G and for each two-vertex subset T of S, there is a cycle of G that contains the vertices of T and avoids the vertices of $S - T$.

3.30 Prove that a graph G of order $n \geq 2k$ is k-connected if and only if for every two disjoint sets V_1 and V_2 of k vertices each, there exist k disjoint paths connecting V_1 and V_2.

3.31 Let G be a k-connected graph and let v be a vertex of G. For a positive integer t, define G_t to be the graph obtained from G by adding t new vertices $u_1, u_2, \ldots, u_t$ and all edges of the form $u_i w$, where $1 \leq i \leq t$ and for which $vw \in E(G)$. Show that G_t is k-connected.

3.32 Show that if G is a k-connected graph with nonempty disjoint subsets S_1 and S_2 of $V(G)$, then there exist k internally disjoint paths $P_1, P_2, \ldots, P_k$ such that P_i is a u_i–v_i path, where $u_i \in S_1$ and $v_i \in S_2$, for $i = 1, 2, \ldots, k$, and $|S_1 \cap V(P_i)| = |S_2 \cap V(P_i)| = 1$.

3.33 Let G be a k-connected graph, $k \geq 3$, and let $v, v_1, v_2, \ldots, v_{k-1}$ be k vertices of G. Show that G has a cycle C containing all of $v_1, v_2, \ldots, v_{k-1}$ but not v and $k - 1$ internally disjoint v–u_i paths

P_i ($1 \leq i \leq k-1$) such that for each i, the vertex u_i is the only vertex of P_i on C.

3.34 Prove Theorem 3.28.

3.35 Prove or disprove: If G is a k-edge-connected graph and $v, v_1, v_2, \ldots, v_k$ are $k+1$ distinct vertices of G, then for $i = 1, 2, \ldots, k$, there exist v–v_i paths P_i such that each path P_i contains exactly one vertex of $\{v_1, v_2, \ldots, v_k\}$, namely v_i, and for $i \neq j$, P_i and P_j are edge-disjoint.

3.36 Prove or disprove: If G is a k-edge-connected graph with nonempty disjoint subsets S_1 and S_2 of $V(G)$, then there exist k edge-disjoint paths $P_1, P_2, \ldots, P_k$ such that P_i is a u_i–v_i path, where $u_i \in S_1$ and $v_i \in S_2$, for $i = 1, 2, \ldots, k$, and $|S_1 \cap V(P_i)| = |S_2 \cap V(P_i)| = 1$.

3.37 Show that $\kappa(Q_n) = \kappa_1(Q_n) = n$ for all positive integers n.

3.38 Assume that G is a *graph* in the proof of Theorem 3.27. Does the proof go through? If not, where does it fail?

3.39 Let G be a graph of order n with $\kappa(G) \geq 1$ Prove that

$$n \geq \kappa(G)[\text{diam } G - 1] + 2.$$

3.5 THE TOUGHNESS OF A GRAPH

The connectivity of a graph is one measure of how strongly connected the graph is, that is, the smaller the connectivity the more vulnerable a graph is. There are other measures of vulnerability. We will discuss the best known of these in the current section.

If G is a noncomplete graph and t is a nonnegative real number such that $t \leq |S|/k(G-S)$ for every vertex-cut S of G, then G is defined to be *t-tough*. If G is a t-tough graph and s is a nonnegative real number such that $s < t$, then G is also *s-tough*. The maximum real number t for which a graph G is a t-tough is called the *toughness* of G and is denoted by $t(G)$. Since complete graphs do not contain vertex-cuts, this definition does not apply to such graphs. Consequently, we define $t(K_n) = +\infty$ for every positive integer n. Certainly, the toughness of a noncomplete graph is a rational number. Also $t(G) = 0$ if and only if G is disconnected. Indeed, it follows that if G is a noncomplete graph, then

$$t(G) = \min |S|/k(G-S), \tag{3.6}$$

where the minimum is taken over all vertex-cuts S of G.

For the graph G of Figure 3.10, $S_1 = \{u, v, w\}$, $S_2 = \{w\}$, and $S_3 = \{u, v\}$ are three (of many) vertex-cuts. Observe that $|S_1|/k(G-S_1) = \frac{3}{8}$, $|S_2|/k(G-S_2) = \frac{1}{3}$ and $|S_3|/k(G-S_3) = \frac{2}{7}$. There is no vertex-cut S of G with $|S|/k(G-S) < \frac{2}{7}$; thus $t(G) = \frac{2}{7}$.

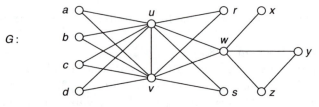

Figure 3.10 A graph of toughness $\frac{2}{7}$.

The toughness of a graph G is then a measure of how tightly the subgraphs of G are held together. Thus the smaller the toughness the more vulnerable the graph is. A 1-tough graph, for example, has the property that breaking the graph into k components (if this is possible) requires the removal of at least k vertices; while breaking a 2-tough graph into k components requires the removal of at least $2k$ vertices.

A parameter that plays an important role in the study of toughness is the independence number. Two vertices that are not adjacent in a graph G are said to be *independent*. A set S of vertices is independent if every two vertices of S are independent. The *vertex independence number* or simply the *independence number* $\beta(G)$ of a graph G is the maximum cardinality among the independent sets of vertices of G. For example, $\beta(K_{r,s}) = \max\{r, s\}$, $\beta(C_n) = \lfloor n/2 \rfloor$ and $\beta(K_n) = 1$.

The independence number is related to toughness in the sense that among all the vertex-cuts S of a noncomplete graph G, the maximum value of $k(G-S)$ is $\beta(G)$, so for every vertex-cut S of G, we have that $\kappa(G) \le |S|$ and $k(G-S) \le \beta(G)$. This leads us to bounds for the toughness of a graph, a result due to Chvátal [C5].

Theorem 3.29

For every noncomplete graph G,

$$\frac{\kappa(G)}{\beta(G)} \le t(G) \le \frac{\kappa(G)}{2}.$$

Proof

According to (3.6),

$$t(G) = \min \frac{|S|}{k(G-S)} \ge \frac{\kappa(G)}{\beta(G)}.$$

Let S' be a vertex-cut with $|S'| = \kappa(G)$ Thus $k(G - S') \geq 2$, so

$$t(G) = \min \frac{|S|}{k(G - S)} \leq \frac{|S'|}{k(G - S')} \leq \frac{\kappa(G)}{2}. \qquad \square$$

Let F be a graph. A graph G is *F-free* if G contains no induced subgraph isomorphic to F. Thus a K_2-free graph is empty. In this context, a graph of particular interest is $K_{1,3}$. A $K_{1,3}$-free graph is also referred to as a *claw-free graph*. The following result by Matthews and Sumner [MS1] provides a class of graphs for which the upper bound given in Theorem 3.29 becomes an equality.

Theorem 3.30

If G is a noncomplete claw-free graph, then $t(G) = \frac{1}{2}\kappa(G)$.

Proof

If G is disconnected, then $t(G) = \kappa(G) = 0$ and the result follows. So we assume that $\kappa(G) = r \geq 1$. Let S be a vertex-cut such that $t(G) = |S|/k(G - S)$. Suppose that $k(G - S) = k$ and that $G_1, G_2, \ldots, G_k$ are the components of $G - S$.

Let $u_i \in V(G_i)$ and $u_j \in V(G_j)$, where $i \neq j$. Since G is r-connected, it follows by Theorem 3.24 that G contains at least r internally disjoint u_i–u_j paths. Each of these paths contains a vertex of S. Consequently, there are at least r edges joining the vertices of S and the vertices of G_i for each i ($1 \leq i \leq k$) such that no two of these edges are incident with the same vertex of S.

Hence there is a set X containing at least kr edges between S and $G - S$ such that any two edges incident with a vertex of S are incident with vertices in distinct components of $G - S$. However, since G is claw-free, no vertex of S is joined to vertices in three components of $G - S$. Therefore,

$$kr = |X| \leq 2|S| = 2kt(G),$$

so $kr \leq 2kt(G)$. Thus $t(G) \geq r/2 = \frac{1}{2}\kappa(G)$. By Theorem 3.29, $t(G) = \frac{1}{2}\kappa(G)$. $\square$

In defining the toughness of a graph we were in some sense fine-tuning the idea of connectivity. For example, if a graph G is 2-connected, then the removal of one vertex from G does not result in a disconnected graph. The removal of two vertices, however, may not only result in a disconnected graph but in fact may result in a graph with many components. If, however, we know that G is 1-tough, then not only is G 2-connected but also the removal of any two vertices of G can result in a graph with at most two components.

EXERCISES 3.5

3.40 Determine the toughness of the complete tripartite graph $K_{r,r,r}$ $(r \geq 2)$.

3.41 Show that if H is a spanning subgraph of a noncomplete graph G, then $t(H) \leq t(G)$.

3.42 Show that if G is a noncomplete graph of order n, then $t(G) \leq (n - \beta(G))/\beta(G)$.

3.43 Show that the order of every noncomplete connected graph G is at least $\beta(G)(1 + t(G))$.

3.44 Show that for positive integers r and s with $r + s \geq 3$, $t(K_{r,s}) = \min\{r, s\}/\max\{r, s\}$.

3.45 Show that every 1-tough graph is 2-connected.

3.46 Show that for every nonnegative rational number r, there exists a graph G with $t(G) = r$.

3.47 Determine a formula for the toughness of a tree. Verify your formula.

4

Eulerian and hamiltonian graphs and digraphs

In this chapter we investigate graphs and digraphs with special circuits and cycles. In particular, we will be concerned with circuits containing every edge of a graph and cycles containing every vertex.

4.1 EULERIAN GRAPHS AND DIGRAPHS

In this section we discuss those trails and circuits in graphs and digraphs which are historically the most famous.

It is difficult to say just when and where graphs originated, but there is justification to the belief that graphs and graph theory may have begun in Switzerland in the early 18th century. In any case, it is evident that the great Swiss mathematician Leonhard Euler [E7] was thinking in graphical terms when he considered the problem of the seven Königsberg bridges.

Figure 4.1 shows a map of Königsberg as it appeared in the 18th century. The river Pregel was crossed by seven bridges, which connected two islands in the river with each other and with the opposite banks. We denote the land regions by the letters *A, B, C* and *D* (as Euler himself did). It is said that the townsfolk of Königsberg amused themselves by trying to devise a route that crossed each bridge just once. (For a more detailed account of the Königsberg Bridge Problem, see Biggs, Lloyd and Wilson [BLW1, p. 1].)

Euler proved that such a route over the bridges of Königsberg is impossible—a fact of which many of the people of Königsberg had already convinced themselves. However, it is probable that Euler's approach to the problem was considerably more sophisticated.

Euler observed that if such a route were possible it could be represented by a sequence of eight letters, each chosen from *A, B, C* and *D*. A term of the sequence would indicate the particular land area to which the route had progressed while two consecutive terms would denote a bridge traversed while proceeding from one land area to another. Since each bridge was to be crossed only once, the letters

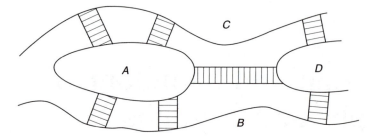

Figure 4.1 The bridges of Königsberg.

A and B would necessarily appear in the sequence as consecutive terms twice, as would A and C. Also, since five bridges lead to region A, Euler saw that the letter A must appear in the sequence a total of three times—twice to indicate an entrance to and exit from land area A, and once to denote either an entrance to A or exit from A. Similarly, each of the letters B, C and D must appear in the sequence twice. However, this implies that nine terms are needed in the sequence, an impossibility; hence the desired route through Königsberg is also impossible.

The Königsberg Bridge Problem has graphical overtones in many ways; indeed, even Euler's representation of a route through Königsberg is essentially that of a walk in a graph. If each land region of Königsberg is represented by a vertex and two vertices are joined by a number of edges equal to the number of bridges joining corresponding land areas, then the resulting structure (Figure 4.2) is a multigraph.

The Königsberg Bridge Problem is then equivalent to the problem of determining whether the multigraph of Figure 4.2 has a trail containing all of its edges.

The Königsberg Bridge Problem suggests the following two concepts. An *eulerian trail* of a graph G is an open trail of G containing all of the edges and vertices of G, while an *eulerian circuit* of G is a circuit containing all of the edges

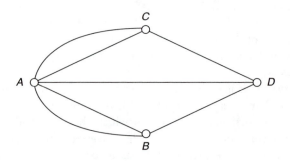

Figure 4.2 The multigraph of Königsberg.

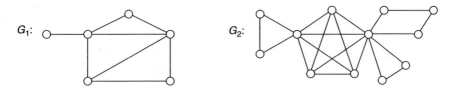

Figure 4.3 Graphs with eulerian trails and eulerian circuits.

and vertices of G. A graph possessing an eulerian circuit is called an *eulerian graph*. Necessarily, then, graphs containing eulerian trails and eulerian circuits are nontrivial connected graphs. The graph G_1 of Figure 4.3 contains an eulerian trail while G_2 is an eulerian graph.

Simple but useful characterizations of both eulerian graphs and graphs with eulerian trails exist; in fact, in each case the characterization was known to Euler [E7]. Complete proofs of these results were not given until 1873, however, in a paper by Hierholzer [H11].

Theorem 4.1

Let G be a nontrivial connected graph. Then G is eulerian if and only if every vertex of G is even.

Proof

Let G be an eulerian graph with eulerian circuit C, and let v be an arbitrary vertex of G. If v is not the initial vertex of C (and therefore not the final vertex either), then each time v is encountered on C, it is entered and left by means of distinct edges. Thus each occurrence of v in C represents a contribution of 2 to the degree of v so that v has even degree. If v is the initial vertex of C, then C begins and ends with v, each term representing a contribution of 1 to its degree while every other occurrence of v indicates an addition of 2 to its degree. This gives an even degree to v. In either case, v is even.

Conversely, let G be a nontrivial connected graph in which every vertex is even. We employ induction on the number m of edges of G. For $m = 3$, the smallest possible value, there is only one such graph, namely K_3, and this graph is eulerian. Assume then that all nontrivial connected graphs having only even vertices and with fewer than m edges, $m \geq 4$, are eulerian, and let G be such a graph with m edges.

Select some vertex u in G, and let W be a u–u circuit of G. Such a circuit exists in G since if W' is any u–v trail of G, where $u \neq v$, then necessarily an odd number of edges of G incident with v are present in W', implying that W' can be extended to a trail W'' containing more edges than that of W'. Hence W' can be extended to a u–u circuit W of G.

If the circuit W contains every edge of G, then W is an eulerian circuit of G and G is eulerian. Otherwise, there are edges of G that are not in W. Remove from G all those edges that are in W together with any resulting isolated vertices, obtaining the graph G'. Since each vertex of W is incident with an even number of edges of W, every vertex of G' is even. Every component of G' is a nontrivial graph with fewer than m edges and is eulerian by hypothesis. Since G is connected, every component of G' has a vertex that also belongs to W. Hence an eulerian circuit of G can be constructed by inserting an eulerian circuit of each component H' of G' at a vertex of H' also belonging to W. □

A graph G is defined to be an *even graph* (*odd graph*) if all of its vertices have even (odd) degree. Thus, by Theorem 4.1, the nontrivial connected even graphs are precisely the eulerian graphs. A characterization of graphs containing eulerian trails can now be presented.

Theorem 4.2

Let G be a nontrivial connected graph. Then G contains an eulerian trail if and only if G has exactly two odd vertices. Furthermore, the trail begins at one of these odd vertices and terminates at the other.

Proof

If G contains an eulerian u–v trail, then, as in the proof of Theorem 4.1, every vertex of G different from u and v is even. It is likewise immediate that each of u and v is odd.

Conversely, let G be a connected graph having exactly two odd vertices u and v. If G does not contain the edge $e = uv$, then the graph $G + e$ is eulerian. If the edge e is deleted from an eulerian circuit of $G + e$, then an eulerian trail of G results. In any case, however, a new vertex w can be added to G together with the edges uw and vw, producing a connected graph H in which every vertex is even. Therefore, H is eulerian and contains an eulerian circuit C. The circuit C necessarily contains uw and vw as consecutive edges so that the deletion of w from C yields an eulerian trail of G. Moreover, this trail begins at u or v and terminates at the other. □

If G is a connected graph with $2k$ odd vertices ($k \geq 1$), then the edge set of G can be partitioned into k subsets, each of which induces a trail connecting odd vertices (Exercise 4.2). However, even more can be said. This result was extended in [CPS1]. (See Exercise 4.3 for a special case of the next theorem.)

Theorem 4.3

If G is a connected graph with $2k$ odd vertices ($k \geq 1$), then $E(G)$ can be partitioned into subsets $E_1, E_2, \ldots, E_k$ so that for each i, $\langle E_i \rangle$ is a trail connecting odd vertices and such that at most one of these trails has odd length.

We note that analogues to Theorems 4.1 and 4.2 exist for multigraphs. It therefore follows that the multigraph of Figure 4.2 contains neither an eulerian trail nor an eulerian circuit. Eulerian graphs have several useful characterizations. The following result, due to Veblen [V2], characterizes eulerian graphs in terms of their cycle structure.

Theorem 4.4

A nontrivial connected graph G is eulerian if and only if $E(G)$ can be partitioned into subsets E_i, $1 \leq i \leq k$, where each subgraph $\langle E_i \rangle$ is a cycle.

Proof

Let G be an eulerian graph. We employ induction on the number m of edges of G. If $m = 3$, then $G = K_3$ and G has the desired property. Assume, then, that the edge set of every eulerian graph with fewer than m edges, $m \geq 4$, can be partitioned into subsets each of which induces a cycle, and let G be an eulerian graph with m edges. Since G is eulerian, G is an even graph and G has at least one cycle C. If $E(G) = E(C)$, then we have the desired (trivial) partition of $E(G)$. Otherwise, there are edges of G not in C. Remove the edges of C to obtain the graph G'. As in the proof of Theorem 4.1, every nontrivial component of G' is a nontrivial connected even graph and so, by Theorem 4.1, is an eulerian graph with fewer than m edges. Thus, by the inductive hypothesis, the edge set of each nontrivial component of G' can be partitioned into subsets, each inducing a cycle. These subsets, together with $E(C)$, give the desired partition of $E(G)$.

For the converse, suppose that the edge set of a nontrivial connected graph G can be partitioned into subsets E_i, $1 \leq i \leq k$, where each subgraph $\langle E_i \rangle$ is a cycle. This implies that G is a nontrivial connected even graph and so, by Theorem 4.1, G is eulerian. □

We next present a characterization of eulerian graphs involving parity and cycle structure. The necessity is due to Toida [T5] and the sufficiency to McKee [M6].

Theorem 4.5

A nontrivial connected graph G is eulerian if and only if every edge of G lies on an odd number of cycles.

Proof

First, let G be an eulerian graph and let $e = uv$ be an edge of G. Then $G - e$ is connected. Consider the set of all u–v trails in $G - e$ for which v appears only once, namely as the terminal vertex. There is an odd number of edges possible for the initial edge of such a trail. Once the initial edge has been chosen and the trail has then proceeded to the next vertex, say w, then again there is an odd number of choices for edges that are incident with w but different from uw. We continue this process until we arrive at vertex v. At each vertex different from v in such a trail, there is an odd number of edges available for a continuation of the trail. Hence there is an odd number of these trails.

Consider the set S of all u–v trails in $G - e$ that are not u–v paths and that contain v only once. Thus, each trail in S contains one or more circuits. Corresponding to each trail T in S and each circuit $C: v_1, v_2, \ldots, v_k, v_1$ in T, there is a unique trail T' in S obtained by traversing C in T in reverse order, that is, by replacing C in T by the circuit $C': v_1, v_k, v_{k-1}, \ldots, v_2, v_1$. Consequently, the trails in S occur in pairs, implying that there is an even number of u–v trails in S. Hence there is an odd number of u–v paths in $G - e$ which, in turn, implies that there is an odd number of cycles containing e.

For the converse, suppose that G is a nontrivial connected graph that is not eulerian. Then G contains a vertex v of odd degree. For each edge e incident with v, denote by $c(e)$ the number of cycles of G containing e. Since each such cycle contains two edges incident with v, it follows that $\Sigma c(e)$ equals twice the number of cycles containing v. Because there is an odd number of terms in this sum, some $c(e)$ is even. □

We now briefly consider the directed analogue of eulerian graphs. An *eulerian trail* of a digraph D is an open trail of D containing all of the arcs and vertices of D, and an *eulerian circuit* is a circuit containing every arc and vertex of D. A digraph that contains an eulerian circuit is called an *eulerian digraph*. The digraph D_1 of Figure 4.4 is eulerian while D_2 has an eulerian trail.

We now present a characterization of eulerian digraphs whose statement and proof are similar to Theorem 4.1.

Theorem 4.6

Let D be a nontrivial connected digraph. Then D is eulerian if and only if od $v =$ id v for every vertex v of D.

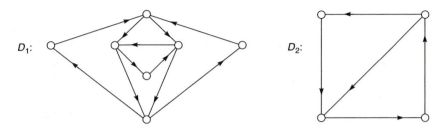

Figure 4.4 Digraphs with eulerian circuits and eulerian trails.

With the aid of Theorem 4.6, it is easy to give a characterization of digraphs containing eulerian trails.

Theorem 4.7

Let D be a nontrivial connected digraph. Then D has an eulerian trail if and only if D contains vertices u and v such that

$$\text{od } u = \text{id } u + 1 \quad \text{and} \quad \text{id } v = \text{od } v + 1$$

and od w = id w for all other vertices w of D. Furthermore, the trail begins at u and ends at v.

According to Theorem 4.4, the edge set of an eulerian graph G can be partitioned into subsets $E_1, E_2, \ldots, E_k$, where $G_i = \langle E_i \rangle$ is a cycle for $i = 1, 2, \ldots, k$. Thus the collection $C = \{G_1, G_1, G_2, G_2, \ldots, G_k, G_k\}$ of cycles of G (where multiplicities are allowed) has the property that every edge of G is in exactly two cycles of C. Such a collection of cycles is called a *cycle double cover* of G. Many graphs that are not eulerian also have cycle double covers. For example, the Petersen graph has a cycle double cover, consisting of five cycles. Szekeres [S8] in 1973 and Seymour [S3] in 1979 independently made the following conjecture.

The Cycle Double Cover Conjecture

Every 2-edge-connected graph has a cycle double cover.

A related conjecture, due to Bondy [B9], bounds the number of cycles in a cycle double cover.

The Small Cycle Double Cover Conjecture

Every 2-edge-connected graph of order $n \geq 3$ has a cycle double cover consisting of fewer than n cycles.

EXERCISES 4.1

4.1 In present-day Königsberg (Kaliningrad), there are two additional bridges, one between regions B and C, and one between regions B and D. Is it now possible to devise a route over all bridges of Königsberg without recrossing any of them?

4.2 Let G be a connected graph with $2k$ odd vertices, $k \geq 1$. Show that $E(G)$ can be partitioned into subsets E_i, $1 \leq i \leq k$, so that $\langle E_i \rangle$ is an open trail for each i. Then show that for $t < k$, $E(G)$ cannot be partitioned into subsets E_i, $1 \leq i \leq t$, so that $\langle E_i \rangle$ is an open trail for each i.

4.3 Show that every nontrivial connected graph G of even size having exactly four odd vertices contains two trails T_1 and T_2 of even length such that $\{E(T_1), E(T_2)\}$ is a partition of $E(G)$.

4.4 Prove Theorem 4.6.

4.5 Prove Theorem 4.7.

4.6 Prove that a nontrivial connected digraph D is eulerian if and only if $E(D)$ can be partitioned into subsets E_i, $1 \leq i \leq k$, where $\langle E_i \rangle$ is a cycle for each i.

4.7 Show that if D is a connected digraph such that $\sum_{v \in V(D)} |\text{od } v - \text{id } v| = 2t$, where $t \geq 1$, then $E(D)$ can be partitioned into subsets E_i, $1 \leq i \leq t$, so that $\langle E_i \rangle$ is an open trail for each i.

4.8 Show that the Petersen graph has a cycle double cover consisting of five cycles.

4.2 HAMILTONIAN GRAPHS AND DIGRAPHS

A graph G is defined to be *hamiltonian* if it has a cycle containing all the vertices of G. The name 'hamiltonian' is derived from Sir William Rowan Hamilton, the well-known Irish mathematician. Surprisingly, though, Hamilton's relationship with the graphs bearing his name is not strictly mathematical (see Biggs, Lloyd and Wilson [BLW1, p. 31]). In 1857, Hamilton introduced a game consisting of a solid regular dodecahedron made of wood, twenty pegs (one inserted at each corner of the dodecahedron), and a supply of string. Each corner was marked with an important city of the time. The aim of the game was to find a route along the edges of the dodecahedron that passed through each city exactly once and that ended at the city where the route began. In order for the player to recall which cities in a route had already been visited, the string was used to connect the appropriate pegs in the appropriate order. There is no indication that the game was ever successful.

The object of Hamilton's game may be described in graphical terms, namely, to determine whether the graph of the dodecahedron has a cycle containing each of its vertices (Figure 4.5). It is from this that we get the term 'hamiltonian'.

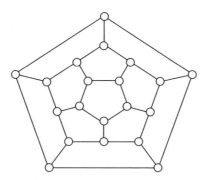

Figure 4.5 The graph of the dodecahedron.

It is interesting to note that in 1855 (two years before Hamilton introduced his game) the English mathematician Thomas P. Kirkman posed the following question in a paper submitted to the Royal Society: Given the graph of a poly-hedron, can one always find a circuit that passes through each vertex once and only once? Thus, Kirkman apparently introduced the general study of 'hamiltonian graphs' although Hamilton's game generated interest in the problem.

A cycle of a graph G containing every vertex of G is called a *hamiltonian cycle* of G; thus, a hamiltonian graph is one that possesses a hamiltonian cycle. Because of the similarity in the definitions of eulerian graphs and hamiltonian graphs, and because a particularly useful characterization of eulerian graphs exists, one might well expect the existence of an analogous criterion for hamiltonian graphs. However, such is not the case; indeed it must be considered one of the major unsolved problems of graph theory to develop an applicable characterization of hamiltonian graphs.

If G is a hamiltonian graph, then certainly G is connected, G contains a hamiltonian cycle and so G has no cut-vertices, and, of course, G has order at least 3; so G is 2-connected. Therefore, a necessary condition for G to be hamiltonian is that G be 2-connected; that is, if G is not 2-connected, then G is not hamiltonian. A less obvious necessary condition is presented next. Recall that the number of components of a graph G is denoted by $k(G)$.

Theorem 4.8

If G is a hamiltonian graph, then for every proper nonempty set S of vertices of G,

$$k(G - S) \leq |S|.$$

Proof

Let S be a proper nonempty subset of $V(G)$, and suppose that $k(G - S) = k \geq 1$, where $G_1, G_2, \ldots, G_k$ are the components of $G - S$. Let C be a hamiltonian cycle of G. When C leaves G_j $(1 \leq j \leq k)$, the next vertex of C belongs to S. Thus $k(G - S) = k \leq |S|$. □

Consequently, for every vertex-cut S of a hamiltonian graph G, it follows that $|S|/k(G - S) \geq 1$.

Corollary 4.9

Every hamiltonian graph is 1-tough.

We now turn our attention to sufficient conditions for a graph to be hamiltonian. Although every hamiltonian graph is 2-connected, the converse is not true. Indeed, no connectivity guarantees that a graph is hamiltonian. For example, let $k \geq 2$ be a positive integer and consider $G = K_{k,k+1}$, which is k-connected. Let S denote the partite set of cardinality k. Then $k(G \pm S) = k + 1 > |S|$, which implies by Theorem 4.8 that G is not hamiltonian.

We have also seen in Corollary 4.9 that every hamiltonian graph is 1-tough. The converse is not true here either.

There have been several sufficient conditions established for a graph to be hamiltonian. We consider some of these in this section. The following result is due to Ore [O1].

Theorem 4.10

If G is a graph of order $n \geq 3$ such that for all distinct nonadjacent vertices u and v,

$$\deg u + \deg v \geq n$$

then G is hamiltonian.

Proof

Assume that the theorem is not true. Hence there exists a maximal nonhamiltonian graph G of order $n \geq 3$ that satisfies the hypothesis of the theorem; that is, G is nonhamiltonian and for every two nonadjacent vertices w_1 and w_2 of G, the graph $G + w_1 w_2$ is hamiltonian. Since G is nonhamiltonian, G is not complete.

Let u and v be two nonadjacent vertices of G. Thus, $G + uv$ is hamiltonian, and, furthermore, every hamiltonian cycle of $G + uv$ contains the edge uv. Thus there is a u–v path $P: u = u_1, u_2, \ldots, u_n = v$ in G containing every vertex of G.

If $u_1, u_i \in E(G)$, $2 \leq i \leq n$, then $u_{i-1} u_n \notin E(G)$; for otherwise,

$$u_1, u_i, u_{i+1}, \ldots, u_n, u_{i-1}, u_{i-2}, \ldots, u_1$$

is a hamiltonian cycle of G. Hence for each vertex of $\{u_2, u_3, \dots, u_n\}$ adjacent to u_1 there is a vertex of $\{u_1, u_2, \dots, u_{n-1}\}$ not adjacent with u_n. Thus, $\deg u_n \le (n-1) - \deg u_1$ so that

$$\deg u + \deg v \le n - 1.$$

This produces a contradiction; so G is hamiltonian. □

If a graph G is hamiltonian, then certainly so is the graph $G + uv$, where u and v are distinct nonadjacent vertices of G. Conversely, suppose that G is a graph of order n with nonadjacent vertices u and v such that $G + uv$ is hamiltonian; furthermore, suppose that $\deg_G u + \deg_G v \ge n$. If G is not hamiltonian, then, as in the proof of Theorem 4.10, we arrive at the contradiction that $\deg_G u + \deg_G v \le n - 1$. Hence we have the following result, which was first observed by Bondy and Chvátal [BC3].

Theorem 4.11

Let u and v be distinct nonadjacent vertices of a graph G of order n such that $\deg u + \deg v \ge n$. Then $G + uv$ is hamiltonian if and only if G is hamiltonian.

Theorem 4.11 motivates our next definition. The *closure* of a graph G of order n, denoted by $C(G)$, is the graph obtained from G by recursively joining pairs of nonadjacent vertices whose degree sum is at least n (in the resulting graph produced at each stage) until no such pair remains. Figure 4.6 illustrates the closure function. That $C(G)$ is well-defined is established next.

Theorem 4.12

If G_1 and G_2 are two graphs obtained from a graph G of order n by recursively joining pairs of nonadjacent vertices whose degree sum is at least n, then $G_1 = G_2$.

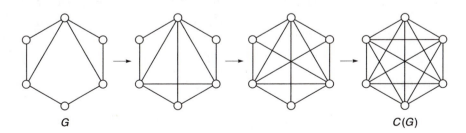

G $C(G)$

Figure 4.6 The closure function.

Proof

Let $e_1, e_2, \ldots, e_j$ and $f_1, f_2, \ldots, f_k$ be the sequences of edges added to G to obtain G_1 and G_2, respectively. It suffices to show that each e_i $(1 \leq i \leq j)$ is an edge of G_2 and that each f_i $(1 \leq i \leq k)$ is an edge of G_1. Assume, to the contrary, that this is not the case. Thus we may assume, without loss of generality, that for some t satisfying $0 \leq t \leq j - 1$, the edge $e_{t+1} = uv$ does not belong to G_2; furthermore, $e_i \in E(G_2)$ for $i \leq t$. Let G_3 be the graph obtained from G by adding the edges $e_1, e_2, \ldots, e_t$. It follows from the definition of G_1 that $\deg_{G_3} u + \deg_{G_3} v \geq n$. This is a contradiction, however, since u and v are nonadjacent vertices of G_2. Thus each edge e_i belongs to G_2 and each edge f_i belongs to G_1; that is, $G_1 = G_2$. □

Our next theorem is a simple consequence of the definition of closure and Theorem 4.11.

Theorem 4.13

A graph is hamiltonian if and only if its closure is hamiltonian.

Since each complete graph with at least three vertices is hamiltonian, we obtain Bondy and Chvátal's [BC3] sufficient condition for a graph to be hamiltonian.

Theorem 4.14

Let G be a graph with at least three vertices. If $C(G)$ is complete, then G is hamiltonian.

If a graph G satisfies the conditions of Theorem 4.10, then $C(G)$ is complete and so, by Theorem 4.14, G is hamiltonian. Thus, Ore's theorem is an immediate corollary of Theorem 4.14 (although chronologically it preceded the theorem of Bondy and Chvátal by several years). Perhaps surprisingly, many well-known sufficient conditions for a graph to be hamiltonian based on vertex degrees can be deduced from Theorem 4.14. Theorem 4.15, due to Chvátal [C4], is an example of one of the strongest of these.

Theorem 4.15

Let G be a graph of order $n \geq 3$, the degrees d_i of whose vertices satisfy $d_1 \leq d_2 \leq \cdots \leq d_n$. If there is no integer $k < n/2$ for which $d_k \leq k$ and $d_{n-k} \leq n - k - 1$, then G is hamiltonian.

Proof

We show that $C(G)$ is complete which, by Theorem 4.14, implies that G is hamiltonian. Assume, to the contrary, that $C(G)$ is not complete. Let u and w be nonadjacent vertices of $C(G)$ for which $\deg_{C(G)} u + \deg_{C(G)} w$ is as large as possible. Since u and w are nonadjacent vertices of $C(G)$, it follows that $\deg_{C(G)} u + \deg_{C(G)} w \le n - 1$. Assume, without loss of generality, that $\deg_{C(G)} u \le \deg_{C(G)} w$. Thus if $k = \deg_{C(G)} u$, we have that $k \le (n-1)/2 < n/2$ and $\deg_{C(G)} w \le n - 1 - k$. Let W denote the vertices other than w that are not adjacent to w in $C(G)$. Then $|W| = n - 1 - \deg_{C(G)} w \ge k$. Also, by the choice of u and w, every vertex $v \in W$ satisfies $\deg_G v \le \deg_{C(G)} v \le \deg_{C(G)} u = k$. Thus, G has at least k vertices of degree at most k and so $d_k \le k$. Similarly, let U denote the vertices other than u that are not adjacent to u in $C(G)$. Then $|U| = n - 1 - \deg_{C(G)} u = n - k - 1$. Every vertex $v \in U$ satisfies $\deg_G v \le \deg_{C(G)} v \le \deg_{C(G)} w \le n - 1 - k$, implying that $d_{n-k-1} \le n - k - 1$. However, $\deg_G u \le \deg_{C(G)} u \le \deg_{C(G)} w \le n - 1 - k$, so $d_{n-k} \le n - k - 1$. This, however, contradicts the hypothesis of the theorem. Thus, $C(G)$ is complete. $\qquad\square$

Perhaps the simplest sufficient condition for a graph to be hamiltonian is due to Dirac [D4]. It is a simple consequence of each of Theorems 4.10, 4.14 and 4.15.

Corollary 4.16

If G is a graph of order $n \ge 3$ such that $\deg v \ge n/2$ for every vertex v of G, then G is hamiltonian.

Each of the sufficient conditions presented so far requires that the graph under consideration contains some vertices of degree at least $n/2$. In the case of regular graphs, however, this situation can be improved. Jackson [J1] showed that every 2-connected r-regular graph of order at most $3r$ is hamiltonian. The Petersen graph, for example, shows that $3r$ cannot be replaced by $3r + 1$.

The sufficient conditions for hamiltonicity that we have presented so far all involve the degrees of the vertices of the graph. Our next result, however, involves the cardinality of independent sets of vertices and the connectivity of the graph. This result is due to Chvátal and Erdös [CE1]. The proof technique has become a standard tool.

Theorem 4.17

Let G be a graph with at least three vertices. If $\kappa(G) \ge \beta(G)$, then G is hamiltonian.

Proof

If $\beta(G) = 1$, then the result follows since G is complete. Hence we assume that $\beta(G) \geq 2$. Let $\kappa(G) = k$. Since $k \geq 2$, G contains at least one cycle. Among all cycles of G, let C be one of maximum length. By Theorem 3.27, there are at least k vertices on C. We show that C is a hamiltonian cycle of G. Assume, to the contrary, that there is a vertex w of G that does not lie on C. Since $|V(C)| \geq k$, we may apply Theorem 3.26 to conclude that there are k paths $P_1, P_2, \ldots, P_k$ having initial vertex w that are pairwise disjoint, except for w, and that share with C only their terminal vertices $v_1, v_2, \ldots, v_k$, respectively. If any two of the vertices v_i are consecutive on C, then there is a cycle containing more vertices than C has. For each $i = 1, 2, \ldots, k$, let u_i be the vertex following v_i in some fixed cyclic ordering of C. No vertex u_i is adjacent to w in G; for otherwise we could replace the edge $v_i u_i$ in C by the v_i–u_i path determined by the path P_i and the edge $u_i w$ to obtain a cycle having length at least $|V(C)| + 1$, which is impossible. Let $S = \{w, u_1, u_2, \ldots, u_k\}$. Since $|S| = k + 1 > \beta(G)$ and $wu_i \notin E(G)$ for $i = 1, 2, \ldots, k$, there are integers j and ℓ such that $u_j u_\ell \in E(G)$. Thus by deleting the edges $v_j u_j$ and $v_l u_\ell$ from C and adding the edge $v_j u_\ell$ together with the paths P_j and P_ℓ, we obtain a cycle of G that is longer than C. This produces a contradiction, so that C is a hamiltonian cycle of G. $\qquad\square$

As we have already remarked, obtaining an applicable characterization of hamiltonian graphs remains an unsolved problem in graph theory. In view of the lack of success in developing such a characterization, it is not surprising that special subclasses of hamiltonian graphs have been singled out for investigation as well as certain classes of nonhamiltonian graphs. We now discuss several types of 'highly hamiltonian' graphs and then briefly consider graphs that are 'nearly hamiltonian'.

A path in a graph G containing every vertex of G is called a *hamiltonian path*. A graph G is *hamiltonian-connected* if for every pair u, v of distinct vertices of G, there exists a hamiltonian u–v path. It is immediate that a hamiltonian-connected graph with at least three vertices is hamiltonian. We define the $(n+1)$-*closure* $C_{n+1}(G)$ of a graph G of order n to be the graph obtained from G by recursively joining pairs of nonadjacent vertices whose degree sum is at least $n+1$ until no such pair remains. We then have the following analogue to Theorem 4.14, also due to Bondy and Chvátal [BC3].

Theorem 4.18

Let G be a graph of order n. If $C_{n+1}(G)$ is complete, then G is hamiltonian-connected.

Proof

If $n = 1$, then the result is obvious; so we assume that $n \geq 2$. Let G be a graph of order n whose $(n + 1)$-closure is complete, and let u and v be any two vertices of G. We show that G contains a hamiltonian u–v path, which will then give us the desired result.

Define the graph H to consist of G together with a new vertex w and the edges uw and vw. Then H has order $n + 1$. Since $C_{n+1}(G)$ is complete, $\langle V(G) \rangle_{C(H)} = K_n$. Thus $\deg_{C(H)} x \geq n - 1$ for $x \in V(G)$; so

$$\deg_{C(H)} x + \deg_{C(H)} w \geq n + 1.$$

Therefore, $C(H) = K_{n+1}$ and by Theorem 4.14, H is hamiltonian. Any hamiltonian cycle of H necessarily contains the edges uw and vw, implying that G has a hamiltonian u–v path. $\qquad\square$

Two immediate corollaries now follow, the first of which is due to Ore [O3].

Corollary 4.19

If G is a graph of order n such that for all distinct nonadjacent vertices u and v,

$$\deg u + \deg v \geq n + 1,$$

then G is hamiltonian-connected.

Corollary 4.20

If G is a graph of order n such that $\deg v \geq (n + 1)/2$ for every vertex v of G, then G is hamiltonian-connected.

A number of other sufficient conditions for a graph to be hamiltonian-connected can be deduced from Theorem 4.18. One of these is the analogue to Theorem 4.15 (see [B7, p. 218]).

Corollary 4.21

Let G be a graph of order $n \geq 3$, the degrees d_i of whose vertices satisfy $d_1 \leq d_2 \leq \cdots \leq d_n$. If there is no integer $k \leq n/2$ for which $d_k \leq k$ and $d_{n-k} \leq n - k$, then G is hamiltonian-connected.

A connected graph G of order n is said to be *panconnected* if for each pair u, v of distinct vertices of G, there exists a u–v path of length ℓ for each ℓ satisfying $d(u, v) \leq \ell \leq n - 1$. If a graph is panconnected, then it is hamiltonian-connected; the next example indicates that these concepts are not equivalent.

For $k \geq 3$, let G_k be that graph such that $V(G_k) = \{v_1, v_2, \ldots, v_{2k}\}$ and

$$E(G_k) = \{v_i v_{i+1} \,|\, i = 1, 2, \ldots, 2k\} \cup \{v_i v_{i+3} \,|\, i = 2, 4, \ldots, 2k - 4\}$$
$$\cup \{v_1 v_3, v_{2k-2} v_{2k}\},$$

where all subscripts are expressed modulo $2k$. Although for each pair u, v of distinct vertices and for each integer ℓ satisfying $k \leq \ell \leq 2k - 1$, the graph G_k contains a u–v path of length ℓ, there is no v_1–v_{2k} path of length ℓ if $1 < \ell < k$. Since $d(v_1, v_{2k}) = 1$, it follows that G_k is not panconnected.

A sufficient condition for a graph G to be panconnected, due to Williamson [W6], can be given in terms of the minimum degree of G.

Theorem 4.22

If G is a graph of order $n \geq 4$ such that $\deg v \geq (n+2)/2$ for every vertex v of G, then G is panconnected.

Proof

If $n = 4$, then $G = K_4$ and the statement is true.

Suppose that the theorem is not true. Thus there exists a graph G of order $n \geq 5$ with $\delta(G) \geq (n+2)/2$ such that G is not panconnected; that is, there are vertices u and v of G and an integer ℓ with $d(u, v) < \ell < n - 1$ such that there is no u–v path of length ℓ. Let $G^* = G - \{u, v\}$. Then G^* has order $n^* = n - 2 \geq 3$ and $\delta(G^*) \geq (n+2)/2 - 2 = n^*/2$. Therefore by Corollary 4.16, the graph G^* contains a hamiltonian cycle $C: v_1, v_2, \ldots, v_{n^*}, v_1$.

If $uv_i \in E(G)$, $1 \leq i \leq n^*$, then $vv_{i+\ell-2} \notin E(G)$, where the subscripts are expressed modulo n^*; for otherwise,

$$u, v_i, v_{i+1}, \ldots, v_{i+\ell-2}, v$$

is a u–v path of length ℓ in G. Thus for each vertex of C that is adjacent with u in G, there is a vertex of C that is not adjacent with v in G. Since $\deg_G u \geq (n+2)/2$, we conclude that u is adjacent with at least $n/2$ vertices of C, so

$$\deg_G v \leq 1 + n^* - \frac{n}{2} = \frac{n}{2} - 1.$$

This, however, produces a contradiction. □

The result presented in Theorem 4.22 cannot be improved in general. Let $n = 2k + 1 \geq 7$, and consider the graph $K_{k,k+1}$ with partite sets V_1 and V_2, where $|V_1| = k$ on $|V_2| = k + 1$. The graph G is obtained from $K_{k,k+1}$ by constructing a path P_{k-1} on $k - 1$ vertices of V_2. Join the remaining two vertices x and y of V_2 by an edge. Then $\deg v \geq (n + 1)/2$ for every vertex v but G is not panconnected since G contains no x–y path of length 3.

A graph G of order $n \geq 3$ is called *pancyclic* if G contains a cycle of length ℓ for each ℓ satisfying $3 \leq \ell \leq n$. We say that G is *vertex-pancyclic* if for each vertex v of G and for every integer ℓ satisfying $3 \leq \ell \leq n$, there is a cycle of G of length ℓ that contains v. Certainly every pancyclic graph is hamiltonian, as is every vertex-pancyclic graph, although the converse is not true. The next theorem, due to Bondy [B8], gives a sufficient condition for a hamiltonian graph to be pancyclic.

Theorem 4.23

If G is a hamiltonian graph of order n and size $m \geq n^2/4$, then either G is pancyclic or n is even and $G = K_{n/2,n/2}$.

If the sum of the degrees of each pair of nonadjacent vertices of a graph G is at least n, where $n = |V(G)| \geq 3$, then by Theorem 4.10, G is hamiltonian. Our next result shows that the condition of Theorem 4.10 actually implies much more about the cycle structure of G.

Corollary 4.24

Let G be a graph of order $n \geq 3$ such that for all distinct nonadjacent vertices u and v,

$$\deg u + \deg v \geq n.$$

Then either G is pancyclic or n is even and $G = K_{n/2,n/2}$.

Proof

Let G have size m. According to Theorem 4.23, we need only show that $m \geq n^2/4$, since G is hamiltonian by Theorem 4.10. Let k be the minimum degree among the vertices of G. If $k \geq n/2$, then clearly $m \geq n^2/4$. Thus we may assume that $k < n/2$.

Let ℓ denote the number of vertices of G of degree k. These ℓ vertices induce a subgraph H that is complete; for if any two vertices of H were not adjacent, then there would exist two nonadjacent vertices the sum of whose degrees would be less than n. This implies that $\ell \leq k + 1$. However, $\ell \neq k + 1$; for otherwise, each vertex of H is adjacent only to vertices of H, which is impossible since G is connected.

Let u be a vertex of degree k. Since $\ell \leq k$, one of the k vertices adjacent to u has degree at least $k + 1$, while each of the other $k - 1$ vertices adjacent to u has degree at least k. If w ($\neq u$) is one of the $n - k - 1$ vertices of G that is not adjacent to u, then $\deg w + \deg u \geq n$, so that $\deg w \geq n - k$. Hence,

$$m = \frac{1}{2} \sum_{v \in V(G)} \deg v \geq \frac{1}{2}[(n - k - 1)(n - k) + k^2 + k + 1]$$

$$= \frac{1}{2}[2k^2 + (2 - 2n)k + (n^2 - n + 1)] \geq \frac{n^2 + 1}{4},$$

the last inequality holding since for $k \leq (n - 1)/2$, the expression $\frac{1}{2}[2k^2 + (2 - 2n)k + (n^2 - n + 1)]$ takes on its minimum value when $k = (n - 1)/2$. □

It is interesting to note that many other conditions that imply that a graph is hamiltonian have been shown to imply that either the graph is pancyclic or else belongs to a simple family of exceptional graphs.

We next briefly consider graphs that are, in certain senses, 'nearly hamiltonian'. Of course, if G is hamiltonian, then G has a hamiltonian path. Sufficient conditions for a graph to possess a hamiltonian path can be obtained from the sufficient conditions for a graph to be hamiltonian. For example, suppose that G is a graph of order $n \geq 2$ such that for all distinct nonadjacent vertices u and v, we have $\deg u + \deg v \geq n - 1$. Then the graph $G + K_1$ satisfies the hypothesis of Theorem 4.10 and so is hamiltonian. This, of course, implies that G contains a hamiltonian path.

We close this section with a brief discussion of hamiltonian digraphs. A digraph D is called *hamiltonian* if it contains a spanning cycle; such a cycle is called a *hamiltonian cycle*. As with hamiltonian graphs, no characterization of hamiltonian digraphs exists. Indeed, if anything, the situation for hamiltonian digraphs is even more complex than it is for hamiltonian graphs. While there are sufficient conditions for a digraph to be hamiltonian, they are analogues of the simpler sufficient conditions for hamiltonian graphs.

Recall that a digraph D is strong if for every two distinct vertices u and v of D, there is both a u–v (directed) path and a v–u path. Clearly, every hamiltonian digraph is strong (though not conversely). The following result of Meyniel [M8] gives a sufficient condition (much like that in Theorem 4.10 for graphs) for a digraph to be hamiltonian.

Theorem 4.25

If D is a nontrivial strong digraph of order n such that $\deg u + \deg v \geq 2n - 1$ for every pair u, v of distinct nonadjacent vertices, then D is hamiltonian.

Theorem 4.25 has a large number of consequences. We consider these now, beginning with a result originally discovered by Woodall [W8].

Corollary 4.26

If D is a nontrivial digraph of order n such that whenever u and v are distinct vertices and $(u, v) \notin E(D)$,

$$\text{od } u + \text{id } v \geq n, \tag{4.1}$$

then D is hamiltonian.

Proof

First we show that condition (4.1) implies that D is strong. Let u and v be any two vertices of D. We show that there is a u–v path in D. If $(u, v) \in E(D)$, then this is obvious. If $(u, v) \notin E(D)$, then by (4.1), there must exist a vertex w in D, with $w \neq u, v$, such that $(u, w), (w, v) \in E(D)$. However, then, u, w, v is the desired u–v path.

To complete the proof we apply Meyniel's theorem. Let u and v be any two nonadjacent vertices of D. Then by (4.1), od u + id $v \geq n$ and od v + id $u \geq n$ so that deg u + deg $v \geq 2n$. Thus, by Theorem 4.25, D is hamiltonian. □

The following well-known theorem is due to Ghouila-Houri [G3]. The proof is an immediate consequence of Theorem 4.25.

Corollary 4.27

If D is a strong digraph of order n such that $\deg v \geq n$ for every vertex v of D, then D is hamiltonian.

This result also has a rather immediate corollary.

Corollary 4.28

If D is a digraph of order n such that

$$\text{od } v \geq n/2 \quad \text{and} \quad \text{id } v \geq n/2$$

for every vertex v of D, then D is hamiltonian.

EXERCISES 4.2

4.9 Show that the graph of the dodecahedron is hamiltonian.

4.10 (a) Show that if G is a 2-connected graph containing a vertex that is adjacent to at least three vertices of degree 2, then G is not hamiltonian.

(b) The *subdivision graph* $S(G)$ of a graph G is that graph obtained from G by replacing each edge uv of G by a vertex w and edges uw and vw. Determine, with proof, all graphs G for which $S(G)$ is hamiltonian.

4.11 Give an example of a 1-tough graph that is not hamiltonian.

4.12 (a) Prove that $K_{r,2r,3r}$ is hamiltonian for every positive integer r.

(b) Prove that $K_{r,2r,3r+1}$ is hamiltonian for no positive integer r.

4.13 (a) Prove that if G and H are hamiltonian graphs, then $G \times H$ is hamiltonian.

(b) Prove that the n-cube Q_n, $n \geq 2$, is hamiltonian.

4.14 Give an independent proof of Corollary 4.16 without using any earlier theorems in the chapter.

4.15 Show that Theorem 4.10 is sharp, that is, show that for infinitely many $n \geq 3$ there are nonhamiltonian graphs G of order n such that $\deg u + \deg v \geq n - 1$ for all distinct nonadjacent vertices u and v.

4.16 Show that Theorem 4.17 is sharp, that is, show that for infinitely many integers $n \geq 3$ there are nonhamiltonian graphs G of order n such that $\kappa(G) \geq \beta(G) - 1$.

4.17 Prove that if G is a graph of order $n \geq 3$ and size $m \geq \binom{n-1}{2} + 2$, then G is hamiltonian.

4.18 Let G be a bipartite graph with partite sets U and W such that $|U| = |W| = k \geq 2$. Prove that if $\deg v > k/2$ for every vertex v of G, then G is hamiltonian.

4.19 Let G be a graph of order $n \geq 2$, the degrees d_i of whose vertices satisfy $d_1 \leq d_2 \leq \cdots \leq d_n$. Show that if there is no integer $k < (n+1)/2$ for which $d_k \leq k - 1$ and $d_{n-k+1} \leq n - k - 1$, then G has a hamiltonian path.

4.20 Show that if G is a graph with at least two vertices for which $\kappa(G) \geq \beta(G) - 1$, then G has a hamiltonian path.

4.21 Show that if G is a graph of order $n \geq 4$ and size $m \geq \binom{n-1}{2} + 3$, then G is hamiltonian-connected.

4.22 Prove that every hamiltonian-connected graph of order 4 or more is 3-connected.

4.23 Give an example of a graph G that is pancyclic but not panconnected.

4.24 Prove or disprove: If G is any graph of order $n \geq 4$ such that for all distinct nonadjacent vertices u and v,

$$\operatorname{od} v \geq n/2 \quad \text{and} \quad \operatorname{id} v \geq n/2$$

then G is panconnected.

4.25 Let G be a connected graph of order n and let k be an integer such that $2 \leq k \leq n - 1$. Show that if $\deg u + \deg v \geq k$ for every pair u, v of nonadjacent vertices of G, then G contains a path of length k.

4.3 LINE GRAPHS AND POWERS OF GRAPHS

The kth *power* G^k of a connected graph G, where $k \geq 1$, is that graph with $V(G^k) = V(G)$ for which $uv \in E(G^k)$ if and only if $1 \leq d_G(u, v) \leq k$. The graphs G^2 and G^3 are also referred to as the *square* and *cube*, respectively, of G. A graph with its square and cube are shown in Figure 4.7.

Since the kth power G^k ($k \geq 2$) of a connected graph G contains G as a subgraph (as a proper subgraph if G is not complete), it follows that G^k is hamiltonian if G is hamiltonian. Whether G is hamiltonian or not, for a connected graph G of order at least 3 and for a sufficiently large integer k, the graph G^k is hamiltonian since G^d is complete if G has diameter d. It is therefore natural to ask for the minimum k for which G^k is hamiltonian. Certainly, for connected graphs in general, $k = 2$ will not suffice since if G is the graph of Figure 4.7, then G^2 is not hamiltonian. The graph G^3 is hamiltonian, however. It is true, in fact, that the cube of every connected graph of order at least 3 is hamiltonian. Indeed, a stronger result exists, discovered independently by Karaganis [K1] and Sekanina [S1].

Theorem 4.29

If G is a connected graph, then G^3 is hamiltonian-connected.

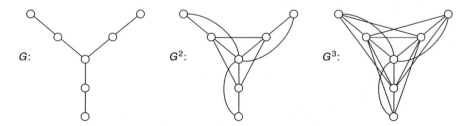

Figure 4.7 A graph whose square is not hamiltonian.

Proof

If H is a spanning tree of G and H^3 is hamiltonian-connected, then G^3 is hamiltonian-connected. Hence it is sufficient to prove that the cube of every tree is hamiltonian-connected. To show this we proceed by induction on n, the order of the tree. For small values of n the result is obvious.

Assume for all trees H of order less than n that H^3 is hamiltonian-connected, and let T be a tree of order n. Let u and v be any two vertices of T. We consider two cases.

Case 1. Suppose that u and v are adjacent in T. Let $e = uv$, and consider the forest $T - e$. This forest has two components, one tree T_u containing u and the other tree T_v containing v. By hypothesis, T_u^3 and T_v^3 are hamiltonian-connected. Let u_1 be any vertex of T_u adjacent to u, and let v_1 be any vertex of T_v adjacent to v. If T_u or T_v is trivial, we define $u_1 = u$ or $v_1 = v$, respectively. Note that u_1 and v_1 are adjacent in T^3 since $d_T(u_1, v_1) \leq 3$. Let P_u be a hamiltonian u–u_1 path (which may be trivial) of T_u^3, and let P_v be a hamiltonian v_1–v path of T_v^3. The path formed by beginning with P_u followed by the edge u_1v_1 and then the path P_v is a hamiltonian u–v path of T^3.

Case 2. Suppose that u and v are not adjacent in T. Since T is a tree, there exists a unique path between every two of its vertices. Let P be the unique u–v path of T, and let $f = uw$ be the edge of P incident with u. The graph $T - f$ consists of two trees, one tree T_u containing u and the other tree T_w containing w. By hypothesis, there exists a hamiltonian w–v path P_w in T_w^3. Let u_1 be a vertex of T_u adjacent to u, or let $u_1 = u$ if T_u is trivial, and let P_u be a hamiltonian u–u_1 path in T_u^3. Because $d_T(u_1, w) \leq 2$, the edge u_1w is present in T^3. Hence the path formed by starting with P_u followed by u_1w and then P_w is a hamiltonian u–v path of T^3. $\square$

It is, of course, an immediate corollary that for any connected graph G of order at least 3, G^3 is hamiltonian.

Although it is not true that the squares of all connected graphs of order at least 3 are hamiltonian, it was conjectured independently by C. Nash-Williams and M. D. Plummer that for 2-connected graphs this is the case. In 1974, Fleischner [F3] proved the conjecture to be correct.

Theorem 4.30

If G is a 2-connected graph, then G^2 is hamiltonian.

A variety of results strengthening (but employing) Fleischner's work have since been obtained; for example, it has been verified [CHJKN1] that the square of a 2-connected graph is hamiltonian-connected.

Theorem 4.31

If G is a 2-connected graph, then G^2 is hamiltonian-connected.

Proof

Since G is 2-connected, G has order at least 3. Let u and v be any two vertices of G. Let $G_1, G_2, \ldots, G_5$ be five distinct copies of G and let u_i and v_i $(i = 1, 2, \ldots, 5)$ be the vertices in G_i corresponding to u and v in G. Form a new graph F by adding to $G_1 \cup G_2 \cup \cdots \cup G_5$ two new vertices w_1 and w_2 and ten new edges $w_1 u_i$ and $w_2 v_i$ $(i = 1, 2, \ldots, 5)$. Clearly, neither w_1 nor w_2 is a cut-vertex of F. Furthermore, since each graph G_i is 2-connected and contains two vertices adjacent to vertices in $V(F) - V(G_i)$, no vertex of G_i is a cut-vertex of F. Hence F is 2-connected.

By Theorem 4.30, F^2 has a hamiltonian cycle C, which, of course, contains w_1 and w_2. At least one of the graphs G_i, say G_1, contains no vertex adjacent to w_1 or w_2 on C. Since u_1 and v_1 are the only vertices of G_1 adjacent on C to vertices not in G_1, it follows that C has a u_1–v_1 path containing all vertices of G_1. Thus $G_1^{\,2}$ has a hamiltonian u_1–v_1 path, which implies that G^2 contains a hamiltonian u–v path. □

We have seen that for every graph G and positive integer k we can determine a new graph, the kth power G^k of G. There are other associated graphs of interest. The *line graph* $L(G)$ of a graph G is that graph whose vertices can be put in one-to-one correspondence with the edges of G in such a way that two vertices of $L(G)$ are adjacent if and only if the corresponding edges of G are adjacent. A graph and its line graph are shown in Figure 4.8.

It is relatively easy to determine the number of vertices and the number of edges in the line graph $L(G)$ of a graph G in terms of easily computed quantities in G. Indeed, if G is a graph of order n and size m with degree sequence $d_1, d_2, \ldots, d_n$

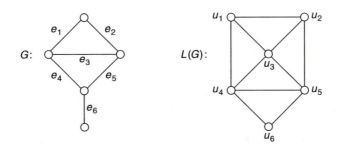

Figure 4.8 A graph and its line graph.

and $L(G)$ is a graph of order n' and size m' graph, then $n' = m$ and

$$m' = \sum_{i=1}^{n} \binom{d_i}{2},$$

since each edge of $L(G)$ corresponds to a pair of adjacent edges of G.

A graph H is called *a line graph* if there exists a graph G such that $H = L(G)$. A natural question to ask is whether a given graph H is a line graph. Several characterizations of line graphs have been obtained, perhaps the best known of which is a 'forbidden subgraph' characterization due to Beineke [B3].

Theorem 4.32

A graph H is a line graph if and only if none of the graphs of Figure 4.9 is an induced subgraph of H.

We turn to the problem of determining relationships between a graph and hamiltonian properties of its line graph. Theorem 4.33, due to Harary and Nash-Williams [HN1], provides a characterization of those graphs having hamiltonian line graphs. A set X of edges in a graph is called a *dominating set* if every edge of G either belongs to X or is adjacent to an edge of X. If $\langle X \rangle$ is a circuit C, then C

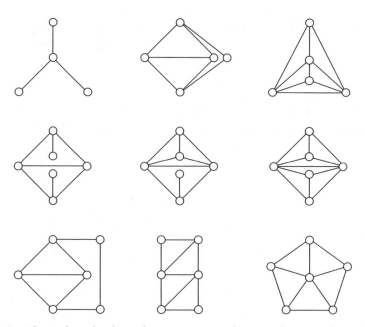

Figure 4.9 The induced subgraphs not contained in any line graph.

is called a *dominating circuit* of G. Equivalently, a circuit C in a graph G is a dominating circuit if every edge of G is incident with a vertex of C.

Theorem 4.33

Let G be a graph without isolated vertices. Then $L(G)$ is hamiltonian if and only if $G = K_{1,\ell}$ for some $\ell \geq 3$ or G contains a dominating circuit.

Proof

If $G = K_{1,\ell}$ for some $\ell \geq 3$, then $L(G)$ is hamiltonian since $L(G) = K_\ell$. Suppose, then, that G contains a dominating circuit

$$C: v_1, v_2 \ldots, v_t, v_1.$$

It suffices to show that there exists an ordering $S: e_1, e_2, \ldots, e_m$ of the m edges of G such that e_i and e_{i+1} are adjacent edges of G, for $1 \leq i \leq m - 1$, as are e_1 and e_m, since such an ordering S corresponds to a hamiltonian cycle of $L(G)$. Begin the ordering S by selecting, in any order, all edges of G incident with v_1 that are not edges of C, followed by the edge $v_1 v_2$. At each successive vertex v_i, $2 \leq i \leq t - 1$, select, in any order, all edges of G incident with v_i that are neither edges of C nor previously selected edges, followed by the edge $v_i v_{i+1}$. This process terminates with the edge $v_{t-1} v_t$. The ordering S is completed by adding the edge $v_t v_1$. Since C is a dominating circuit of G, every edge of G appears exactly once in S. Furthermore, consecutive edges of S as well as the first and last edges of S are adjacent in G.

Conversely, suppose that G is not a star but $L(G)$ is hamiltonian. We show that G contains a dominating circuit. Since $L(G)$ is hamiltonian, there is an ordering S: $e_1, e_2, \ldots, e_m$ of the m edges of G such that e_i and e_{i+1} are adjacent edges of G, for $1 \leq i \leq m - 1$, as are e_1 and e_m. For $1 \leq i \leq m - 1$, let v_i be the vertex of G incident with both e_i and e_{i+1}. (Note that $1 \leq k \neq q \leq m - 1$ does not necessarily imply that $v_k \neq v_q$.) Since G is not a star, there is a smallest integer j_1 exceeding 1 such that $v_{j_1} \neq v_1$. The vertex $v_{j_1 - 1}$ is incident with e_{j_1}, the vertex v_{j_1} is incident with e_{j_1}, and $v_{j_1 - 1} = v_1$. Thus, $e_{j_1} = v_1 v_{j_1}$. Next, let j_2 (if it exists) be the smallest integer exceeding j_1 such that $v_{j_2} \neq v_{j_1}$. The vertex $v_{j_2 - 1}$ is incident with e_{j_2}, the vertex v_{j_2} is incident with e_{j_2}, and $v_{j_2 - 1} = v_{j_1}$. Thus, $e_{j_2} = v_{j_1} v_{j_2}$. Continuing in this fashion, we finally arrive at a vertex v_{j_t} such that $e_{j_t} = v_{j_t - 1} v_{j_t}$ where $v_{j_t} = v_{m-1}$. Since every edge of G appears exactly once in S and since $1 < j_1 < j_2 < \cdots < j_t \leq m - 1$, this construction yields a trail

$$T: v_1, e_{j_1}, v_{j_1}, e_{j_2}, v_{j_2}, \ldots, v_{j_{t-1}}, e_{j_t}, v_{j_t} = v_{m-1}$$

in G with the properties that (i) every edge of G is incident with a vertex of T and (ii) neither e_1 nor e_m is an edge of T.

Let w be the vertex of G incident with both e_1 and e_m. We consider four possible cases.

Case 1. Suppose that $w = v_1 = v_{m-1}$. Then T itself is a dominating circuit of G.

Case 2. Suppose that $w = v_1$ and $w \neq v_{m-1}$. Since e_m is incident with both w and v_{m-1}, it follows that $e_m = v_{m-1}w = v_{m-1}v_1$. Thus $C: T, e_m, v_1$ is a dominating circuit of G.

Case 3. Suppose that $w = v_{m-1}$ and $w \neq v_1$. Since e_1 is incident with both w and v_1, we have that $e_1 = wv_1 = v_{m-1}v_1$. Thus $C: T, e_1, v_1$ is a dominating circuit of G.

Case 4. Suppose that $w \neq v_{m-1}$ and $w \neq v_1$. Since e_m is incident with both w and v_{m-1}, it follows that $e_m = v_{m-1} w$. Since e_1 is incident with both w and v_1, we have that $e_1 = wv_1$. Thus $v_1 \neq v_{m-1}$, and $C: T, e_m, w, e_1, v_1$ is a dominating circuit of G. $\qquad \square$

It follows from Theorem 4.33 that if G is either eulerian or hamiltonian, then $L(G)$ is hamiltonian.

EXERCISES 4.3

4.26 Show that the graph G^2 of Figure 4.7 is not hamiltonian.

4.27 Prove that if v is any vertex of a connected graph G of order at least 4, then $G^3 - v$ is hamiltonian.

4.28 Prove that if G is a self-complementary graph of order at least 5, then G^2 is hamiltonian-connected.

4.29 Determine a formula for the number of triangles in the line graph $L(G)$ in terms of quantities in G.

4.30 Prove that $L(G)$ is eulerian if G is eulerian.

4.31 Find a necessary and sufficient condition for a graph G to have the property that $G = L(G)$.

4.32 For each of the following, prove or disprove.

    (a) If G is hamiltonian, then G^2 is hamiltonian-connected.

    (b) If G is connected and $L(G)$ is eulerian, then G is eulerian.

    (c) If G is hamiltonian, then $L(G)$ is hamiltonian-connected.

    (d) If G has a dominating circuit, then $L(G)$ has a dominating circuit.

5

Directed graphs

We return to digraphs, first considering graphs having an orientation that produces a strong digraph. The main emphasis here, however, is the study of tournaments.

5.1 STRONG DIGRAPHS

In Chapter 1 we described two types of connectedness that a digraph may possess, namely, weakly connected (or more simply, connected) and strongly connected (or, more simply, strong). In this section we explore strong digraphs in more detail. Recall that a digraph D is strong if for every pair u, v of vertices of D, there is both a (directed) u–v path and a (directed) v–u path. Strong digraphs are characterized in the following theorem.

Theorem 5.1

A digraph D is strong if and only if D contains a closed spanning walk.

Proof

Assume that $W: u_1, u_2, \ldots, u_k, u_1$ is a closed spanning walk in D. Let $u, v \in V(D)$. Then $u = u_i$ and $v = u_j$ for some i, j with $1 \leq i, j \leq k$ and $i \neq j$. Without loss of generality, we assume that $i < j$. Then $W_1: u_i, u_{i+1}, \ldots, u_j$ is a u_i–u_j walk in D and $W_2: u_j, u_{j+1}, \ldots, u_k, u_1, \ldots, u_i$ is a u_j–u_i walk in D. Consequently, D contains both a u_i–u_j path and a u_j–u_i path in D, and so D is strong.

Conversely, assume that D is a nontrivial strong digraph. We show that D contains a closed spanning walk. Suppose that this is not the case, and let W be a closed walk containing a maximum number of vertices of D. Let x be a vertex of D that is not on W, and let v be a vertex on W. Since D is strong, D contains a v–x path P_1 and an x–v path P_2. When v is encountered on W, we insert P_1 followed by P_2. This results in a closed walk W' containing more vertices of D than W, which produces a contradiction. Thus D contains a closed spanning walk. $\qquad\square$

Recall that an orientation of a graph G is a digraph obtained by assigning a direction to each edge of G. We are now interested in those graphs having a strong orientation. Certainly, if G has a strong orientation, then G must be connected. Also, if G has a bridge, then it is impossible to produce a strong orientation of G. On the other hand, if G is a bridgeless connected graph, then G always has a strong orientation. This observation was first made by Robbins [R9].

Theorem 5.2

A nontrivial graph G has a strong orientation if and only if G is 2-edge-connected.

Proof

We have already observed that if a graph G has a strong orientation, then G is 2-edge-connected. Suppose that the converse is false. Then there exists a 2-edge-connected graph G that has no strong orientation. Among the subgraphs of G, let H be one of maximum order that has a strong orientation; such a subgraph exists since for each $v \in V(G)$, the subgraph $\langle \{v\} \rangle$ trivially has a strong orientation. Thus $|V(H)| < |V(G)|$, since, by assumption, G has no strong orientation.

Assign directions to the edges of H so that the resulting digraph D is strong, but assign no directions to the edges of $G - E(H)$. Let $u \in V(H)$ and let $v \in V(G) - V(H)$. Since G is 2-edge-connected, there exist two edge-disjoint (graphical) u–v paths in G. Let P be one of these u–v paths and let Q be the v–u path that results from the other u–v path. Further, let u_1 be the last vertex of P that belongs to H, and let v_1 be the first vertex of Q belonging to H. Next, let P_1 be the u_1–v subpath of P and let Q_1 be the v–v_1 subpath of Q. Direct the edges of P_1 from u_1 toward v, producing the directed path P_1', and direct the edges of Q_1 from v toward v_1, producing the directed path Q_1'.

Define the digraph D' by $V(D') = V(H) \cup V(P_1') \cup V(Q_1')$ and $E(D') = E(H) \cup E(P_1') \cup E(Q_1')$. Since H is strong, so is D', which contradicts the choice of H. □

EXERCISES 5.1

5.1 Let $e = uv$ be an edge of a graph G. Show that if G has a strong orientation, then G has a strong orientation in which u is adjacent to v and a strong orientation in which v is adjacent to u.

5.2 Let G be a connected graph with cut-vertices. Show that an orientation D of G is strong if and only if the subdigraph of D induced by the vertices of each block of G is strong.

5.3 Prove that a graph G has an eulerian orientation if and only if G is eulerian.

5.2 TOURNAMENTS

The oriented graphs that have received the greatest attention are the tournaments, that is, those digraphs obtained by orienting the edges of complete graphs.

The number of nonisomorphic tournaments increases sharply with order. For example, there is only one tournament of order 1 and one of order 2. There are two tournaments of order 3, namely the tournaments T_1 and T_2 shown in Figure 5.1. There are four tournaments of order 4, 12 of order 5, 56 of order 6, and over 154 billion of order 12.

If T is a tournament of order n, then its size is $\binom{n}{2}$ and so,

$$\sum_{v \in V(T)} \text{od } v = \sum_{v \in V(T)} \text{id } v = \binom{n}{2}.$$

A tournament T is *transitive* if whenever (u, v) and (v, w) are arcs of T, then (u, w) is also an arc of T. The tournament T_2 of Figure 5.1 is transitive while T_1 is not. The following result gives an elementary property of transitive tournaments.

Theorem 5.3

A tournament is transitive if and only if it is acyclic.

Proof

Let T be an acyclic tournament and suppose that (u, v) and (v, w) are arcs of T. Since T is acyclic, $(w, u) \notin E(T)$. Therefore, $(u, w) \in E(T)$ and T is transitive.

Conversely, suppose that T is a transitive tournament and assume that T contains a cycle, say $C: v_1, v_2, \ldots, v_k, v_1$ (where $k \geq 3$ since T is asymmetric). Since

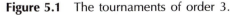

Figure 5.1 The tournaments of order 3.

(v_1, v_2) and (v_2, v_3) are arcs of the transitive tournament T, (v_1, v_3) is an arc of T. Similarly, (v_1, v_4), (v_1, v_5), ..., (v_1, v_k) are arcs of T. However, this contradicts the fact that (v_k, v_1) is an arc of T. Thus, T is acyclic. $\square$

Every tournament of order n can be thought of as representing or modeling a round robin tournament involving competition among n teams. In a round robin tournament, each team plays every other team exactly once and ties are not permitted. Let $v_1, v_2, \ldots, v_n$ represent the teams as well as the vertices of the corresponding tournament T. If, in the competition between v_i and v_j, $i \neq j$, team v_i defeats team v_j, then (v_i, v_j) is an arc of T. The number of victories by team v_i is the outdegree of v_i. For this reason, the outdegree of the vertex v_i in a tournament is also referred to as the *score* of v_i.

A sequence $s_1, s_2, \ldots, s_n$ of nonnegative integers is called a *score sequence* (of a tournament) if there exists a tournament T of order n whose vertices can be labeled $v_1, v_2, \ldots, v_n$ such that od $v_i = s_i$ for $i = 1, 2, \ldots, n$. The following result describes precisely which sequences are score sequences of transitive tournaments.

Theorem 5.4

A nondecreasing sequence S of n nonnegative integers is a score sequence of a transitive tournament if and only if S is the sequence $0, 1, \ldots, n-1$.

Proof

First we show that S: $0, 1, \ldots, n-1$ is a score sequence of a transitive tournament. Let T be the tournament with vertex set $V(T) = \{v_1, v_2, \ldots, v_n\}$ and arc set $E(T) = \{(v_i, v_j) \mid 1 \leq i < j \leq n\}$. We claim that T is transitive. Let (v_i, v_j) and (v_j, v_k) be arcs of T. Then $i < j < k$. Since $i < k$, (v_i, v_k) is an arc of T and so T is transitive. For $1 \leq i \leq n$, od $v_i = n - i$. Therefore, a score sequence of T is S: $0, 1, \ldots, n-1$.

Next, we show that if T is a transitive tournament of order n, then $0, 1, \ldots, n-1$ is a score sequence of T. This is equivalent to showing that every two vertices of T have distinct scores. Let u and w be two vertices of T. Assume, without loss of generality, that (u, w) is an arc of T. Let W be the set of vertices of T to which w is adjacent. Therefore od $w = |W|$. For each $x \in W$, (w, x) is an arc of T. Since T is transitive, (u, x) is also an arc of T. Thus, od $u \geq |W| + 1$ and so od $u \neq$ od w. $\square$

The proof of Theorem 5.4 shows that the structure of a transitive tournament is uniquely determined.

Corollary 5.5

For every positive integer n, there is exactly one transitive tournament of order n.

Combining this corollary with Theorem 5.3, we arrive at yet another corollary.

Corollary 5.6

For every positive integer n, there is exactly one acyclic tournament of order n.

Although there is only one transitive tournament of each order n, in a certain sense that we now explore, every tournament has the structure of a transitive tournament. Let T be a tournament. We define a relation on $V(T)$ by u is related to v if there is both a u–v path and a v–u path in T. This relation is an equivalence relation and, as such, this relation partitions $V(T)$ into equivalence classes $V_1, V_2, \ldots, V_k$ ($k \geq 1$). Let $S_i = \langle V_i \rangle$ for $i = 1, 2, \ldots, k$. Then each S_i is a strong subdigraph and, indeed, is maximal with respect to the property of being strong. The subdigraphs $S_1, S_2, \ldots, S_k$ are called the *strong components* of T. So the vertex sets of the strong components of T produce a partition of $V(T)$.

Let T be a tournament with strong components $S_1, S_2, \ldots, S_k$, and let $\tilde{T}$ denote that digraph whose vertices $u_1, u_2, \ldots, u_k$ are in one-to-one correspondence with the strong components (u_i corresponds to S_i, $i = 1, 2, \ldots, k$) such that (u_i, u_j) is an arc of $\tilde{T}$, $i \neq j$, if and only if some vertex of S_i is adjacent to at least one vertex of S_j. Since S_i and S_j are distinct strong components of T, it follows that *every* vertex of S_i is adjacent to *every* vertex of S_j. Hence, $\tilde{T}$ is obtained by identifying the vertices of S_i for $i = 1, 2, \ldots, k$. A tournament T and associated digraph $\tilde{T}$ are shown in Figure 5.2.

Observe that for the tournament T of Figure 5.2, $\tilde{T}$ is necessarily a tournament and is, in fact, a transitive tournament. That this always occurs follows from Theorem 5.7 (Exercise 5.6).

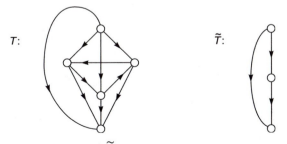

T: $\tilde{T}$:

Figure 5.2 A tournament T and $\tilde{T}$.

Theorem 5.7

If T is a tournament with (exactly) k strong components, then $\widetilde{T}$ is the transitive tournament of order k.

Since for every tournament T, the tournament $\widetilde{T}$ is transitive, it follows that if T is a tournament that is not strong, then $V(T)$ can be partitioned as $V_1 \cup V_2 \cup \cdots \cup V_k$ $(k \geq 2)$ such that $\langle V_i \rangle$ is a strong tournament for each i, and if $v_i \in V_i$ and $v_j \in V_j$, where $i > j$, then $(v_i, v_j) \in E(T)$. This decomposition is often useful when studying the properties of tournaments that are not strong.

We already noted that there are four tournaments of order 4. Of course, one of these is transitive, which consists of four trivial strong components S_1, S_2, S_3, S_4, where the vertex of S_j is adjacent to the vertex of S_i if and only if $j > i$. There are two tournaments of order 4 containing two strong components S_1 and S_2, depending on whether S_1 or S_2 is the strong component of order 3. (No strong component has order 2.) Since there are four tournaments of order 4, there is exactly one strong tournament of order 4. These tournaments are depicted in Figure 5.3. The arcs not drawn in T_1, T_2 and T_3 are all directed downward, as indicated by the 'double arrow'.

We also stated that there are 12 tournaments of order 5. There are six tournaments of order 5 that are not strong, shown in Figure 5.4. Again all arcs that are not drawn are directed downward. Thus there are six strong tournaments of order 5.

Theorem 5.4 characterizes score sequences of transitive tournaments. We next investigate score sequences in more generality. We begin with a theorem similar to Theorem 1.4.

Theorem 5.8

A nondecreasing sequence S: $s_1, s_2, \ldots, s_n$ $(n \geq 2)$ of nonnegative integers is a score sequence if and only if the sequence S_1: $s_1, s_2, \ldots, s_{s_n}, s_{s_{n+1}} - 1, \ldots, s_{n-1} - 1$ is a score sequence.

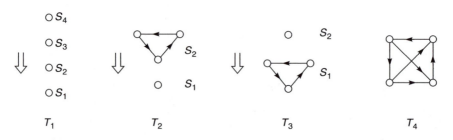

Figure 5.3 The four tournaments of order 4.

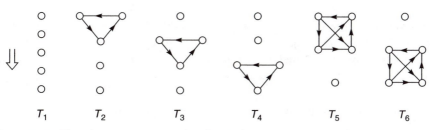

Figure 5.4   The six tournaments of order 5 that are not strong.

Proof

Assume that S_1 is a score sequence. Then there exists a tournament T_1 of order $n - 1$ such that S_1 is a score sequence of T_1. Hence the vertices of T_1 can be labeled as $v_1, v_2, \ldots, v_{n-1}$ such that

$$\text{od } v_i = \begin{cases} s_i & \text{for } 1 \leq i \leq s_n \\ s_i - 1 & \text{for } i > s_n. \end{cases}$$

we construct a tournament T by adding a vertex v_n to T_1. Furthermore, for $1 \leq i \leq n$, v_n is adjacent to v_i if $1 \leq i \leq s_n$, and v_n is adjacent from v_i otherwise. The tournament T then has S as a score sequence.

For the converse, we assume that S is a score sequence. Hence there exist tournaments of order n whose score sequence is S. Among all such tournaments, let T be one such that $V(T) = \{v_1, v_2, \ldots, v_n\}$, od $v_i = s_i$ for $i = 1, 2, \ldots, n$, and the sum of the scores of the vertices adjacent from v_n is minimum. We claim that v_n is adjacent to vertices having scores $s_1, s_2, \ldots, s_{s_n}$. Assume, to the contrary, that v_n is not adjacent to vertices having scores $s_1, s_2, \ldots, s_{s_n}$. Necessarily, then, there exist vertices v_j and v_k with $j < k$ and $s_j < s_k$ such that v_n is adjacent to v_k, and v_n is adjacent from v_j. Since the score of v_k exceeds the score of v_j, there exists a vertex v_t such that v_k is adjacent to v_t, and v_t is adjacent from v_j (Figure 5.5[a]). Thus, a 4-cycle C: v_n, v_k, v_t, v_j, v_n is produced. If we reverse the directions of the arcs of C, a tournament T' is obtained also having S as a score sequence (Figure 5.5[b]).

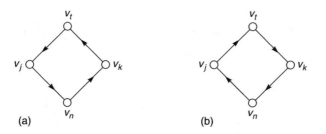

(a)   (b)

Figure 5.5   A step in the proof of Theorem 5.8.

However, in T', the vertex v_n is adjacent to v_j rather than v_k. Hence the sum of the scores of the vertices adjacent from v_n is smaller in T' than in T, which is impossible. Thus, as claimed, v_n is adjacent to vertices having scores $s_1, s_2, \ldots, s_{s_n}$. Then $T - v_n$ is a tournament having score sequence S. □

As an illustration of Theorem 5.8, we consider the sequence

S: 1, 2, 2, 3, 3, 4.

In this case, s_n (actually s_6) has the value 4; thus, we delete the last term, repeat the first $s_n = 4$ terms, and subtract 1 from the remaining terms, obtaining

S_1': 1, 2, 2, 3, 2.

Rearranging, we have

S_1: 1, 2, 2, 2, 3.

Repeating this process twice more, we have

S_2': 1, 2, 2, 1

S_2: 1, 1, 2, 2

S_3: 1, 1, 1.

The sequence S_3 is clearly a score sequence. We can use this information to construct a tournament with score sequence S. The sequence S_3 is the score sequence of the tournament T_3 of Figure 5.6. Proceeding from S_3 to S_2, we add a

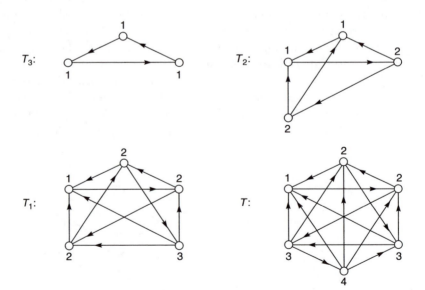

Figure 5.6 Construction of a tournament with a given score sequence.

new vertex to T_3 and join it *to* two vertices of T_3 and *from* the other, producing a tournament T_2 with score sequence S_2. To proceed from S_2 to S_1, we add a new vertex to T_2 and join it *to* vertices having scores 1, 2 and 2, and *from* the remaining vertex of T_2, producing a tournament T_1 with score sequence S_1. Continuing in the same fashion, we finally produce a desired tournament T with score sequence S by adding a new vertex to T_1 and joining it *to* vertices having scores 1, 2, 2 and 3, and joining it *from* the other vertex.

The following theorem by Landau [L1] gives a nonconstructive criterion for a sequence to be a score sequence. There are many proofs of this result; the one we give is due to Thomassen [T2].

Theorem 5.9

A nondecreasing sequence S: $s_1, s_2, \ldots, s_n$ of nonnegative integers is a score sequence if and only if for each k $(1 \leq k \leq n)$,

$$\sum_{i=1}^{k} s_i \geq \binom{k}{2}, \tag{5.1}$$

with equality holding when $k = n$.

Proof

Assume that S: $s_1, s_2, \ldots, s_n$ is a score sequence. Then there exists a tournament T of order n with $V(T) = \{v_1, v_2, \ldots, v_n\}$ such that $\text{od}_T v_i = s_i$ for $i = 1, 2, \ldots, n$. Let k be an integer with $1 \leq k \leq n$. Then $T_1 = \langle \{v_1, v_2, \ldots, v_k\} \rangle$ is a tournament of order k and size $\binom{k}{2}$. Since $\text{od}_T v_i \geq \text{od}_{T_1} v_i$ for $1 \leq i \leq k$, it follows that

$$\sum_{i=1}^{k} s_i = \sum_{i=1}^{k} \text{od}_T v_i \geq \sum_{i=1}^{k} \text{od}_{T_1} v_i = \binom{k}{2},$$

with equality holding when $k = n$.

We prove the converse by contradiction. Assume that S: $s_1, s_2, \ldots, s_n$ is a counterexample to the theorem, chosen so that n is as small as possible and so that s_1 is as small as possible among all these counterexamples.

Suppose first that there exists an integer k with $1 \leq k \leq n - 1$ such that

$$\sum_{i=1}^{k} s_i = \binom{k}{2}. \tag{5.2}$$

Thus the sequence S_1: $s_1, s_2, \ldots, s_k$ satisfies (5.1) and so, by the minimality of n, there exists a tournament T_1 of order k having score sequence S_1.

Consider the sequence $\mathcal{T}: t_1, t_2, \ldots, t_{n-k}$, where $t_i = s_{k+i} - k$ for $i = 1, 2, \ldots, n - k$. Since

$$\sum_{i=1}^{k+1} s_i \geq \binom{k+1}{2},$$

it follows from (5.2) that

$$s_{k+1} = \sum_{i=1}^{k+1} s_i - \sum_{i=1}^{k} s_i \geq \binom{k+1}{2} - \binom{k}{2} = k.$$

Thus, since S is a nondecreasing sequence,

$$t_i = s_{k+i} - k \geq s_{k+1} - k \geq 0$$

for $i = 1, 2, \ldots, n - k$, and so $\mathcal{T}$ is a nondecreasing sequence of nonnegative integers. We show that $\mathcal{T}$ satisfies (5.1).

For each r satisfying $1 \leq r \leq n - k$, we have

$$\sum_{i=1}^{r} t_i = \sum_{i=1}^{r} (s_{k+i} - k) = \sum_{i=1}^{r} s_{k+i} - rk = \sum_{i=1}^{r+k} s_i - \sum_{i=1}^{r} s_i - rk.$$

Since

$$\sum_{i=1}^{r+k} s_i \geq \binom{r+k}{2}$$

and

$$\sum_{i=1}^{k} s_i = \binom{k}{2},$$

it follows that

$$\sum_{i=1}^{r} t_i \geq \binom{r+k}{2} - \binom{k}{2} - rk = \binom{r}{2},$$

with equality holding for $r = n - k$. Thus, $\mathcal{T}$ satisfies (5.1) and so, by the minimality of n, there exists a tournament T_2 of order $n - k$ having score sequence $\mathcal{T}$.

Let T be a tournament with $V(T) = V(T_1) \cup V(T_2)$ and

$$E(T) = E(T_1) \cup E(T_2) \cup \{(u, v) \mid u \in V(T_2), v \in V(T_1)\}.$$

Then S is a score sequence for T, contrary to assumption. Thus for $k = 1$, $2, \ldots, n - 1$,

$$\sum_{i=1}^{k} s_i > \binom{k}{2}.$$

In particular, $s_1 > 0$.

Consider the sequence S': $s_1 - 1, s_2, s_3, \ldots, s_{n-1}, s_n + 1$. Clearly S' is a nondecreasing sequence of nonnegative integers that satisfies (5.1). By the minimality of s_1, then, there exists a tournament T' of order n having score sequence S'. Let x and y be vertices of T' such that $\operatorname{od}_{T'} x = s_n + 1$ and $\operatorname{od}_{T'} y = s_1 - 1$. Since $\operatorname{od}_{T'} x \geq \operatorname{od}_{T'} y + 2$, there is a vertex $w \neq x, y$ such that $(x, w) \in E(T')$ and $(y, w) \notin E(T')$. Thus, P: x, w, y is a path in T'.

Let T be the tournament obtained from T' by reversing the directions of the arcs of P. Then S is a score sequence for T, again producing a contradiction and completing the proof. $\qquad\square$

With a slight alteration in the hypothesis of the preceding theorem, we obtain a necessary and sufficient condition for a score sequence of a *strong* tournament. This result is due to L. Moser (see Harary and Moser [HM1]).

Theorem 5.10

A nondecreasing sequence S: $s_1, s_2, \ldots, s_n$ *of nonnegative integers is a score sequence of a strong tournament if and only if*

$$\sum_{i=1}^{k} s_i > \binom{k}{2}$$

for $1 \leq k \leq n - 1$ *and*

$$\sum_{i=1}^{n} s_i = \binom{n}{2}.$$

Furthermore, if S is a score sequence of a strong tournament, then every tournament with S as a score sequence is strong.

We close this section with a brief discussion involving distance in a tournament. Recall that if u and v are vertices of a digraph D, and D contains at least one (directed) u–v path, then the length of a shortest u–v path is called the directed distance from u to v and is denoted by $d(u, v)$.

Theorem 5.11

Let v be a vertex of maximum score in a nontrivial tournament T. If u is a vertex of T different from v, then $\mathrm{d}(v, u) \leq 2$.

Proof

Assume that od $v = k$. Necessarily, $k \geq 1$. Let $v_1, v_2, \ldots, v_k$ denote the vertices of T adjacent from v. Then $d(v, v_i) = 1$ for $i = 1, 2, \ldots, k$. If $V(T) = \{v, v_1, v_2, \ldots, v_k\}$, then the proof is complete.

Assume, then, that $V(T) - \{v, v_1, v_2, \ldots, v_k\}$ is nonempty, and let $u \in V(T) - \{v, v_1, v_2, \ldots, v_k\}$. If u is adjacent from some vertex v_i, $1 \leq i \leq k$, then $d(v, u) = 2$, producing the desired result. Suppose that this is not the case. Then u is adjacent to all of the vertices $v_1, v_2, \ldots, v_k$, as well as to v, so od $u \geq 1 + k = 1 + $ od v. However, this contradicts the fact that v is a vertex of maximum score. □

Theorem 5.11 was first discovered by the sociologist Landau [L1] during a study of pecking orders and domination among chickens. In the case of chickens, the theorem says that if chicken c pecks the largest number of chickens, then for every other chicken d, either c pecks d, or c pecks some chicken that pecks d. Thus c dominates every other chicken either directly or indirectly in two steps.

Let D be a strong digraph. Recall that the *eccentricity* $e(v)$ of a vertex v of D is defined as $e(v) = \max_{w \in V(D)} d(v, w)$. The *radius* of D is rad $D = \min_{v \in V(D)} e(v)$ and the *center* Cen(D) of D is defined as $\langle\{v \mid e(v) = \text{rad } D\}\rangle$

Theorem 5.11 provides an immediate result dealing with the radius of a strong tournament.

Corollary 5.12

Every nontrivial strong tournament has radius 2.

We conclude this section with a result on the center of a strong tournament.

Theorem 5.13

The center of every nontrivial strong tournament contains at least three vertices.

Proof

Let T be a nontrivial strong tournament. By Corollary 5.12, rad $T = 2$. Let w be a vertex having eccentricity 2. Since T is strong, there are vertices adjacent to w; let v be one of these having maximum score. Among the vertices adjacent to v, let u be one of maximum score. We show that both u and v have eccentricity 2, which will complete the proof.

Assume, to the contrary, that one of the vertices u and v does *not* have eccentricity 2. Suppose, then, that $x \in \{u, v\}$ and $e(x) \geq 3$. Hence, there exists a vertex y in T such that $d(x, y) \geq 3$. Thus, y is adjacent to x. Moreover, y is adjacent to every vertex adjacent from x. These observations imply that od $y >$ od x.

Suppose that $x = v$. Since x is adjacent to w, it follows that y is adjacent to w. However, od $y >$ od v, which contradicts the defining property of v. Therefore, $x = u$. Here x is adjacent to v so that y is adjacent to v, but od $y >$ od u. Hence, $x \neq u$ and the proof is complete. □

EXERCISES 5.2

5.4 Draw all four (nonisomorphic) tournaments of order 4.

5.5 Give an example of two nonisomorphic regular tournaments of the same order.

5.6 Prove Theorem 5.7.

5.7 Determine those positive integers n for which there exist regular tournaments of order n.

5.8 Show that if two vertices u and v have the same score in a tournament T, then u and v belong to the same strong component of T.

5.9 Which of the following sequences are score sequences? Which are score sequences of strong tournaments? For each sequence that is a score sequence, construct a tournament having the given sequence as a score sequence.

  (a) 0, 1, 1, 4, 4
  (b) 1, 1, 1, 4, 4, 4
  (c) 1, 3, 3, 3, 3, 3, 5
  (d) 2, 3, 3, 4, 4, 4, 4, 4

5.10 What can be said about a tournament T with score sequence $s_1, s_2, \ldots, s_n$ such that equality holds in (5.1) for every k, $1 \leq k \leq n$?

5.11 Show that if S: $s_1, s_2, \ldots, s_n$ is a score sequence of a tournament, then S_1: $n - 1 - s_1, n - 1 - s_2, \ldots, n - 1 - s_n$ is a score sequence of a tournament.

5.12 Prove that every regular tournament is strong.

5.13 Prove that every two vertices in a nontrivial regular tournament lie on a 3-cycle.

5.14 Prove that if T is a nontrivial regular tournament, then diam $T = 2$.

5.15 Prove that every vertex of a nontrivial strong tournament lies on a 3-cycle.

5.16 Prove Corollary 5.12.

5.17 (a) A vertex v of a tournament T is called a *winner* if $d(v, u) \leq 2$ for every $u \in V(T)$. Show that no tournament has exactly two winners.

   (b) Show that if n is a positive integer, $n \neq 2, 4$, then there is a tournament of order n in which every vertex is a winner.

5.3 HAMILTONIAN TOURNAMENTS

The large number of arcs that a tournament has often produces a variety of paths and cycles. In this section we investigate these types of subdigraphs in tournaments. We begin with perhaps the most basic result of this type, a property of tournaments first observed by Rédei [R2].

Theorem 5.14

Every tournament contains a hamiltonian path.

Proof

Let T be a tournament of order n, and let $P: v_1, v_2, \ldots, v_k$ be a longest path in T. If P is not a hamiltonian path of T, then $1 \leq k < n$ and there is a vertex v of T not on P. Since P is a longest path, $(v, v_1), (v_k, v) \notin E(T)$, and so $(v_1, v), (v, v_k) \in E(T)$. This implies that there is an integer i $(1 \leq i < k)$ such that $(v_i, v) \in E(T)$ and $(v, v_{i+1}) \in E(T)$. But then

$$v_1, v_2, \ldots, v_i, v, v_{i+1}, \ldots, v_k$$

is a path whose length exceeds that of P, producing a contradiction. □

A simple but useful consequence of Theorem 5.14 concerns transitive tournaments.

Corollary 5.15

Every transitive tournament contains exactly one hamiltonian path.

The preceding corollary is a special case of a result by Szele [S10], who showed that every tournament contains an odd number of hamiltonian paths.

While not every tournament is hamiltonain, such is the case for strong tournaments, a fact discovered by Camion [C1]. It is perhaps surprising that if a tournament is hamiltonian, then it must possess significantly stronger properties. A digraph D of order $n \geq 3$ is *pancyclic* if it contains a cycle of length ℓ for each $\ell = 3, 4, \ldots, n$ and is *vertex-pancyclic* if each vertex v of D lies on a cycle of length ℓ for each $\ell = 3, 4, \ldots, n$. Harary and Moser [HM1] showed that every nontrivial strong tournament is pancyclic. The following result was discovered by Moon [M9]. The proof here is due to C. Thomassen.

Theorem 5.16

Every nontrivial strong tournament is vertex-pancyclic.

Proof

Let T be a strong tournament of order $n \geq 3$, and let v_1 be a vertex of T. We show that v_1 lies on an ℓ-cycle for each $\ell = 3, 4, \ldots, n$. We proceed by induction on ℓ.

Since T is strong, it follows from Exercise 5.15 that v_1 lies on a 3-cycle. Assume that v_1 lies on an ℓ-cycle $v_1, v_2, \ldots, v_\ell, v_1$, where $3 \leq \ell \leq n - 1$. We prove that v_1 lies on an $(\ell + 1)$-cycle.

Case 1. Suppose that there is a vertex v not on C that is adjacent to at least one vertex of C and is adjacent from at least one vertex of C. This implies that for some $i\,(1 \leq i \leq \ell)$, both (v_i, v) and (v, v_{i+1}) are arcs of T (where all subscripts are expressed modulo ℓ). Thus, v_1 lies on the $(\ell + 1)$-cycle.

$$v_1, v_2, \ldots, v_i, v, v_{i+1}, \ldots, v_\ell, v_1.$$

Case 2. Suppose that no vertex v exists as in Case 1. Let A denote the set of all vertices in $V(T) - V(C)$ that are adjacent to every vertex of C, and let B be the set of all vertices in $V(T) - V(C)$ that are adjacent from every vertex of C. Then $A \cup B = V(T) - V(C)$. Since T is strong, neither A nor B is empty. Furthermore, there is a vertex b in B and a vertex a in A such that $(b, a) \in E(T)$. Thus, v_1 lies on the $(\ell + 1)$-cycle

$$a, v_1, v_2, \ldots, v_{\ell-1}, b, a. \qquad \square$$

Corollary 5.17

Every nontrivial strong tournament is pancyclic.

EXERCISES 5.3

5.18 Prove that if T is a tournament that is not transitive, then T has at least three hamiltonian paths.

5.19 Use Corollary 5.15 to give an alternative proof of Theorem 5.4.

5.20 Prove or disprove: Every arc of a nontrivial strong tournament T lies on a hamiltonian cycle of T.

5.21 Prove or disprove: Every vertex-pancyclic tournament is hamiltonian-connected.

5.22 Show that if a tournament T has an ℓ-cycle, then T has an s-cycle for $s = 3, 4, \ldots \ell$.

6

Planar graphs

We now consider graphs that can be drawn in the plane without their edges crossing. A result obtained by Euler plays a central role in the study of these 'planar' graphs. We describe two characterizations of planar graphs. A necessary condition for a planar graph to be hamiltonian is discussed. Two parameters associated with nonplanar graphs are then considered.

6.1   THE EULER IDENTITY

A graph G of order n and size m is said to be *realizable* or *embeddable* on a surface S if it is possible to distinguish a collection of n distinct points of S that correspond to the vertices of G and a collection of m curves, pairwise disjoint except possibly for endpoints, on S that correspond to the edges of G such that if a curve A corresponds to the edge $e = uv$, then only the endpoints of A correspond to vertices of G, namely u and v. Intuitively, G is embeddable on S if G can be drawn on S so that edges (more precisely, the curves corresponding to edges) intersect only at a vertex (that is, a point corresponding to a vertex) mutually incident with them. In this chapter we are concerned exclusively with the case in which S is a plane or sphere.

A graph is *planar* if it can be embedded in the plane. Embedding a graph in the plane is equivalent to embedding it on the sphere. In order to see this, we perform what is called a *stereographic projection*. Let S be a sphere tangent to a plane π, where P is the point of S diametrically opposite to the point of tangency. If a graph G is embedded on S in such a way that no vertex of G is P and no edge of G passes through P, then G may be projected onto π to produce an embedding of G on π. The inverse of this projection shows that any graph that can be embedded in the plane can also be embedded on the sphere.

If a planar graph is embedded in the plane, then it is called a *plane graph*. The graph $G_1 = K_{2,3}$ of Figure 6.1 is planar, although, as drawn, it is not plane; however, $G_2 = K_{2,3}$ is both planar and plane. The graph $G_3 = K_{3,3}$ is nonplanar. This last statement will be proved presently.

Given a plane graph G, a *region* of G is a maximal portion of the plane for which any two points may be connected by a curve A such that each point of A

127

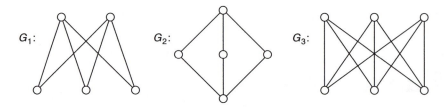

Figure 6.1   Planar, plane and nonplanar graphs.

neither corresponds to a vertex of G nor lies on any curve corresponding to an edge of G. Intuitively, the regions of G are the connected portions of the plane remaining after all curves and points corresponding, respectively, to edges and vertices of G have been deleted. For a plane graph G, the *boundary* of a region R consists of all those points x corresponding to vertices and edges of G having the property that x can be connected to a point of R by a curve, all of whose points different from x belong to R. Every plane graph G contains an unbounded region called the *exterior region* of G. If G is embedded on the sphere, then no region of G can be regarded as being exterior. On the other hand, it is equally clear that a plane graph G can always be embedded in the plane so that a given region of G becomes the exterior region. Hence a plane graph G can always be realized in the plane so that any vertex or edge lies on the boundary of its exterior region. The plane graph G_2 of Figure 6.1 has three regions, and the boundary of each is a 4-cycle.

The order, size and number of regions of any connected plane graph are related by a theorem discovered by Euler [E8].

Theorem 6.1 (The Euler Identity)

If G is a connected plane graph with n vertices, m edges and r regions, then

$$n - m + r = 2.$$

Proof

We employ induction on m, the result being obvious for $m = 0$ since in this case $G = K_1$ and so $n = 1$ and $r = 1$. Assume that the result is true for all connected plane graphs with $m - 1$ edges, where $m \geq 1$. Let G be a connected plane graph with m edges. If G is a tree, then $n = m + 1$ and $r = 1$ so the desired formula follows. On the other hand, if G is not a tree, let e be a cycle edge of G and consider $G - e$. The connected plane graph $G - e$ has n vertices, $m - 1$ edges, and $r - 1$ regions so that by the inductive hypothesis, $n - (m - 1) + (r - 1) = 2$, which implies that $n - m + r = 2$. □

From the preceding theorem, it follows that every two embeddings of a connected planar graph in the plane result in plane graphs having the same number of regions; thus one can speak of the number of regions of a connected planar graph. For planar graphs in general, we have the following result.

Corollary 6.2

If G is a plane graph with n vertices, m edges and r regions, then $n - m + r = 1 + k(G)$.

A planar graph G is called *maximal planar* if, for every pair u, v of non-adjacent vertices of G, the graph $G + uv$ is nonplanar. Thus in any embedding of a maximal planar graph G having order $n \geq 3$, the boundary of every region of G is a triangle. For this reason, maximal planar graphs are also referred to as *triangulated planar graphs*; triangulated plane graphs are often called simply *triangulations*.

On a given number n of vertices, a planar graph is quite limited as to how large its size m can be. A bound on m follows from our next result.

Theorem 6.3

If G is a maximal planar graph of order $n \geq 3$ and size m, then

$$m = 3n - 6.$$

Proof

Embed G in the plane, resulting in a plane graph with r regions. The boundary of every region is a triangle, and each edge is on the boundary of two regions. Therefore, if the number of edges on the boundary of a region is summed over all regions, the result is $3r$. On the other hand, such a sum counts each edge twice, so $3r = 2m$. Applying Theorem 6.1, we obtain $m = 3n - 6$. □

Corollary 6.4

If G is a planar graph of order $n \geq 3$ and size m, then

$$m \leq 3n - 6.$$

Proof

Add to G sufficiently many edges so that the resulting graph of order n' and size m' is maximal planar. Clearly, $n = n'$ and $m \leq m'$. By Theorem 6.3, $m' = 3n - 6$ and so $m \leq 3n - 6$. □

An immediate but important consequence of Corollary 6.4 is given next.

Corollary 6.5

Every planar graph contains a vertex of degree at most 5.

Proof

Let G be a planar graph of order n and size m with $V(G) = \{v_1, v_2, \ldots, v_n\}$. If $n \leq 6$, then the result is obvious. Otherwise, $m \leq 3n - 6$ implies that

$$\sum_{i=1}^{n} \deg v_i = 2m \leq 6n - 12.$$

Not all vertices of G have degree 6 or more, for then $2m \geq 6n$. Thus G contains a vertex of degree 5 or less. □

We next consider another corollary involving degrees. In it we make use of the fact that the minimum degree is at least 3 in a maximal planar graph of order at least 4. In order to see this, let G be a maximal planar graph of order $n \geq 4$ and size m. By Theorem 6.3, $m = 3n - 6$. Surely, G contains no vertices of degree 0 or 1. It remains only to show that G contains no vertex of degree 2. Suppose that G does contain a vertex v with deg $v = 2$. Then $G - v$ is planar of order $n - 1$ and size $m - 2$. Since $m = 3n - 6$, it follows that

$$m - 2 = 3n - 8 = 3(n - 1) - 5 > 3(n - 1) - 6,$$

which contradicts Corollary 6.4.

Corollary 6.6

Let G be a maximal planar graph of order $n \geq 4$, and let n_i denote the number of vertices of degree i in G for $i = 3, 4, \ldots, k = \Delta(G)$. Then

$$3n_3 + 2n_4 + n_5 = n_7 + 2n_8 + \cdots + (k - 6)n_k + 12.$$

Proof

Let G have size m. Then, by Theorem 6.3, $m = 3n - 6$. Since

$$n = \sum_{i=3}^{k} n_i \quad \text{and} \quad 2m = \sum_{i=3}^{k} in_i,$$

it follows that

$$\sum_{i=3}^{k} in_i = 6 \sum_{i=3}^{k} n_i - 12$$

and, consequently,

$$3n_3 + 2n_4 + n_5 = n_7 + 2n_8 + \cdots + (k-6)n_k + 12. \qquad \square$$

An interesting feature of planar graphs is that they can be embedded in the plane so that every edge is a straight line segment. This result was proved independently by Fáry [F1] and Wagner [W1].

The theory of planar graphs is very closely allied with the study of polyhedra; in fact, with every polyhedron P is associated a connected planar graph $G(P)$ whose vertices and edges are the vertices and edges of P. Necessarily, then, every vertex of $G(P)$ has degree at least 3. Moreover, if $G(P)$ is a plane graph, then the faces of P are the regions of $G(P)$ and every edge of $G(P)$ is on the boundary of two regions. A polyhedron and its associated plane graph are shown in Figure 6.2.

It is customary to denote the number of vertices, edges and faces of a polyhedron P by V, E and F, respectively. However, these are the number of vertices, number of edges, and number of regions of a connected planar graph, namely $G(P)$. According to Theorem 6.1, V, E and F are related. In this form, the statement of this result is known as the Euler Polyhedron Formula.

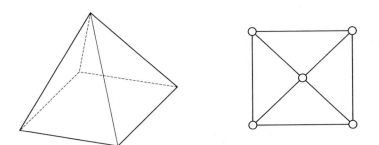

Figure 6.2 A polyhedron and its associated graph.

Theorem 6.7 (Euler Polyhedron Formula)

If V, E and F are the number of vertices, edges and faces of a polyhedron, then

$$V - E + F = 2.$$

When dealing with a polyhedron P (as well as the graph $G(P)$), it is customary to represent the number of vertices of degree k by V_k and number of faces (regions) bounded by a k-cycle by F_k. It follows then that

$$2E = \sum_{k \geq 3} kV_k = \sum_{k \geq 3} kF_k. \tag{6.1}$$

By Corollary 6.5, every polyhedron has at least one vertex of degree 3, 4 or 5. As an analogue to this result, we have the following.

Theorem 6.8

At least one face of every polyhedron is bounded by a k-cycle for some $k = 3, 4, 5$.

Proof

Assume, to the contrary, that $F_3 = F_4 = F_5 = 0$. By equation (6.1),

$$2E = \sum_{k \geq 6} kF_k \geq \sum_{k \geq 6} 6F_k = 6 \sum_{k \geq 6} F_k = 6F.$$

Hence $E \geq 3F$. Also,

$$2E = \sum_{k \geq 3} kV_k \geq \sum_{k \geq 3} 3V_k = 3 \sum_{k \geq 3} V_k = 3V.$$

By Theorem 6.3, $V - E + F = 2$ and so $3V - 3E + 3F = 6$. Hence $6 = 3V - 3E + 3F \leq 2E - 3E + E = 0$, a contradiction. $\qquad\square$

A *regular polyhedron* is a polyhedron whose faces are bounded by congruent regular polygons and whose polyhedral angles are congruent. In particular, for a regular polyhedron, $F = F_s$ for some s and $V = V_t$ for some t. For example, a cube is a regular polyhedron with $V = V_3$ and $F = F_4$. There are only four other regular polyhedra. These five regular polyhedra are also called *platonic solids*. The Greeks were aware, over two thousand years ago, that there are only five such polyhedra.

Theorem 6.9

There are exactly five regular polyhedra.

Proof

Let P be a regular polyhedron and let $G(P)$ be an associated planar graph. Then $V - E + F = 2$, where V, E and F denote the number of vertices, edges and faces of P and $G(P)$. Therefore,

$$
\begin{aligned}
-8 &= 4E - 4V - 4F \\
&= 2E + 2E - 4V - 4F \\
&= \sum_{k \geq 3} kF_k + \sum_{k \geq 3} kV_k - 4\sum_{k \geq 3} V_k - 4\sum_{k \geq 3} F_k \\
&= \sum_{k \geq 3} (k-4)F_k + \sum_{k \geq 3} (k-4)V_k.
\end{aligned}
$$

Since P is regular, there exist integers $s\,(\geq 3)$ and $t\,(\geq 3)$ such that $F = F_s$ and $V = V_t$. Hence $-8 = (s-4)F_s + (t-4)V_t$. Moreover, we note that $3 \leq s \leq 5$, $3 \leq t \leq 5$, and $sF_s = 2E = tV_t$. This gives us nine cases to consider.

Case 1. Assume that $s = 3$ and $t = 3$. Here we have

$$
-8 = -F_3 - V_3 \quad \text{and} \quad 3F_3 = 3V_3,
$$

so $F_3 = V_3 = 4$. Thus P is the *tetrahedron*. (That the tetrahedron is the only regular polyhedron with $V_3 = F_3 = 4$ follows from geometric considerations.)

Case 2. Assume that $s = 3$ and $t = 4$. Therefore

$$
-8 = -F_3 \quad \text{and} \quad 3F_3 = 4V_4.
$$

Hence $F_3 = 8$ and $V_4 = 6$, implying that P is the *octahedron*.

Case 3. Assume that $s = 3$ and $t = 5$. In this case,

$$
-8 = -F_3 + V_5 \quad \text{and} \quad 3F_3 = 5V_5,
$$

so $F_3 = 20$, $V_5 = 12$ and P is the *icosahedron*.

Case 4. Assume that $s = 4$ and $t = 3$. We find here that

$$
-8 = -V_3 \quad \text{and} \quad 4F_4 = 3V_3.
$$

Thus $V_3 = 8$, $F_4 = 6$ and P is the *cube*.

Case 5. Assume that $s = 4$ and $t = 4$. This is impossible since $-8 \neq 0$.

Case 6. Assume that $s = 4$ and $t = 5$. This case, too, cannot occur, for otherwise $-8 = V_5$.

Case 7. Assume that $s = 5$ and $t = 3$. For these values,

$$-8 = F_5 - V_3 \quad \text{and} \quad 5F_5 = 3V_3.$$

Solving for F_5 and V_3, we find that $F_5 = 12$ and $V_3 = 20$, so P is the *dodecahedron*.

Case 8. Assume that $s = 5$ and $t = 4$. Here $-8 = F_5$, which is impossible.

Case 9. Assume that $s = 5$ and $t = 5$. This, too, is impossible since $-8 \neq F_5 + V_5$. This completes the proof. □

The graphs of the five regular polyhedra are shown in Figure 6.3.

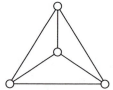

tetrahedron

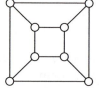

cube

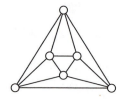

octahedron

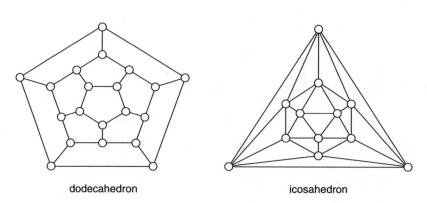

dodecahedron   icosahedron

Figure 6.3 The graphs of the regular polyhedra.

EXERCISES 6.1

6.1 Give an example of a planar graph that contains no vertex of degree less than 5.

6.2 Show that every planar graph of order $n \geq 4$ has at least four vertices of degree less than or equal to 5.

6.3 Prove Corollary 6.2.

6.4 Prove that a planar graph of order $n \geq 3$ and size m is maximal planar if and only if $m = 3n - 6$.

6.5 Prove that there exists only one 4-regular maximal planar graph.

6.6 Let $k \geq 3$ be an integer, and let G be a plane graph of order n ($\geq k$) and size m.

    (a) If the length of every cycle is at least k, then determine an upper bound B for m in terms of n and k.

    (b) Show that the bound B obtained in (a) is sharp by determining, for arbitrary $k \geq 3$, a plane graph G of order n and size B, every cycle of which has length at least k.

6.7 If the boundary of every interior region of a plane graph G of order n and size m is a triangle and the boundary of the exterior region is a cycle ($k \geq 3$), express m in terms of n.

6.2 CHARACTERIZATIONS OF PLANAR GRAPHS

There are two graphs, namely K_5 and $K_{3,3}$ (Figure 6.4), that play an important role in the study of planar graphs.

Theorem 6.10

The graphs K_5 and $K_{3,3}$ are nonplanar.

K_5: 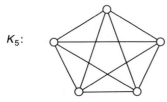 $K_{3,3}$: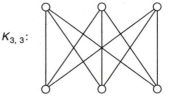

Figure 6.4 The nonplanar graphs K_5 and $K_{3,3}$.

Proof

Assume, to the contrary, that K_5 is a planar graph. Since K_5 has $n = 5$ vertices and $m = 10$ edges,

$$10 = m > 3n - 6 = 9,$$

which contradicts Corollary 6.4. Thus K_5 is nonplanar.

Suppose next that $K_{3,3}$ is a planar graph, and consider any plane embedding of it. Since $K_{3,3}$ is bipartite, it has no triangles; thus each of its regions is bounded by at least four edges. Let the number of edges bounding a region be summed over all r regions of $K_{3,3}$, denoting the result by N. Thus, $N \geq 4r$. Since the sum N counts each edge twice and $K_{3,3}$ contains $m = 9$ edges, $N = 18$ so that $r \leq \frac{9}{2}$. However, by Theorem 6.1, $r = 5$, and this is a contradiction. Hence $K_{3,3}$ is nonplanar. □

For the purpose of presenting two useful, interesting criteria for graphs to be planar, we describe two relations on graphs in this section.

An *elementary subdivision* of a nonempty graph G is a graph obtained from G by removing some edge $e = uv$ and adding a new vertex w and edges uw and vw. A *subdivision* of G is a graph obtained from G by a succession of one or more elementary subdivisions. In Figure 6.5 the graphs G_1 and G_2 are elementary subdivisions of G_3.

It should be clear that any subdivision of a graph G is planar or nonplanar according to whether G is planar or nonplanar. Also it is an elementary observation that if a graph G contains a nonplanar subgraph, then G is nonplanar. Combining these facts with our preceding results, we obtain the following.

Theorem 6.11

If a graph G contains a subgraph that is isomorphic to or a subdivision of either K_5 or $K_{3,3}$, then G is nonplanar.

The remarkable property of Theorem 6.11 is that its converse is also true. These two results provide a characterization of planar graphs that is undoubtedly one of the best known theorems in the theory of graphs. Before presenting a proof

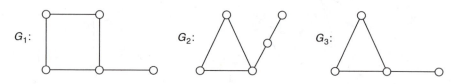

Figure 6.5 Subdivision.

of this result, first discovered by Kuratowski [K11], we need one additional fact about planar graphs.

Theorem 6.12

A graph is planar if and only if each of its blocks is planar.

Proof

Certainly, a graph G is planar if and only if each of its components is planar, so we may assume G to be connected. It is equally clear that if G is planar, then each block of G is planar. For the converse, we employ induction on the number of blocks of G. If G has only one block and this block is planar, then, of course, G is planar. Assume that every graph with fewer than $k \geq 2$ blocks, each of which is planar, is a planar graph, and suppose that G has k blocks, all of which are planar. Let B be an end-block of G, and denote by v the cut-vertex of G common to B. Delete from G all vertices of B different from v, calling the resulting graph G'. By the inductive hypothesis, G' is a planar graph. Since the block B is planar, it may be embedded in the plane so that v lies on the exterior region. In any region of a plane embedding of G' containing v, the plane block B may now be suitably placed so that the two vertices of G' and B labeled v are 'identified'. The result is a plane graph of G; hence G is planar. □

We can now give a characterization of planar graphs. The proof of this result (Theorem 6.15), known as Kuratowski's theorem [K11], is based on a proof by Dirac and Schuster [DS1]. First, we present two lemmas.

Lemma 6.13

If G is a 2-connected graph of order at least 4, then either G contains a vertex v such that $G - v$ is also 2-connected or G contains a vertex of degree 2.

Proof

Suppose that G contains no vertex v such that $G - v$ is 2-connected. Then, for each vertex x of G, there is a vertex y of $G - x$ such that $G - x - y$ is disconnected. Among all such pairs x, y of vertices G, let u, v be a pair such that $G - u - v$ is disconnected and contains a component G_1 of minimum order k. Necessarily, each of u and v is adjacent to at least one vertex in every component of $G - u - v$. If $k = 1$, then the vertex of G_1 can be adjacent only to u and v and consequently has degree 2 in G. Thus we may assume that $k \geq 2$. Let G_2 be the graph that is the union of the components of $G - u - v$ that are different from G_1. Now let $H = \langle V(G_1) \cup \{u, v\}\rangle$.

Let $w_1 \in V(G_1)$. Then there exists a vertex w_2 in $G - w_1$ such that $G - w_1 - w_2$ is disconnected. The vertex w_2 either belongs to H or to G_2. We consider these two cases.

Case 1. Assume $w_2 \in V(H)$. Since each of $\langle V(G_2) \cup \{u\}\rangle$, $\langle V(G_2) \cup \{v\}\rangle$, and $\langle V(G_2) \cup \{u, v\}\rangle$ is connected, some component of $G - w_1 - w_2$ has order less than k, which is impossible.

Case 2. Assume $w_2 \in V(G_2)$. Since $G - w_1 - w_2$ is disconnected, the vertices u and v must be nonadjacent and belong to distinct components in $G - w_1 - w_2$. This implies that $H - w_1$ has exactly two components, namely a component H_u containing u and a component H_v containing v. If H_u is trivial, and consequently contains only u, then u is adjacent in G only to w_1 and w_2 and $\deg_G u = 2$. Similarly, $\deg_G v = 2$ if H_v is trivial. On the other hand, if both H_u and H_v are nontrivial, then $G - w_1 - u$ and $G - w_1 - v$ are disconnected graphs containing a component whose order is less than k, again a contradiction. □

Figure 6.6 shows a 2-connected graph G_1 containing no vertex v for which $G_1 - v$ is also 2-connected. Thus, according to Lemma 6.13, G_1 contains a vertex of degree 2, which, of course, is the case. On the other hand, G_1 does contain an edge, namely $w_1 x_1$, whose removal from G_1 results in a 2-connected graph. The graph G_2 in Figure 6.6, however, contains no such edge. However, G_2 does contain a vertex, namely w_2, whose removal from G_2 results in a 2-connected graph. There is a result analogous to Lemma 6.13 concerning edges. Indeed, it is a consequence of Lemma 6.13.

Lemma 6.14

If G is a 2-connected graph of order at least 4, then either G contains an edge e such that $G - e$ is also 2-connected or G contains a vertex of degree 2.

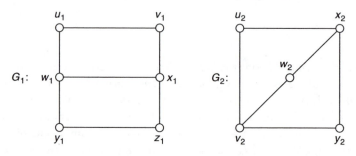

Figure 6.6 2-Connected graphs.

Proof

Suppose that G contains no edge e such that $G - e$ is 2-connected. Further, assume, to the contrary, that G contains no vertex of degree 2. By Lemma 6.13, G contains a vertex v such that $G - v$ is 2-connected. Let e be an edge incident with v. By hypothesis, $G - e$ is not 2-connected. Consequently, $G - e$ contains a cut-vertex u that is necessarily distinct from v.

Hence $G - e - u = G - u - e$ is disconnected, which implies that e is a bridge in the graph $G - u$. Since $G - u - v = G - v - u$ is connected, the edge e is a pendant edge in $G - u$ and so v is an end-vertex of $G - u$. However, then, v has degree 1 in $G - u$ and so has degree 2 in G, producing a contradiction. ☐

We are now prepared to present a proof of Kuratowski's theorem.

Theorem 6.15

A graph G is planar if and only if G contains no subgraph isomorphic to K_5 or $K_{3,3}$ or a subdivision of one of these.

Proof

The necessity is a consequence of Theorem 6.11; thus we need only consider the sufficiency. In view of Theorem 6.12, the proof reduces to showing that if a 2-connected graph does not contain K_5 or $K_{3,3}$ or a subdivision of one of these as a subgraph, then it must be planar. Assume, to the contrary, that there is a nonplanar 2-connected graph that fails to contain K_5, $K_{3,3}$, or a subdivision of one of these as a subgraph. Among all such nonplanar 2-connected graphs, let G be one of a minimum size.

We claim that $\delta(G) \geq 3$. Since G is 2-connected, $\delta(G) \geq 2$. Suppose that G contains a vertex v with $\deg_G v = 2$, where v is adjacent to u and w. We consider two cases, according to whether $uw \in E(G)$ or $uw \notin E(G)$.

Case 1. Assume $uw \in E(G)$. Then $G - v$ is a 2-connected graph that also fails to contain K_5, $K_{3,3}$, or a subdivision of one of these as a subgraph. Since the size of $G - v$ is less than that of G, it follows that $G - v$ is planar. In a planar embedding of $G - v$, the vertex v and the edges uv and vw may be inserted in a region of $G - v$ whose boundary includes the edge uw such that the resulting graph G is plane, but this is impossible.

Case 2. Assume $uw \notin E(G)$. Then $G' = G - v + uw$ is a 2-connected graph whose size is less than that of G. We show that G' does not contain K_5, $K_{3,3}$, or a subdivision of one of these as a subgraph; for assume, to the contrary, that G' does contain such a subgraph F. Necessarily, F contains the edge uw. If we replace uw by the vertex v and the edges uv and vw, then the resulting graph F' is a subdivision of F and thus a subdivision of K_5 or $K_{3,3}$. However, then F' is a subgraph of G, which is impossible. Thus G' is a 2-connected graph that contains neither K_5, $K_{3,3}$, nor a subdivision of one of these as a subgraph, and the size of G' is less than that of G. Hence G' is planar. Since G is a subdivision of G', it follows that G too is planar, which produces a contradiction.

Since a contradiction is produced in both cases, no vertex of G has degree 2 and so $\delta(G) \geq 3$, as claimed. Because G is a 2-connected graph of order at least 4 with $\delta(G) \geq 3$, it follows by Lemma 6.14 that G contains an edge $e = uv$ such that $H = G - e$ is also 2-connected.

Since H does not contain K_5, $K_{3,3}$, or a subdivision of one of these as a subgraph and the size of H is smaller than that of G, the graph H is planar. Now, since H is 2-connected, H has a cycle containing u and v by Theorem 2.5. Among all planar embeddings of H, let H be embedded in the plane so that H has a cycle C containing u and v for which the number of regions interior to C is maximum. We may further assume that $C: u = v_0, v_1, \ldots, v_i = v, \ldots, v_k = u$, where $2 \leq i \leq k - 2$.

Several observations regarding the plane graph H can now be made. In order to do this, it is convenient to define two special subgraphs of H. By the *exterior subgraph* (*interior subgraph*) of H, we mean the subgraph of G induced by those edges lying exterior (interior) to the cycle C. First, since the graph G is nonplanar, both the exterior and interior subgraphs exist, for otherwise, the edge e could be added to H (either exterior to C or interior to C) so that the resulting graph, namely G, is planar.

We note further that no two distinct vertices of the set $\{v_0, v_1, \ldots, v_i\}$ are connected by a path in the exterior subgraph of H, for otherwise this would contradict the choice of C as being a cycle containing u and v having the maximum number of regions interior to it. A similar statement can be made regarding the set $\{v_i, v_{i+1}, \ldots, v_k\}$. These remarks in connection with the fact that $H + e$ is nonplanar imply the existence of a v_s–v_t path P, $0 < s < i < t < k$, in the exterior subgraph of H such that no vertex of P different from v_s and v_t belongs to C. This structure is illustrated in Figure 6.7. We further note that no vertex of P different from v_s and v_t is adjacent to a vertex of C other than v_s or v_t, and, moreover, any path connecting a vertex of P with a vertex of C must contain at least one of v_s and v_t.

Let H_1 be the component of $H - \{v_r \mid 0 \leq r < k, r \neq s, t\}$ containing P. By the choice of C, the subgraph H_1 cannot be inserted in the interior of C in a plane manner. This, together with the assumption that G is nonplanar, implies that the interior subgraph of H must contain one of the following:

1. A v_a–v_b path Q, $0 < a < s$, $i < b < t$ (or, equivalently, $s < a < i$ and $t < b < k$), none of whose vertices different from v_a and v_b belong to C.

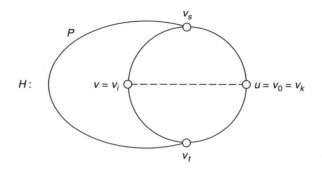

Figure 6.7 Structure of the graph H of Theorem 6.15.

2. A vertex w not on C that is connected to C by three internally disjoint paths such that the end-vertex of one such path P' is one of v_0, v_s, v_i and v_t. If P' ends at v_0, the end-vertices of the other paths are v_a and v_b, where $s \le a < i$ and $i < b \le t$ but not both $a = s$ and $b = t$ hold. If P' ends at any of v_s, v_i or v_t, there are three analogous cases.

3. A vertex w not on C that is connected to C by three internally disjoint paths P_1, P_2, P_3 such that the end-vertices of the path (different from w) are three of the four vertices v_0, v_s, v_i, v_t, say v_0, v_i, v_s, respectively, together with a v_c–v_t path P_4 ($v_c \ne v_0, v_i, w$) where v_c is on P_1 or P_2, and P_4 is disjoint from P_1, P_2 and C except for v_c and v_t. The remaining choices for P_1, P_2 and P_3 produce three analogous cases.

4. A vertex w not on C that is connected to the vertices v_0, v_s, v_i, v_t by four internally disjoint paths.

These four cases exhaust the possibilities. In the first three cases, the graph G has a subgraph that contain $K_{3,3}$ or a subdivision of $K_{3,3}$ as a subgraph, while in the fourth case, G contains K_5 or a subdivision of K_5 as a subgraph. This, however, is contrary to our assumption. □

Thus the Petersen graph (Figure 6.8[a]) is nonplanar since it contains the subgraph of Figure 6.8(b) that is a subdivision of $K_{3,3}$. Despite its resemblance to the complete graph K_5, the Petersen graph does *not* contain a subgraph that is a subdivision of K_5.

For graphs G_1 and G_2, a mapping ϕ from $V(G_1)$ onto $V(G_2)$ is called an *elementary contraction* if there exist adjacent vertices u and v of G_1 such that

(i) $\phi u = \phi v$,
(ii) $\{u_1, v_1\} \ne \{u, v\}$ implies that $\phi u_1 \ne \phi v_1$;
(iii) $\{u_1, v_1\} \cap \{u, v\} = \emptyset$ implies that $u_1 v_1 \in E(G_1)$ if and only if $\phi u_1 \phi v_1 \in E(G_2)$; and
(iv) for $w \in V(G_1)$, $w \ne u, v$, then $uw \in E(G_1)$ or $vw \in E(G_1)$ if and only if $\phi u \phi w \in E(G_2)$.

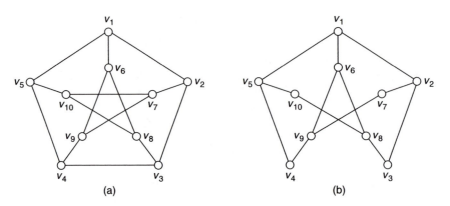

Figure 6.8 The Petersen graph and a subgraph that is a subdivision of $K_{3,3}$.

The graph G_2 is then said to be obtained from G_1 by the identification of the adjacent vertices u and v.

For graphs G and H, the graph G is a *contraction* of H if there exists a mapping from $V(H)$ to $V(G)$ that is either an isomorphism or a composition of elementary contractions.

In Figure 6.9, the graph G is a contraction of H. Letting $H = G_1$, we see that G_2 is obtained from G_1 by the identification of the adjacent vertices u and v. Then $G = G_3$ is obtained from G_2 by identifying x and t. Hence there is a mapping from $V(H)$ to $V(G)$ which is the composition of two elementary contradictions.

If a graph G is a contraction of a graph H, then we also say that H *contracts* to G and H is *contractible* to G. A *subcontraction* of a graph H is a contraction of a subgraph of H.

There is an alternate definition of contraction that is more intuitive. A graph G is a *contraction* of a graph H if there exists a one-to-one correspondence between

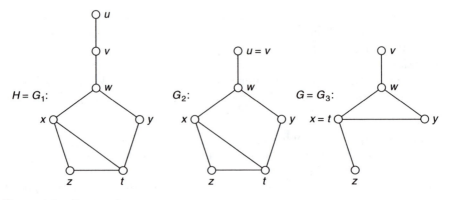

Figure 6.9 Contraction.

$V(G)$ and the elements of a partition of $V(H)$, where each element induces a connected subgraph of H, and two vertices of G are adjacent if the subgraph of H induced by the union of the corresponding elements is connected. For graphs G and H of Figure 6.9, the mapping ϕ defined by

$$\phi(v) = \{u, v\}, \quad \phi(w) = \{u\}, \quad \phi(x) = \{x, t\}, \quad \phi(y) = \{y\}, \quad \phi(z) = \{z\}$$

has the desired properties.

Theorem 6.16

If a graph H is a subdivision of a graph G, then G is a contraction of H.

Proof

If $G = H$, then clearly G is a contraction of H. Hence we may assume that H is obtained from G by a sequence of elementary subdivisions. Suppose that G' is an elementary subdivision of G; then G' is obtained from G by removing some edge uv and adding a vertex w together with the edges uw and vw. However, then, G' is contractible to G by an elementary contraction ϕ, which fixes every element of $V(G)$ and $\phi w = \phi u$. Hence G can be obtained from H by a mapping that is a composition of finitely many elementary contractions so that G is a contraction of H. ☐

Corollary 6.17 will actually prove to be of more use than the theorem itself.

Corollary 6.17

If a graph H contains a subgraph that is a subdivision of a connected nontrivial graph G, then G is a subcontraction of H.

We can now present our second characterization of planar graphs (Halin [H4], Wagner [W2] and Harary and Tutte [HT1]), which is often referred to as Wagner's theorem.

Theorem 6.18

A graph G is planar if and only if neither K_5 nor $K_{3,3}$ is a subcontraction of G.

Proof

Let G be a nonplanar graph. By Theorem 6.15, G contains a subgraph that is isomorphic to K_5 or $K_{3,3}$, or a subdivision of one of these. Thus by Corollary 6.17, K_5 or $K_{3,3}$ is a subcontraction of G.

In order to verify the converse, we first suppose that G is a graph such that $H = K_{3,3}$ is a subcontraction of G. We show, in this case, that G contains a subgraph that is isomorphic to or is a subdivision of $K_{3,3}$, implying that G is nonplanar.

Denote the vertices of H by u_i and u'_i, $1 \leq i \leq 3$, such that every edge of H is of the type $u_i u'_j$. Taking the alternate definition of contraction, we let G_i, $1 \leq i \leq 3$, be the connected subgraph of G corresponding to u_i and let G'_i correspond to u'_i. Since $u_i u'_j \in E(H)$ for $1 \leq i \leq 3$, $1 \leq j \leq 3$, in the graph G there exists a vertex v_{ij} of G_i adjacent with a vertex v'_{ij} of G'_j. Among the vertices v_{i1}, v_{i2}, v_{i3} of G_i, two or possibly all three may actually represent the same vertex. If $v_{i1} = v_{i2} = v_{i3}$, we set each $v_{ij} = v_i$; otherwise, we define v_i to be a vertex of G_i connected to the distinct elements of $\{v_{i1}, v_{i2}, v_{i3}\}$ with internally disjoint paths in G_i. (It is possible that $v_i = v_{ij}$ for some j.) We now proceed as above with the subgraphs G'_i, thereby obtaining vertices v'_i. The subgraph of G induced by the nine edges $v_{ij} v'_{ij}$ together with the edge sets of any necessary aforementioned paths from a vertex v_i or v'_i is isomorphic to or is a subdivision of $K_{3,3}$.

Assume now that $H = K_5$ is a subcontraction of G. Let $V(H) = \{u_i \mid \leq i \leq 5\}$, and suppose that G_i is the connected subgraph of G that corresponds to u_i. As before, there exists a vertex v_{ij} of G_i adjacent with a vertex v_{ji} of G_j, $i \neq j$, $1 \leq i$, $j \leq 5$. For a fixed i, $1 \leq i \leq 5$, we consider the vertices v_{ij}, $j \neq i$. If the vertices v_{ij} represent the same vertex, we denote this vertex by v_i. If the vertices v_{ij} are distinct and there exists a vertex (possibly some v_{ij}) from which there are internally disjoint paths (one of which may be trivial) to the v_{ij}, then denote this vertex by v_i. If three of the vertices v_{ij} are the same vertex, call this vertex v_i. If two vertices v_{ij} are the same while the other two are distinct, then denote the two coinciding vertices by v_i if there exist internally disjoint paths to the other two vertices. Hence in several instances we have defined a vertex v_i, for $1 \leq i \leq 5$. Should v_i exist for each $i = 1, 2, \ldots, 5$, then G contains a subgraph that is isomorphic to or is a subdivision of K_5.

Otherwise, for some i, there exist distinct vertices w_i and w'_i of G_i, each of which is connected to two of the v_{ij} by internally disjoint (possibly trivial) paths of G_i while w_i and w'_i are connected by a path of G_i, none of whose internal vertices are the vertices v_{ij}. If two vertices v_{ij} coincide, then this vertex is w_i. If the other two vertices v_{ij} should also coincide, then this vertex is w'_i. Without loss of generality, we assume that $i = 1$ and that w_1 is connected to v_{12} and v_{13}, while w'_1 is connected to v_{14} and v_{15} as described above.

Denote the edge set of these five paths of G_1 by E_1. We now turn to G_2. If $v_{21} = v_{24} = v_{25}$, we set $E_2 = \emptyset$; otherwise, there is a vertex w_2 of G_2 (which may coincide with v_{21}, v_{24} or v_{25}) connected by internally disjoint (possibly trivial) paths in G_2 to the distinct elements of $\{v_{21}, v_{24}, v_{25}\}$. We then let E_2 denote the edge sets of these paths. In an analogous manner, we define accordingly the sets E_3, E_4 and E_5 with the aid of the sets $\{v_{31}, v_{34}, v_{35}\}$, $\{v_{41}, v_{42}, v_{43}\}$ and $\{v_{51}, v_{52}, v_{53}\}$, respectively. The subgraph induced by the union of the sets E_i and the edges $v_{ij}v_{ji}$ contains a subgraph F that is isomorphic to or is a subdivision of $K_{3,3}$ such that the vertices of degree 3 of F are w_1, w_1' and the vertices w_i, $i = 2, 3, 4, 5$. In either case, G is nonplanar. □

As an application of this theorem, we again note the nonplanarity of the Petersen graph of Figure 6.8(a). The Petersen graph contains K_5 as a subcontraction, which follows by considering the partition V_1, V_2, V_3, V_4, V_5 of its vertex set, where $V_i = \{v_i, v_{i+5}\}$.

EXERCISES 6.2

6.8 Show that the converse of Theorem 6.16 is not, in general, true.
6.9 Show that the Petersen graph of Figure 6.8(a) is nonplanar by

   (a) showing that it has $K_{3,3}$ as a subcontraction, and
   (b) using Exercise 6.6(a).

6.10 Let T be a tree of order at least 4, and let $e_1, e_2, e_3 \in E(\overline{T})$. Prove that $T + e_1 + e_2 + e_3$ is planar.
6.11 A graph G is *outerplanar* if it can be embedded in the plane so that every vertex of G lies on the boundary of the exterior region. Prove the following:

   (a) A graph G is outerplanar if and only if $G + K_1$ is planar.
   (b) A graph is outerplanar if and only if it contains no subgraph that is a subdivision of either K_4 or $K_{2,3}$.
   (c) If G is on outerplanar graph of order $n \geq 2$ and size m, then $m \leq 2n - 3$.

6.3 HAMILTONIAN PLANAR GRAPHS

We have encountered many sufficient conditions for a graph to be hamiltonian but only two necessary conditions. In this section we reverse our point of view and consider a necessary condition for a planar graph to be hamiltonian.

Let G be a hamiltonian plane graph of order n and let C be a fixed hamiltonian cycle in G. With respect to this cycle, a *chord* is an edge of G that does not lie on C. Let r_i ($i = 3, 4, \ldots, n$) denote the number of regions of G in the interior of C whose boundary contains exactly i edges. Similarly, let r_i' denote the number of

G :

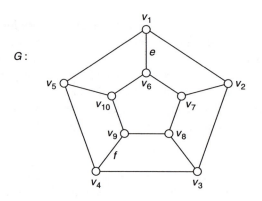

Figure 6.10 A hamiltonian plane graph.

regions of G in the exterior of C whose boundary contains i edges. To illustrate these definitions, let G be the plane graph of Figure 6.10 with hamiltonian cycle C: v_1, v_6, v_7, v_8, v_9, v_{10}, v_5, v_4, v_3, v_2, v_1. Then $r_i = 0$ if $i \neq 4$ and $r_4 = 4$. Also, $r'_i = 0$ if $i \neq 4, 5$ while $r'_4 = 1$ and $r'_5 = 2$.

Using the notation of the previous paragraph, we have the following necessary condition, due to Grinberg [G6], for a plane graph to be hamiltonian.

Theorem 6.19

Let G be a plane graph of order n with hamiltonian cycle C. Then with respect to this cycle C,

$$\sum_{i=3}^{n} (i - 2)(r_i - r'_i) = 0.$$

Proof

We first consider the interior of C. If d denotes the number of chords of G in the interior of C, then exactly $d + 1$ regions of G lie inside C. Therefore,

$$\sum_{i=3}^{n} r_i = d + 1,$$

implying that

$$d = \left(\sum_{i=3}^{n} r_i \right) - 1. \tag{6.2}$$

Let the number of edges bounding a region interior to C be summed over all $d+1$ such regions, denoting the result by N. Hence $N = \sum_{i=3}^{n} ir_i$. However, N counts each interior chord twice and each edge of C once, so that $N = 2d + n$. Thus,

$$\sum_{i=3}^{n} ir_i = 2d + n. \tag{6.3}$$

Substituting (6.2) into (6.3), we obtain

$$\sum_{i=3}^{n} ir_i = 2 \sum_{i=3}^{n} r_i - 2 + n,$$

so

$$\sum_{i=3}^{n} (i-2)r_i = n - 2. \tag{6.4}$$

By considering the exterior of C, we conclude in a similar fashion that

$$\sum_{i=3}^{n} (i-2)r_i' = n - 2. \tag{6.5}$$

It follows from (6.4) and (6.5) that

$$\sum_{i=3}^{n} (i-2)(r_i - r_i') = 0. \qquad \square$$

The following observations often prove quite useful in applying Theorem 6.19. Let G be a plane graph with hamiltonian cycle C. Furthermore, suppose that the edge e of G is on the boundary of two regions R_1 and R_2 of G. If e is an edge of C, then one of R_1 and R_2 is in the interior of C and the other is in the exterior of C. If, on the other hand, e is not an edge of C, then R_1 and R_2 are either both in the interior of C or both in the exterior of C.

In 1880, the English mathematician P. G. Tait conjectured that every 3-connected cubic planar graph is hamiltonian. This conjecture was disproved in 1946 by Tutte [T8], who produced the graph G in Figure 6.11 as a counterexample. In addition to disproving Tait's conjecture, Tutte [T11] proved that every 4-connected planar graph is hamiltonian. This result was later extended by Thomassen [T3].

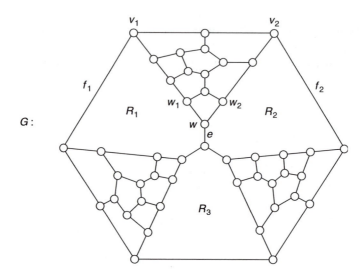

Figure 6.11 The Tutte graph.

Theorem 6.20

Every 4-connected planar graph is hamiltonian-connected.

As an illustration of Grinberg's theorem, we now verify that the Tutte graph (Figure 6.11) is not hamiltonian. Assume, to the contrary, that the Tutte graph G, which has order 46, contains a hamiltonian cycle C. Observe that C must contain exactly two of the edges e, f_1 and f_2.

Consider the regions R_1, R_2 and R_3 of G. Suppose that two of them, say R_1 and R_2, lie in the exterior of C. Then the edges f_1 and f_2 do not belong to C since the unbounded region of G also lies in the exterior of C. This, however, is impossible; thus at most one of the regions R_1, R_2 and R_3 lies in the exterior of C. We conclude that at least two of these regions, say R_1 and R_2, lie in the interior of C. This, of course, implies that their common boundary edge e does not belong to C. Therefore, f_1 and f_2 are edges of C. Now let G_1 denote the component of $G - \{e, f_1, f_2\}$ containing w. Then the cycle C contains a v_1–v_2 subpath P that is a hamiltonian path of G_1. Consider the graph $G_2 = G_1 + v_1 v_2$. Then G_2 has a hamiltonian cycle C_2 consisting of P together with the edge $v_1 v_2$.

An application of Theorem 6.19 to G_2 and C_2 yields

$$1(r_3 - r_3') + 2(r_4 - r_4') + 3(r_5 - r_5') + 6(r_8 - r_8') = 0. \tag{6.6}$$

Since $v_1 v_2$ is an edge of C_2 and since the unbounded region of G_2 lies in the exterior of C_2, we have that

$$r_3 - r_3' = 1 - 0 = 1 \quad \text{and} \quad r_8 - r_8' = 0 - 1 = -1.$$

Therefore, from (6.6) we obtain

$$2(r_4 - r_4') + 3(r_5 - r_5') = 5.$$

Since $\deg_{G_2} w = 2$, both ww_1 and ww_2 are edges of C_2. This implies that $r_4 \geq 1$, so

$$r_4 - r_4' = 1 - 1 = 0 \quad \text{or} \quad r_4 - r_4' = 2 - 0 = 2.$$

If $r_4 - r_4' = 0$, then $3(r_5 - r_5') = 5$, which is impossible. If, on the other hand, $r_4 - r_4' = 2$, then $3(r_5 - r_5') = 1$, again impossible. We conclude that Tutte's graph is not hamiltonian.

For many years, Tutte's graph was the only known example of a 3-connected cubic planar graph that was not hamiltonian. Much later, however, other such graphs have been found; for example, Grinberg himself provided the graph in Exercise 6.12 as another counterexample to Tait's conjecture.

EXERCISES 6.3

6.12 Show, by applying Theorem 6.19, that the Grinberg graph is non-hamiltonian.

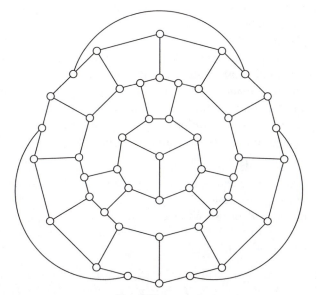

The Grinberg graph

6.13 Show, by applying Theorem 6.19, that the Herschel graph is non-hamiltonian.

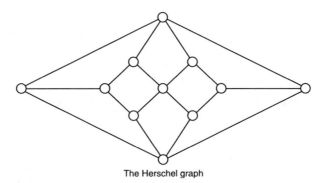

The Herschel graph

6.14 Show, by applying Theorem 6.19, that no hamiltonian cycle in the graph of Figure 6.10 contains both the edges e and f.

6.4 CROSSING NUMBER AND THICKNESS

There are several ways of measuring how nonplanar a graph is. In this section, we discuss two of these measures.

Nonplanar graphs cannot, of course, be embedded in the plane. Hence, whenever a nonplanar graph is 'drawn' in the plane, some of its edges must cross. This rather simple observation suggests our next concept.

The *crossing number* $\nu(G)$ of a graph G is the minimum number of crossings (of its edges) among the drawings of G in the plane. Before proceeding further, we comment on the assumptions we are making regarding the idea of 'drawings'. In all drawings under consideration, we assume that

- adjacent edges never cross
- two nonadjacent edges cross at most once
- no edge crosses itself
- no more than two edges cross at a point of the plane and
- the (open) arc in the plane corresponding to an edge of the graph contains no vertex of the graph.

A few observations will prove useful. Clearly a graph G is planar if and only if $\nu(G) = 0$. Further, if $G \subseteq H$, then $\nu(G) \leq \nu(H)$; while if H is a subdivision of G, then $\nu(G) = \nu(H)$. For very few classes of graphs is the crossing number known. It has been shown by Blažek and Koman [BK1] and Guy [G8], among others, that for complete graphs,

$$v(K_n) \leq \frac{1}{4} \left\lfloor \frac{n}{2} \right\rfloor \left\lfloor \frac{n-1}{2} \right\rfloor \left\lfloor \frac{n-2}{2} \right\rfloor \left\lfloor \frac{n-3}{2} \right\rfloor, \tag{6.7}$$

and Guy has conjectured that equality holds in (6.7) for all n. As far as exact results are concerned, the best obtained is the following (Guy [G9]).

Theorem 6.21

For $1 \leq n \leq 10$,

$$v(K_n) = \frac{1}{4} \left\lfloor \frac{n}{2} \right\rfloor \left\lfloor \frac{n-1}{2} \right\rfloor \left\lfloor \frac{n-2}{2} \right\rfloor \left\lfloor \frac{n-3}{2} \right\rfloor. \tag{6.8}$$

Since K_n is planar for $1 \leq n \leq 4$, Theorem 6.21 is obvious for $1 \leq n \leq 4$. Further, K_5 is nonplanar; thus, $v(K_5) \geq 1$. On the other hand, there exists a drawing (Figure 6.12) of K_5 in the plane with one crossing so that $v(K_5) = 1$.

The inequality $v(K_6) \leq 3$ follows from Figure 6.13, where a drawing of K_6 with three crossings is shown. We now verify that $v(K_6) \geq 3$, completing the proof that $v(K_6) = 3$. Let there be given a drawing of K_6 in the plane with $c = v(K_6)$ crossings, where, of course, $c \geq 1$. At each crossing we introduce a new vertex, producing a connected plane graph G of order $6 + c$ and size $15 + 2c$. By Corollary 6.4,

$$15 + 2c \leq 3(6 + c) - 6,$$

so that $c \geq 3$ and, consequently, $v(K_6) \geq 3$.

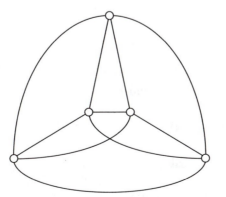

Figure 6.12   A drawing of K_5 with one crossing.

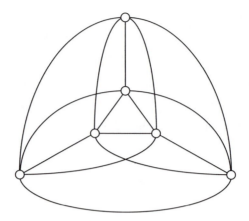

Figure 6.13 A drawing of K_6 with three crossings.

Considerably more specialized techniques are required to verify Theorem 6.21 for $7 \leq n \leq 10$.

It was mentioned in Section 6.1 that every planar graph can be embedded in the plane so that each edge is a straight line segment. Thus, if a graph G has crossing number 0, this fact can be realized by considering only drawings in the plane in which the edges are straight line segments. One may very well ask if, in general, it is sufficient to consider only drawings of graphs in which edges are straight line segments in determining crossing numbers. With this question in mind, we introduce a variation of the crossing number.

The *rectilinear crossing number* $\bar{v}(G)$ of a graph G is the minimum number of crossings among all those drawings of G in the plane in which each edge is a straight line segment. Since the crossing number $v(G)$ considers *all* drawings of G in the plane (not just those for which edges are straight line segments), we have the obvious inequality

$$v(G) \leq \bar{v}(G). \tag{6.9}$$

As previously stated, $v(G) = \bar{v}(G)$ for every planar graph G. It has also been verified that $v(K_n) = \bar{v}(K_n)$ for $1 \leq n \leq 7$ and $n = 9$; however,

$$v(K_8) = 18 \quad \text{and} \quad \bar{v}(K_8) = 19$$

(Guy [G9]), so strict inequality in (6.9) is indeed a possibility.

We return to our chief interest, namely the crossing number, and consider the complete bipartite graphs. The problem of determining $v(K_{s,t})$ has a rather curious history. It is sometimes referred to as *Turán's Brick-Factory Problem* (named for

Paul Turán). We quote from Turán [T7]:

> We worked near Budapest, in a brick factory. There were some kilns where
> the bricks were made and some open storage yards where the bricks were
> stored. All the kilns were connected by rail with all the storage yards. The
> bricks were carried on small wheeled trucks to the storage yards. All we had
> to do was to put the bricks on the trucks at the kilns, push the trucks to the
> storage yards, and unload them there. We had a reasonable piece rate for
> the trucks, and the work itself was not difficult; the trouble was only at the
> crossings. The trucks generally jumped the rails there, and the bricks fell out
> of them; in short this caused a lot of trouble and loss of time which was
> precious to all of us. We were all sweating and cursing at such occasions, I
> too; but *nolens volens* the idea occurred to me that this loss of time could have
> been minimized if the number of crossings of the rails had been minimized.
> But what is the minimum number of crossings? I realized after several days
> that the actual situation could have been improved, but the exact solution of
> the general problem with *s* kilns and *t* storage yards seemed to be very
> difficult . . . the problem occurred to me again . . . at my first visit to Poland
> where I met Zarankiewicz. I mentioned to him my 'brick-factory'-
> problem . . . and Zarankiewicz thought to have solved (it). But Ringel found
> a gap in his published proof, which nobody has been able to fill so far—in
> spite of much effort. This problem has also become a notoriously difficult
> unsolved problem . . . .

Zarankiewicz [Z2] thus thought that he had proved

$$v(K_{s,t}) = \left\lfloor \frac{s}{2} \right\rfloor \left\lfloor \frac{s-1}{2} \right\rfloor \left\lfloor \frac{t}{2} \right\rfloor \left\lfloor \frac{t-1}{2} \right\rfloor \tag{6.10}$$

but, in actuality, he only verified that the right hand expression of (6.10) is an
upper bound for $v(K_{s,t})$. As it turned out, both P. C. Kainen and G. Ringel found
flaws in Zarankiewicz's proof. Hence, (6.10) remains only a conjecture. It is further
conjectured that $v(K_{s,t}) = \bar{v}(K_{s,t})$. The best general result on crossing number of
complete bipartite graphs is the following, due to the combined work of Kleitman
[K7] and Woodall [W9].

Theorem 6.22

If s and t are integers $(s \leq t)$ and either $s \leq 6$ or $s = 7$ and $t \leq 10$, then

$$v(K_{s,t}) = \left\lfloor \frac{s}{2} \right\rfloor \left\lfloor \frac{s-1}{2} \right\rfloor \left\lfloor \frac{t}{2} \right\rfloor \left\lfloor \frac{t-1}{2} \right\rfloor.$$

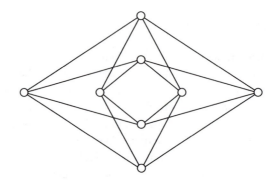

Figure 6.14 A drawing of $K_{4,4}$ with four crossings.

It follows, therefore, from Theorem 6.22 that

$$v(K_{3,t}) = \left\lfloor \frac{t}{2} \right\rfloor \left\lfloor \frac{t-1}{2} \right\rfloor, \quad v(K_{4,t}) = 2 \left\lfloor \frac{t}{2} \right\rfloor \left\lfloor \frac{t-1}{2} \right\rfloor,$$

$$v(K_{5,t}) = 4 \left\lfloor \frac{t}{2} \right\rfloor \left\lfloor \frac{t-1}{2} \right\rfloor \quad \text{and} \quad v(K_{6,t}) = 6 \left\lfloor \frac{t}{2} \right\rfloor \left\lfloor \frac{t-1}{2} \right\rfloor$$

for all t. For example, $v(K_{3,3}) = 1$, $v(K_{4,4}) = 4$, $v(K_{5,5}) = 16$, $v(K_{6,6}) = 36$ and $v(K_{7,7}) = 81$. A drawing of $K_{4,4}$ with four crossings is shown in Figure 6.14.

As would be expected, the situation regarding crossing numbers of complete k-partite graphs, $k \geq 3$, is even more complicated. For the most part, only bounds and highly specific results have been obtained in these cases. On the other hand, some of the proof techniques employed have been enlightening. As an example, we establish the crossing number of $K_{2,2,3}$ (see White [W4], p. 67).

Theorem 6.23

The crossing number of $K_{2,2,3}$ is $v(K_{2,2,3}) = 2$.

Proof

Let $v(K_{2,2,3}) = c$. Since $K_{3,3}$ is nonplanar and $K_{3,3} \subseteq K_{2,2,3}$, it follows that $K_{2,2,3}$ is nonplanar so that $c \geq 1$. Let there be given a drawing of $K_{2,2,3}$ in the plane with c crossings. At each crossing we introduce a new vertex, producing a connected plane graph G of order $n = 7 + c$ and size $m = 16 + 2c$. By Corollary 6.4, $m \leq 3n - 6$.

Let $u_1 u_2$ and $v_1 v_2$ be two (nonadjacent) edges of $K_{2,2,3}$ that cross in the given drawing, giving rise to a new vertex. If G is a triangulation, then $C: u_1, v_1, u_2, v_2, u_1$, is a cycle of G, implying that the induced subgraph $\langle \{u_1, u_2, v_1, v_2\} \rangle$ in $K_{2,2,3}$ is

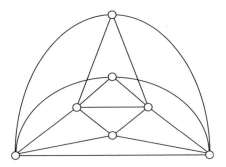

Figure 6.15 A drawing of $K_{2,2,3}$ with two crossings.

isomorphic to K_4. However, $K_{2,2,3}$ contains no such subgraph; thus, G is not a triangulation so that $m < 3n - 6$. We have

$$16 + 2c < 3(7 + c) - 6,$$

from which it follows that $c \geq 2$. The inequality $c \leq 2$ follows from the fact that there exists a drawing of $K_{2,2,3}$ with two crossings (Figure 6.15).   □

Other graphs whose crossing numbers have been investigated with little success are the n-cubes Q_n. Since Q_n is planar for $n = 1, 2, 3$, of course, $v(Q_n) = 0$ for each such n. Eggleton and Guy [EG1] have shown that $v(Q_4) = 8$ but $v(Q_n)$ is unknown for $n \geq 5$. One might observe that

$$Q_4 = K_2 \times K_2 \times K_2 \times K_2 = C_4 \times C_4,$$

so that $v(C_4 \times C_4) = 8$. This raises the problem of determining $v(C_s \times C_t)$ for $s, t \geq 3$. For the case $s = t = 3$, Harary, Kainen and Schwenk [HKS1] showed that $v(C_3 \times C_3) = 3$. Their proof consisted of the following three steps:

Step 1. Exhibiting a drawing of $C_3 \times C_3$ with three crossings so that $v(C_3 \times C_3) \leq 3$.
Step 2. Showing that $C_3 \times C_3 - e$ is nonplanar for every edge e of $C_3 \times C_3$ so that $v(C_3 \times C_3) \geq 2$.
Step 3. Showing, by case exhaustion, that it is impossible to have a drawing of $C_3 \times C_3$ with exactly two crossings so that $v(C_3 \times C_3) \geq 3$ (Exercise 6.19).

Ringeisen and Beineke [RB1] then extended this result significantly by determining $v(C_3 \times C_t)$ for all integers $t \geq 3$.

Theorem 6.24

For all $t \geq 3$,

$$v(C_3 \times C_t) = t.$$

Proof

We label the vertices of $C_3 \times C_t$ by the $3t$ ordered pairs $(0,j)$, $(1,j)$ and $(2,j)$, where $j = 0, 1, \ldots, t-1$, and, for convenience, we let

$$u_j = (0,j), v_j = (1,j) \quad \text{and} \quad w_j = (2,j).$$

First, we note that $\nu(C_3 \times C_t) \le t$. This observation follows from the fact that there exists a drawing of $C_3 \times C_t$ with t crossings. A drawing of $C_3 \times C_4$ with four crossings is shown in Figure 6.16. Drawings of $C_3 \times C_t$ with t crossings for other values of t can be given similarly.

To complete the proof, we show that $\nu(C_3 \times C_t) \ge t$. We verify this by induction on $t \ge 3$. For $t = 3$, we recall the previously mentioned result $\nu(C_3 \times C_3) = 3$.

Assume that $\nu(C_3 \times C_k) \ge k$, where $k \ge 3$, and consider the graph $C_3 \times C_{k+1}$. We show that $\nu(C_3 \times C_{k+1}) \ge k+1$. Let there be given a drawing of $C_3 \times C_{k+1}$ with $\nu(C_3 \times C_{k+1})$ crossings. We consider two cases.

Case 1. Suppose that no edge of any triangle $T_j = \langle\{u_j, v_j, w_j, \}\rangle$, $j = 0, 1, \ldots, k$, is crossed. For $j = 0, 1, \ldots, k$, define

$$H_j = \langle\{u_j, v_j, w_j, u_{j+1}, v_{j+1}, w_{j+1}\}\rangle,$$

where the subscripts are expressed modulo $k+1$. We show that for each $j = 0$, $1, \ldots, k$, the number of times edges of H_j are crossed totals at least two. Since, by

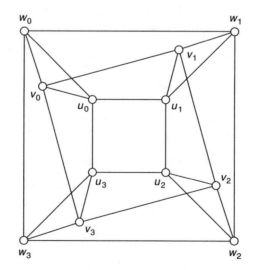

Figure 6.16 A drawing of $C_3 \times C_4$ with four crossings.

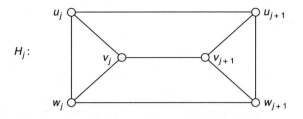

Figure 6.17 The subgraph H_j in the proof of Theorem 6.24.

assumption, no triangle T_j has an edge crossed and since every edge not in any T_j belongs to exactly one subgraph H_j, it will follow that there are at least $k+1$ crossings in the drawing because then every crossing of an edge in H_j involves either two edges of H_j or an edge of H_j and an edge of H_i for some $i \neq j$.

If two of the edges $u_j u_{j+1}$, $v_j v_{j+1}$ and $w_j w_{j+1}$ cross each other, then two edges of H_j are crossed. Assume then that no two edges of H_j cross each other. Thus, H_j is a plane subgraph in the drawing of $C_3 \times C_{k+1}$ (Figure 6.17). The triangle T_{j+2} must lie within some region of H_j. If T_{j+2} lies in a region of H_j bounded by a triangle, say T_j, then at least one edge of the cycle $u_0, u_1, \dots, u_k, u_0$, for example, must cross an edge of T_j, contradicting our assumption. Thus, T_{j+2} must lie in a region of H_j bounded by a 4-cycle, say $u_j, u_{j+1}, w_{j+1}, w_j, u_j$, without loss of generality. However, then edges of the cycle $v_0, v_1, \dots, v_k, v_0$ must cross edges of the cycle $u_j, u_{j+1}, w_{j+1}, w_j, u_j$ at least twice and hence edges of H_j at least twice, as asserted.

Case 2. Assume that some triangle, say T_0, has at least one of its edges crossed. Suppose that $\nu(C_3 \times C_{k+1}) < k+1$. Then the graph $C_3 \times C_{k+1} - E(T_0)$, which is a subdivision of $C_3 \times C_k$, is drawn with fewer than k crossings, contradicting the inductive hypothesis. □

The only other result giving the crossing number of graphs $C_s \times C_t$ is the following formula by Beineke and Ringeisen [BR1]. This theorem and the succeeding theorems in this chapter are stated only to illustrate the types of results obtained in this area.

Theorem 6.25

For all $t \geq 4$,

$$\nu(C_4 \times C_t) = 2t.$$

Beineke and Ringeisen [BR1] have also found a formula for $\nu(K_4 \times C_t)$.

Theorem 6.26

For all $t \geq 3$,

$$\nu(K_4 \times C_t) = 3t.$$

In addition to the crossing number, another parameter that is interesting for nonplanar graphs only is the thickness. The *edge-thickness* or simply the *thickness* $\theta_1(G)$ of a nonempty graph G is the minimum number of pairwise edge-disjoint planar spanning subgraphs of G whose edge sets is a partition of $E(G)$. This provides another measure of the nonplanarity of a graph. Once again, it is the complete graphs, complete bipartite graphs and n-cubes that have received the most attention.

A formula for the thickness of the complete graphs was established primarily due to the efforts of Beineke [B2], Beineke and Harary [BH2], Vasak [V1] and Alekseev and Gonchakov [AG1].

Theorem 6.27

The thickness of K_n is given by

$$\theta_1(K_n) = \begin{cases} \left\lfloor \dfrac{n+7}{6} \right\rfloor & n \neq 9, 10 \\ 3 & n = 9, 10. \end{cases}$$

Although only partial results exist for the thickness of complete bipartite graphs (Beineke, Harary and Moon [BHM1]), a formula is known for the thickness of the n-cubes, due to Kleinert [K6].

Theorem 6.28

The thickness of Q_n is given by

$$\theta_1(Q_n) = \left\lceil \frac{n+1}{4} \right\rceil.$$

EXERCISES 6.4

6.15 Draw K_7 in the plane with nine crossings.

6.16 Determine $\nu(K_{3,3})$ without using Theorem 6.22.

6.17 Show that $\nu(K_{7,7}) \leq 81$.

6.18 Determine $\nu(K_{2,2,2})$.

6.19 Determine $v(K_{1,2,3})$.

6.20 Show that $2 \leq v(C_3 \times C_3) \leq 3$.

6.21 Prove that $\bar{v}(C_3 \times C_t) = t$ for $t \geq 3$.

6.22 (a) It is known that $v(W_4 \times K_2) = 2$, where W_4 is the wheel $C_4 + K_1$ of order 5. Draw $W_4 \times K_2$ in the plane with two crossings.

     (b) Prove or disprove: If G is a nonplanar graph containing an edge e such that $G - e$ is planar, then $v(G) = 1$.

6.23 Prove that $\theta_1(K_n) \geq \lfloor (n+7)/6 \rfloor$ for all positive integers n.

6.24 Verify that $\theta_1(K_n) = \lfloor (n+7)/6 \rfloor$ for $n = 4, 5, 6, 7, 8$.

7

Graph embeddings

In Chapter 6 the emphasis was on embeddings of graphs in the plane. Here this notion is extended to embeddings of graphs on other surfaces.

7.1 THE GENUS OF A GRAPH

We now introduce the best known parameter involving nonplanar graphs. A compact orientable 2-manifold is a surface that may be thought of as a sphere on which has been placed a number of 'handles' or, equivalently, a sphere in which has been inserted a number of 'holes'. The number of handles (or holes) is referred to as the *genus of the surface*. By the *genus* $\gamma(G)$ *of a graph* G is meant the smallest genus of all surfaces (compact orientable 2-manifolds) on which G can be embedded. Every graph has a genus; in fact, it is a relatively simple observation that a graph of size m can be embedded on a surface of genus m.

Since the embedding of graphs on spheres and planes is equivalent, the graphs of genus 0 are precisely the planar graphs. The graphs with genus 1 are therefore the nonplanar graphs that are embeddable on the torus. The (nonplanar) graphs K_5 and $K_{3,3}$ have genus 1. Embeddings of $K_{3,3}$ on the torus and on the surface of genus 2 are shown in Figures 7.1(a) and (b).

Not only is K_5 embeddable on the torus, but so are K_6 and K_7. (The graph K_8 is not embeddable on the torus.) Figure 7.2 gives an embedding of K_7 on the torus. The torus is obtained by identifying opposite sides of the square. The vertices of K_7 are labeled $v_0, v_1, \ldots, v_6$. Thus note that the 'vertices' located at the corners of the square actually represent the same vertex of K_7, namely the one labeled v_5.

For graphs embedded on surfaces of positive genus, the regions and the boundaries of the regions are defined in entirely the same manner as for embeddings in the plane. Thus, if G is embedded on a surface S, then the components of $S - G$ are the regions of the embedding. In Figure 7.1(a) there are three regions, in Figure 7.1(b) there are two regions and in Figure 7.2 there are 14 regions.

A region is called a *2-cell* if any simple closed curve in that region can be continuously deformed or contracted in that region to a single point. Equivalently, a region is a 2-cell if it is topologically homeomorphic to 2-dimensional Euclidean

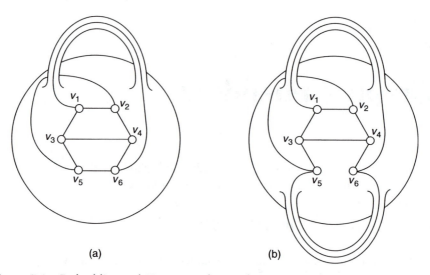

Figure 7.1 Embeddings of $K_{3,3}$ on surfaces of genus 1 and 2.

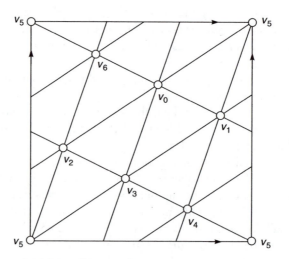

Figure 7.2 An embedding of K_7 on the torus.

space. Although every region of a connected graph embedded on the sphere is necessarily a 2-cell, this need not be the case for connected graphs embedded on surfaces of positive genus. Of the two regions determined by the embedding of $K_{3,3}$ on the 'double torus' in Figure 7.1(b), one is a 2-cell and the other is not. The boundary of the 2-cell is a 4-cycle while the boundary of the other region consists of all vertices and edges of $K_{3,3}$. Indeed, in the closed walk bounding the second region, five of the edges are encountered twice.

An embedding of a graph G on a surface S is called a *2-cell embedding* of G on S if all the regions so determined are 2-cells. The embeddings in Figure 7.1(a) and Figure 7.2 are both 2-cell embeddings.

Theorem 7.1

Let G be a connected multigraph of order n and size m with a 2-cell embedding on the surface of genus g and having r regions. Then

$$n - m + r = 2 - 2g. \tag{7.1}$$

Proof

The proof is by induction on g. For $g = 0$, the formula holds for connected graphs by Theorem 6.1. If G is a connected multigraph of order n and size m (which is not a graph) embedded in the plane and having r regions, then a plane graph H is obtained by deleting from G all loops and all but one edge in any set of multiple edges joining the same two vertices. If H has order n_1, size m_1 and r_1 regions, then $n_1 - m_1 + r_1 = 2$ by Theorem 6.1. We now add back the deleted edges to form the originally embedded multigraph G. Note that the addition of each such edge increases the number of regions by 1. If G has k more edges than does H, then $n = n_1$, $m = m_1 + k$ and $r = r_1 + k$ so that

$$n - m + r = n_1 - (m_1 + k) + (r_1 + k) = n_1 - m_1 - r_1 = 2,$$

producing the desired result for $g = 0$.

Assume the theorem to be true for all connected multigraphs that are 2-cell embedded on the surface of genus $g - 1$, where $g > 0$, and let G be a connected multigraph of order n and size m that is 2-cell embedded on the surface S of genus g and having r regions. We verify that (7.1) holds.

Since the surface S has genus g and $g > 0$, S has handles. Draw a curve C around a handle of S such that C contains no vertices of G. Necessarily, C will cross edges of G; for otherwise C lies in a region of G and cannot be contracted in that region to a single point, contradicting the fact that the embedding on S is a 2-cell embedding. By re-embedding G on S, if necessary, we may assume that the total number of intersections of C with edges of G is finite, say k, where $k > 0$. If $e_1, e_2, \ldots, e_t$ are the edges of G that are crossed by C, then $1 \le t \le k$ (Figure 7.3). Moreover, if edge e_i, $1 \le i \le t$, is crossed by C a total of ℓ_i times, then $\sum_{i=1}^{t} \ell_i = k$.

At each of the k intersections of C with the edges of G we add a new vertex; further, each subset of C lying between consecutive new vertices is specified as a new edge. Moreover, each edge of G that is crossed by C, say a total of ℓ times, is subdivided into $\ell + 1$ new edges.

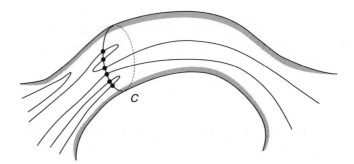

Figure 7.3 A curve C drawn on a handle of the surface S.

Let the new multigraph so formed be denoted by G'; further, suppose G' has order n', size m' and r' regions. Since k new vertices have been introduced in forming G', it follows that $n' = n + k$. The curve C has resulted in an increase of k in the number of edges. Also, each edge e_i, $1 \leq i \leq t$, has given rise to an increase of ℓ_i edges and since $\sum_{i=1}^{t} \ell_i = k$, the total increase in size from G to G' is $2k$; that is, $m' = m + 2k$.

Each portion of C that became an edge of G' is in a region of G. Thus, the addition of such an edge divides that region into two regions. Since there exist k such edges, $r' = r + k$. Because every region of G is a 2-cell, it follows that every region of G' is a 2-cell.

We now make a 'cut' in the handle along C, separating the handle into two pieces (as shown in Figure 7.4). The two resulting holes are 'patched' or 'capped', producing a new (2-cell) region in each case. (This is called a *'capping' operation*.)

In the process of performing this capping operation, several changes have occurred. First, the surface S has been transformed into a new surface S'. The two capped pieces of the handle of S are now part of the sphere of S'. Hence S' has one less handle than S so that S' has genus $g - 1$. Furthermore, the multigraph G' itself has been altered. The vertices and edges resulting from the curve C have been divided into two copies, one copy on each of the two pieces of the capped handle. If G'' denotes this new multigraph, then G'' has order $n'' = n' + k = n + 2k$ and size $m'' = m' + k = m + 3k$. Also, the number r'' of regions satisfies $r'' = r' + 2 = r + k + 2$. Since each of these r'' regions in the connected multigraph G'' is a 2-cell, the inductive hypothesis applies so that $n'' - m'' + r'' = 2 - 2(g - 1)$ or

$$(n + 2k) - (m + 3k) + (r + k + 2) = 2 - 2(g - 1);$$

thus,

$$n - m + r = 2 - 2g,$$

giving the desired result. □

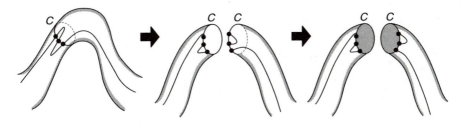

Figure 7.4   Capping a cut handle.

Restating Theorem 7.1 for graphs, we have the following.

Corollary 7.2

Let G be a connected graph of order n and size m with a 2-cell embedding on the surface of genus g and having r regions. Then

$$n - m + r = 2 - 2g.$$

Theorem 7.3

If G is a connected graph embedded on the surface of genus $\gamma(G)$, then every region of G is a 2-cell.

Corollary 7.2 and Theorem 7.3 now immediately imply the following.

Theorem 7.4

If G is a connected graph of order n and size m embedded on the surface of genus $\gamma(G)$ and having r regions, then

$$n - m + r = 2 - 2\,\gamma(G).$$

An important conclusion, which can be reached with the aid of Theorem 7.4, is that every two embeddings of a connected graph G on the surface of genus $\gamma(G)$ results in the same number of regions. With the theorems obtained thus far, we can now provide a lower bound for the genus of a connected graph in terms of its order and size.

Theorem 7.5

If G is a connected graph of order n ≥ 3 and size m, then

$$\gamma(G) \geq \frac{m}{6} - \frac{n}{2} + 1.$$

Proof

The result is immediate for $n = 3$, so we assume that $n \geq 4$. Let G be embedded on the surface of genus $\gamma(G)$. By Theorem 7.4, $n - m + r = 2 - 2\gamma(G)$, where r is the number of regions of G. (Necessarily, each of these regions is a 2-cell by Theorem 7.3.) Since the boundary of every region contains at least three edges and every edge is on the boundary of at most two regions, $3r \leq 2m$. Thus,

$$2 - 2\gamma(G) = n - m + r \leq n - m + \frac{2m}{3},$$

and the desired result follows. □

The lower bound for $\gamma(G)$ presented in Theorem 7.5 can be improved when more information on cycle lengths in G is available.

Theorem 7.6

If G is a connected graph of order n and size m whose smallest cycle has length k, then

$$\gamma(G) \geq \frac{m}{2}\left(1 - \frac{2}{k}\right) - \frac{n}{2} + 1.$$

A special case of Theorem 7.6 that includes bipartite graphs is of special interest. A graph is called *triangle-free* if it contains no triangles.

Corollary 7.7

If G is a connected, triangle-free graph of order n ≥ 3 and size m, then

$$\gamma(G) \geq \frac{m}{4} - \frac{n}{2} + 1.$$

As one might have deduced by now, no general formula for the genus of an arbitrary graph is known. Indeed, it is unlikely that such a formula will ever be developed in terms of quantities that are easily calculable. On the other hand, the following result by Battle, Harary, Kodama and Youngs [BHKY1] implies that, as far as genus formulas are concerned, one need only investigate blocks.

Theorem 7.8

If G is a graph having blocks $B_1, B_2, \ldots, B_k$, then

$$\gamma(G) = \sum_{i=1}^{k} \gamma(B_i).$$

The following corollary is a consequence of the preceding result.

Corollary 7.9

If G is a graph with components $G_1, G_2, \ldots, G_k$, then

$$\gamma(G) = \sum_{i=1}^{k} \gamma(G_i).$$

As is often the case, when no general formula exists for the value of a parameter for an arbitrary graph, formulas (or partial formulas) are established for certain families of graphs. Ordinarily the first classes to be considered are the complete graphs, the complete bipartite graphs, and the *n*-cubes. The genus offers no exception to this rule.

In 1968, Ringel and Youngs [RY1] completed a proof of a result that has a remarkable history. They solved a problem that became known as the *Heawood Map Coloring Problem*; this problem will be discussed in Chapter 8. The solution involved the verification of a conjectured formula for the genus of a complete graph; the proof can be found in (and, in fact, *is*) the book by Ringel [R8].

Theorem 7.10

The genus of the complete graph is given by

$$\gamma(K_n) = \left\lceil \frac{(n-3)(n-4)}{12} \right\rceil, \quad n \geq 3.$$

A formula for the genus of the complete bipartite graph was discovered by Ringel [R7].

Theorem 7.11

The genus of the complete bipartite graph is given by

$$\gamma(K_{r,s}) = \left\lceil \frac{(r-2)(s-2)}{4} \right\rceil, \quad r, s \geq 2.$$

A formula for the genus of the *n*-cube was found by Ringel [R5] and by Beineke and Harary [BH1]. We prove this result to illustrate some of the techniques involved. We omit the obvious equality $\gamma(Q_1) = 0$.

Theorem 7.12

For $n \geq 2$, the genus of the n-cube is given by

$$\gamma(Q_n) = (n-4) \cdot 2^{n-3} + 1.$$

Proof

The *n*-cube is a triangle-free graph of order 2^n and size $n \cdot 2^{n-1}$; thus, by Corollary 7.7,

$$\gamma(Q_n) \geq (n-4) \cdot 2^{n-3} + 1.$$

To verify the reverse inequality, we employ induction on *n*. For $n \geq 2$, define the statement $A(n)$ as follows: The graph Q_n can be embedded on the surface of genus $(n-4) \cdot 2^{n-3} + 1$ such that the boundary of every region is a 4-cycle and such that there exist 2^{n-2} regions with pairwise disjoint boundaries. That the statements $A(2)$ and $A(3)$ are true is trivial. Assume $A(k-1)$ to be true, $k \geq 4$, and, accordingly, let S be the surface of genus $(k-5) \cdot 2^{k-4} + 1$ on which Q_{k-1} is embedded such that the boundary of each region is a 4-cycle and such that there exist 2^{k-3} regions with pairwise disjoint boundaries. We note that since Q_{k-1} has order 2^{k-1}, each vertex of Q_{k-1} belongs to the boundary of precisely one of the aforementioned 2^{k-3} regions. Now let Q_{k-1} be embedded on another copy S' of the surface of genus $(k-5) \cdot 2^{k-4} + 1$ such that the embedding of Q_{k-1} on S' is a 'mirror image' of the embedding of Q_{k-1} on S (that is, if v_1, v_2, v_3, v_4 are the vertices of the boundary of a region of Q_{k-1} on S, where the vertices are listed clockwise about the 4-cycle, then there is a region on S', with the vertices v_1, v_2, v_3, v_4 on its boundary listed counterclockwise). We now consider the 2^{k-3} distinguished regions of S together with the corresponding regions of S', and join each pair of associated regions by a handle. The addition of the first handle produces the surface of genus $2[(k-5) \cdot 2^{k-4}+1]$ while the addition of each of the other $2^{k-3}-1$ handles results in an increase of one to the genus. Thus, the surface just constructed has genus $(k-4) \cdot 2^{k-3}+1$. Now each set of four vertices on the boundary of a distinguished region can be joined to the corresponding four vertices on the boundary of the associated region so that the four edges are embedded on the handle joining the regions. It is now immediate that the resulting graph is isomorphic to Q_k and that every region is bounded by a 4-cycle. Furthermore, each added handle gives rise to four regions, 'opposite' ones of which have disjoint boundaries, so there exist 2^{k-2} regions of Q_k that are pairwise disjoint.

Thus, $A(n)$ is true for all $n \geq 2$, proving the result. $\qquad \square$

EXERCISES 7.1

7.1 Determine $g = \gamma(K_{4,4})$ without using Theorem 7.11 and label the regions in a 2-cell embedding of $K_{4,4}$ on the surface of genus g.

7.2 (a) Show that $\gamma(G) \leq \nu(G)$ for every graph G.
(b) Prove that for every positive integer k, there exists a graph G such that $\gamma(G) = 1$ and $\nu(G) = k$.

7.3 Prove Theorem 7.6.

7.4 Use Theorem 7.8 to prove Corollary 7.9.

7.5 Show that

$$\gamma(K_n) \geq \left\lceil \frac{(n-3)(n-4)}{12} \right\rceil \quad \text{for } n \geq 3.$$

7.6 Show that

$$\gamma(K_{r,s}) \geq \left\lceil \frac{(r-2)(s-2)}{4} \right\rceil \quad \text{for } r, s \geq 2.$$

7.7 (a) Find a lower bound for $\gamma(K_{3,3} + \bar{K}_n)$.
(b) Determine $\gamma(K_{3,3} + \bar{K}_n)$ exactly for $n = 1, 2, 3$.

7.8 Determine $\gamma(K_2 \times C_4 \times C_6)$.

7.9 Prove, for every positive integer g, that there exists a connected graph G of genus g.

7.10 Prove, for each positive integer k, that there exists a connected planar graph G such that $\gamma(G \times K_2) \geq k$.

7.2 2-CELL EMBEDDINGS OF GRAPHS

In the preceding section we saw that every graph G has a genus; that is, there exists a surface (a compact orientable 2-manifold) of minimum genus on which G can be embedded. Indeed, by Theorem 7.3 if G is a connected graph that is embedded on the surface of genus $\gamma(G)$, then the embedding is necessarily a 2-cell embedding. On the other hand, if G is disconnected, then no embedding of G is a 2-cell embedding.

Our primary interest lies with embeddings of (connected) graphs that are 2-cell embeddings. In this section, we investigate graphs and the surfaces on which they can be 2-cell embedded. It is convenient to denote the surface of genus k by S_k. Thus, S_0 represents the sphere (or plane), S_1 represents the torus, and S_2 represents the double torus (or sphere with two handles).

We have already mentioned that the torus can be represented as a square with opposite sides identified. More generally, the surface S_k ($k > 0$) can be represented as a regular $4k$-gon whose $4k$ sides can be listed in clockwise order as

$$a_1 b_1 a_1^{-1} b_1^{-1} a_2 b_2 a_2^{-1} b_2^{-1} \ldots a_k b_k a_k^{-1} b_k^{-1}, \tag{7.2}$$

where, for example, a_1 is a side directed clockwise and a_1^{-1} is a side also labeled a_1 *but* directed counterclockwise. These two sides are then identified in a manner consistent with their directions. Thus, the double torus can be represented as shown in Figure 7.5. The 'two' points labeled X are actually the same point on S_2 while the 'eight' points labeled Y are, in fact, a single point.

Although it is probably obvious that there exist a variety of graphs that can be embedded on the surface S_k of a given nonnegative integer k, it may not be entirely obvious that there always exist graphs for which a 2-cell embedding on S_k exists.

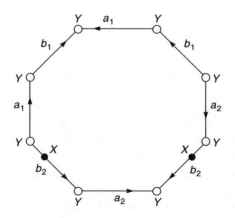

Figure 7.5 A representation of the double torus.

Theorem 7.13

For every nonnegative integer k, there exists a connected graph that has a 2-cell embedding on S_k.

Proof

For $k = 0$, every connected planar graph has the desired property; thus, we assume that $k > 0$.

We represent S_k as a regular $4k$-gon whose $4k$ sides are described and identified as in (7.2). First, we define a multigraph H as follows. At each vertex of the $4k$-gon, let there be a vertex of H. Actually, the identification process associated with the $4k$-gon implies that there is only one vertex of H. Let each side of the $4k$-gon represent an edge of H. The identification produces $2k$ distinct edges, each of which is a loop. This completes the construction of H. Hence, the multigraph H has order 1 and size $2k$. Furthermore, there is only one region, namely the interior of the polygon; this region is clearly a 2-cell. Therefore, there exists a 2-cell embedding of H on S_K.

To convert the multigraph H into a graph, we subdivide each loop twice, producing a graph G having order $4k + 1$, size $6k$, and again a single 2-cell region. □

Figure 7.6 illustrates the construction given in the proof of Theorem 7.13 in the case of the torus S_1. The graph G so constructed is shown in Figure 7.6(a). In Figures 7.6(b)–(e) we see a variety of ways of visualizing the embedding. In Figure 7.6(b), a 3-dimensional embedding is described. In Figures 7.6(c) and (d), the torus is represented as a rectangle with opposite sides identified. (Figure 7.6(c) is the actual drawing described in the proof of the theorem.) In Figure 7.6(e), a portion of G is drawn in the plane, then two circular holes are made in the plane and a tube (or handle) is placed over the plane joining the two holes. The edge uv is then drawn over the handle, completing the 2-cell embedding.

The graphs G constructed in the proof of Theorem 7.13 are planar. Hence, for every nonnegative integer k, there exist planar graphs that can be 2-cell embedded on S_k. It is also true that for every planar graph G and *positive* integer k, there exists an embedding of G on S_k that is *not* a 2-cell embedding. In general, for a given graph G and positive integer k with $k > \gamma(G)$, there always exists an embedding of G on S_k that is not a 2-cell embedding, which can be obtained from an embedding of G on $S_{\gamma(G)}$ by adding $k - \gamma(G)$ handles to the interior of some region of G. If $k = \gamma(G)$ and G is connected, then by Theorem 7.3 every embedding of G on S_k is a 2-cell embedding while, of course, if $k < \gamma(G)$, there is no embedding whatsoever of G on S_k.

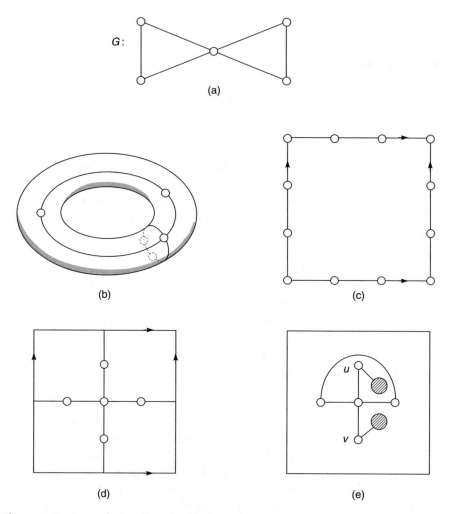

Figure 7.6 A graph 2-cell embedded on the torus.

Thus far, whenever we have described a 2-cell embedding (or, in fact, any embedding) of a graph G on a surface S_k, we have resorted to a geometric description, such as the ones shown in Figure 7.6. There is a far more useful method, algebraic in nature, which we shall now discuss.

Consider the 2-cell embedding of K_5 on S_1 shown in Figure 7.7, with the vertices of K_5 labeled as indicated. Observe that in this embedding the edges incident with v_1 are arranged cyclically counterclockwise about v_1 in the order $v_1v_2, v_1v_3, v_1v_4, v_1v_5$ (or, equivalently, $v_1v_3, v_1v_4, v_1v_5, v_1v_2$, and so on). This induces a cyclic permutation π_1 of the subscripts of the vertices adjacent with v_1, namely $\pi_1 = (2\ 3\ 4\ 5)$, expressed as a permutation cycle. Similarly, this embedding induces a cyclic permutation π_2 of the subscripts of the vertices adjacent with v_2;

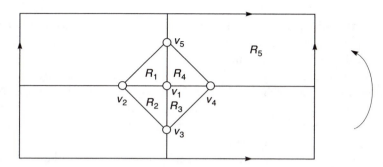

Figure 7.7 A 2-cell embedding of K_5 on the torus.

in particular, $\pi_2 = (1\,5\,4\,3)$. In fact, for each vertex $v_i\,(1 \le i \le 5)$, one can associate a cyclic permutation π_i with v_i. In this case, we have

$$\pi_1 = (2\,3\,4\,5),$$
$$\pi_2 = (1\,5\,4\,3),$$
$$\pi_3 = (1\,2\,5\,4),$$
$$\pi_4 = (1\,3\,2\,5),$$
$$\pi_5 = (1\,4\,3\,2).$$

In the 2-cell embedding of K_5 on S_1 shown in Figure 7.7, there are five regions, labeled $R_1, R_2, \ldots, R_5$. Each region R_i $(1 \le i \le 5)$ is, of course, a 2-cell. The boundary of the region R_1 consists of the vertices v_1, v_2 and v_5 and the edges v_1v_2, v_2v_5, and v_5v_1. If we trace out the edges of the boundary in a clockwise direction, that is, keeping the boundary at our left and the region to our right (Figure 7.8), beginning with the edge v_1v_2, we have v_1v_2, followed by v_2v_5, and finally v_5v_1. This information can also be obtained from the cyclic permutations

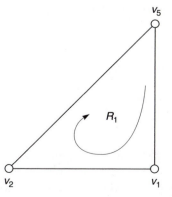

Figure 7.8 Tracing out a region.

$\pi_1, \pi_2, \ldots, \pi_5$; indeed, the edge following $v_1 v_2 = v_2 v_1$ as we trace the boundary edges of R_1 in a clockwise direction is precisely the edge incident with v_2 that follows $v_2 v_1$ if one proceeds counterclockwise about v_2; that is, the edge following $v_1 v_2$ in the boundary of R_1 is $v_2 v_{\pi_2(1)} = v_2 v_5$ Similarly, the edge following $v_2 v_5 = v_5 v_2$ as we trace out the edges of the boundary of R_1 in a clockwise direction is $v_5 v_{\pi_5(2)} = v_5 v_1$ Hence with the aid of the cyclic permutations $\pi_1, \pi_2, \ldots, \pi_5$, we can trace out the edges of the boundary of R_1. In a like manner, the boundary of every region of the embedding can be so described.

Since the direction (namely, clockwise) in which the edges of the boundary of a region are traced in the above description is of utmost importance, it is convenient to regard each edge of K_5 as a symmetric pair of arcs and, thus, to interpret K_5 itself as a digraph D. With this interpretation, the boundary of the region R_1 and thus R_1 itself can be described, starting at v_1, as

$$(v_1, v_2), (v_2, v_{\pi_2(1)}), (v_5, v_{\pi_5(2)})$$

or

$$(v_1, v_2), (v_2, v_5), (v_5, v_1). \tag{7.3}$$

We now define a mapping $\pi \colon E(D) \to E(D)$ as follows. Let $a \in E(D)$, where $a = (v_i, v_j)$. Then

$$\pi(a) = \pi((v_i, v_j)) = \pi(v_i, v_j) = (v_j, v_{\pi_j(i)}).$$

The mapping π is one-to-one and so is a permutation of $E(D)$. Thus, π can be expressed as a product of disjoint permutation cycles. In this context, each permutation cycle of π is referred to as an 'orbit' of π. Hence (7.3) corresponds to an orbit of π and is often denoted more compactly as $v_1 - v_2 - v_5 - v_1$. (Although this orbit corresponds to a cycle in the graph, this is not always the case for an arbitrary orbit in a graph that is 2-cell embedded). For the embedding of K_5 on S_1 shown in Figure 7.7, the list of all five orbits (one for each region) is given below:

$R_1 \colon v_1 - v_2 - v_5 - v_1,$

$R_2 \colon v_1 - v_3 - v_2 - v_1,$

$R_3 \colon v_1 - v_4 - v_3 - v_1,$

$R_4 \colon v_1 - v_5 - v_4 - v_1,$

$R_5 \colon v_2 - v_3 - v_5 - v_2 - v_4 - v_5 - v_3 - v_4 - v_2.$

The orbits of π form a partition of $E(D)$ and, as such, each arc of D appears in exactly one orbit of π. Since D is the digraph obtained by replacing each edge of K_5 by a symmetric pair of arcs, each edge of K_5 appears twice among the orbits of π, once for each of the two possible directions that are assigned to the edge.

The 2-cell embedding of K_5 on S_1 shown in Figure 7.7 uniquely determines the collection $\{\pi_1, \pi_2, \ldots, \pi_5\}$ of permutations of the subscripts of the vertices adjacent to the vertices of K_5. This set of permutations, in turn, completely describes the embedding of K_5 on S_1 shown in Figure 7.7.

This method of describing an embedding is referred to as the *Rotational Embedding Scheme*. Such a scheme was observed and used by Dyck [D10] in 1888 and by Heffter [H9] in 1891. It was formalized by Edmonds [E1] in 1960 and discussed in more detail by Youngs [Y1] in 1963.

We now describe the Rotational Embedding Scheme in a more general setting. Let G be a nontrivial connected graph with $V(G) = \{v_1, v_2, \ldots, v_n\}$. Let

$$V(i) = \{j \mid v_j \in N(v_i)\}.$$

For each i ($1 \leq i \leq n$), let π_i: $V(i) \to V(i)$ be a cyclic permutation (or rotation) of $V(i)$. Thus, each permutation π_i can be represented by a (permutation) cycle of length $|V(i)| = |N(v_i)| = \deg v_i$. The Rotational Embedding Scheme states that there is a one-to-one correspondence between the 2-cell embeddings of G (on all possible surfaces) and the n-tuples $(\pi_1, \pi_2, \ldots, \pi_n)$ of cycle permutations.

Theorem 7.14 (The Rotational Embedding Scheme)

Let G be a nontrivial connected graph with $V(G) = \{v_1, v_2, \ldots, v_n\}$. For each 2-cell embedding of G on a surface, there exists a unique n-tuple $(\pi_1, \pi_2, \ldots, \pi_n)$, where for $i = 1, 2, \ldots, n$, π_i: $V(i) \to V(i)$ is a cyclic permutation that describes the subscripts of the vertices adjacent to v_i in counterclockwise order about v_i. Conversely, for each such n-tuple $(\pi_1, \pi_2, \ldots, \pi_n)$, there exists a 2-cell embedding of G on some surface such that for $i = 1, 2, \ldots, n$, the subscripts of the vertices adjacent to v_i and in counterclockwise order about v_i are given by π_i.

Proof

Let there be given a 2-cell embedding of G on some surface. For each vertex v_i of G, define π_i: $V(i) \to V(i)$ as follows: If $v_i v_j \in E(G)$ and $v_i v_t$ (possibly $t = j$) is the next edge encountered after $v_i v_j$ as we proceed counterclockwise about v_i, then we define $\pi_i(j) = t$. Each π_i so defined is a cyclic permutation.

Conversely, assume that we are given an n-tuple $(\pi_1, \pi_2, \ldots, \pi_n)$ such that for each i ($1 \leq i \leq n$), π_i: $V(i) \to V(i)$ is a cyclic permutation. We show that this determines a 2-cell embedding of G on some surface. (By necessity, this proof requires the use of properties of compact orientable 2-manifolds.)

Let D denote the digraph obtained from G by replacing each edge of G by a symmetric pair of arcs. We define a mapping $\pi \colon E(D) \to E(D)$ by

$$\pi((v_i, v_j)) = \pi(v_i, v_j) = (v_j, v_{\pi_j(i)}).$$

The mapping π is one-to-one and, thus, is a permutation of $E(D)$. Hence, π can be expressed as a product of disjoint permutation cycles. Each of these permutation cycles is called an *orbit* of π. Thus, the orbits partition the set $E(D)$. Assume that

$$R \colon ((v_i, v_j)(v_j, v_t)\ldots(v_\ell, v_i))$$

is an orbit of π, which we also write as

$$R \colon v_i - v_j - v_t - \cdots - v_\ell - v_i.$$

Hence, this implies that in the desired embedding, if we begin at v_i and proceed along (v_i, v_j) to v_j, then the next arc we must encounter after (v_i, v_j) in a counterclockwise direction about v_j is $(v_j, v_{\pi_j(i)}) = (v_j, v_t)$. Continuing in this manner, we must finally arrive at the arc (v_ℓ, v_i) and return to v_i, in the process describing the boundary of a (2-cell) region (considered as a subset of the plane) corresponding to the orbit R. Therefore, each orbit of π gives rise to a 2-cell region in the desired embedding.

To obtain the surface S on which G is 2-cell embedded, pairs of regions, with their boundaries, are 'pasted' along certain arcs; in particular, if (v_i, v_j) is an arc on the boundary of R_1 and (v_j, v_i) is an arc on the boundary of R_2, then (v_i, v_j) is identified with (v_j, v_i) as shown in Figure 7.9. The properties of compact orientable 2-manifolds imply that S is indeed an appropriate surface.

In order to determine the genus of S, one needs only to observe that the number r of regions equals the number of orbits. Thus, if G has order n and size m, then by Corollary 7.2, $S = S_k$ where k is the nonnegative integer satisfying the equation $n - m + r = 2 - 2k$. □

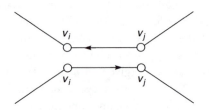

Figure 7.9 A step in the proof of Theorem 7.14.

As an illustration of the Rotational Embedding Scheme, we once again consider the complete graph K_5, with $V(K_5) = \{v_1, v_2, v_3, v_4, v_5\}$. Let there be given the 5-tuple $(\pi_1, \pi_2, \pi_3, \pi_4, \pi_5)$, where

$$\pi_1 = (2\,3\,4\,5),$$
$$\pi_2 = (1\,3\,4\,5),$$
$$\pi_3 = (1\,2\,4\,5),$$
$$\pi_4 = (1\,2\,3\,5),$$
$$\pi_5 = (1\,2\,3\,4).$$

Thus, by Theorem 7.14, this 5-tuple describes a 2-cell embedding of K_5 on some surface S_k. To evaluate k, we consider the digraph D obtained by replacing each edge of K_5 by a symmetric pair of arcs and determine the orbits of the permutation $\pi \colon E(D) \to E(D)$ defined in the proof of Theorem 7.14. The orbits are

$$R_1\colon v_1 - v_2 - v_3 - v_4 - v_5 - v_1,$$
$$R_2\colon v_1 - v_3 - v_2 - v_4 - v_3 - v_5 - v_4 - v_1 - v_5 - v_2 - v_1,$$
$$R_3\colon v_1 - v_4 - v_2 - v_5 - v_3 - v_1;$$

and each orbit corresponds to a 2-cell region. Thus, the number of regions in the embedding is $r = 3$. Since K_5 has order $n = 5$ and size $m = 10$, and since $n - m + r = -2 = 2 - 2k$, it follows that $k = 2$, so that the given 5-tuple describes an embedding of K_5 on S_2.

Given an n-tuple of cyclic permutations as we have described, it is not necessarily an easy problem to present a geometric description of the embedding, particularly on surfaces of high genus. For the example just presented, however, we give two geometric descriptions in Figure 7.10. In Figure 7.10(a), a portion of K_5 is drawn in the plane. Two handles are then inserted over the plane, as indicated, and the remainder of K_5 is drawn along these handles. The edge $e_1 = v_2 v_5$ is drawn along the handle H_1, the edge $e_2 = v_3 v_5$ is drawn along H_2 while $e_3 = v_1 v_3$ is drawn along both H_1 and H_2. The three 2-cell regions produced are denoted by R_1, R_2 and R_3.

In Figure 7.10(b), this 2-cell embedding of K_5 on S_2 is shown on the regular octagon. The labeling of the eight sides (as in [7.2]) indicates the identification used in producing S_2.

As a more general illustration of Theorem 7.14 we determine the genus of the complete bipartite graph $K_{2a,2b}$. According to Theorem 7.11, $\gamma(K_{2a,2b}) = (a-1)(b-1)$. That $(a-1)(b-1)$ is a lower bound for $\gamma(K_{2a,\,2b})$ follows from Exercise 7.6. We use Theorem 7.14 to show that $K_{2a,2b}$ is 2-cell embeddable on

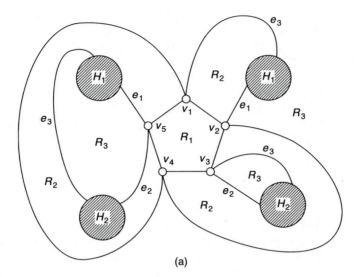

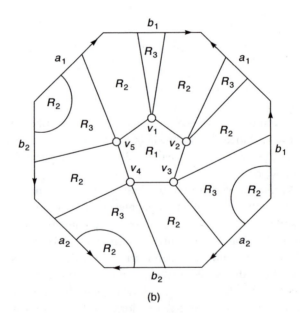

Figure 7.10 A 2-cell embedding of K_5 on the double torus.

$S_{(a-1)(b-1)}$, thereby proving that $\gamma(K_{2a,2b}) \leq (a-1)(b-1)$ and completing the argument.

Denote the partite sets of $K_{2a,2b}$ by U and W, where $|U| = 2a$ and $|W| = 2b$. Further, label the vertices so that

$$U = \{v_1, v_3, v_5, \ldots, v_{4a-1}\} \quad \text{and} \quad W = \{v_2, v_4, v_6, \ldots, v_{4b}\}.$$

Let there be given the $(2a + 2b)$-tuple (assuming that $a \le b$)

$$(\pi_1, \pi_2, \ldots, \pi_{4a-1}, \pi_{4a}, \pi_{4a+2}, \pi_{4a+4}, \ldots, \pi_{4b}),$$

where

$$\pi_1 = \pi_5 = \cdots = \pi_{4a-3} = (2\,4\,6\,\ldots\,4b),$$
$$\pi_3 = \pi_7 = \cdots = \pi_{4a-1} = (4b\,\ldots\,6\,4\,2),$$
$$\pi_2 = \pi_6 = \cdots = \pi_{4b-2} = (1\,3\,5\,\ldots\,4a - 1),$$
$$\pi_4 = \pi_8 = \cdots = \pi_{4b} = (4a - 1\,\ldots\,5\,3\,1).$$

By Theorem 7.14, then, this $(2a + 2b)$-tuple describes a 2-cell embedding of $K_{2a, 2b}$ on some surface S_h. In order to evaluate h, we let D denote the digraph obtained by replacing each edge of $K_{2a, 2b}$ by a symmetric pair of arcs and determine the orbits of the permutation $\pi: E(D) \to E(D)$ defined in the proof of Theorem 7.14.

Every orbit of π contains an arc of the type (v_s, v_t), where $v_s \in U$ and $v_t \in W$. If $s \equiv 1 \pmod 4$ and $t \equiv 2 \pmod 4$, then the resulting orbit R containing (v_s, v_t) is

$$R: v_s - v_t - v_{s+2} - v_{t-2} - v_s,$$

with $s + 2$ expressed modulo $4a$ and $t - 2$ expressed modulo $4b$. Note that R also contains the arc (v_{s+2}, v_{t-2}), where, then, $s + 2 \equiv 3 \pmod 4$ and $t - 2 \equiv 0 \pmod 4$. If $s \equiv 1 \pmod 4$ and $t \equiv 0 \pmod 4$, then the orbit R' containing (v_s, v_t) is

$$R': v_s - v_t - v_{s-2} - v_{t-2} - v_s,$$

where, again, $s - 2$ is expressed modulo $4a$ and $t - 2$ is expressed modulo $4b$. The orbit R' also contains the arc (v_{s-2}, v_{t-2}), where, $s - 2 \equiv 3 \pmod 4$ and $t - 2 \equiv 2 \pmod 4$. Thus every orbit of π is either of the type R (where $s \equiv 1 \pmod 4$ and $t \equiv 2 \pmod 4$) or the type R' (where $s \equiv 1 \pmod 4$ and $t \equiv 0 \pmod 4$). Since there are a choices for s and b choices for t in each case, the total number of orbits is $2ab$; therefore, the number of regions in this embedding is $r = 2ab$.

Since $K_{2a, 2b}$ has order $n = 2a + 2b$ and size $m = 4ab$ and because $n - m + r = 2 - 2h$, we have

$$(2a + 2b) - 4ab + 2ab = 2 - 2h,$$

so that $h = (a - 1)(b - 1)$. Hence there is a 2-cell embedding of $K_{2a,2b}$ on $S_{(a-1)(b-1)}$, as we wished to show.

As a theoretical application of Theorem 7.14, we present a result that is referred to as the Ringeisen–White Edge-Adding Lemma (see Ringeisen [R4]).

Theorem 7.15

Let G be a connected graph with $V(G) = \{v_1, v_2, \ldots, v_n\}$ such that v_i and v_j are distinct nonadjacent vertices. Suppose that there exists a 2-cell embedding of G on some surface S_h with r regions such that v_i is on the boundary of region R_i and v_j is on the boundary of region R_j. Let $H = G + v_i v_j$. Then

(i) *if $R_i \neq R_j$, there exists a 2-cell embedding of H on S_{h+1} with $r-1$ regions in which v_i and v_j are on the boundary of the same region;*

(ii) *if $R_i = R_j$, there exists a 2-cell embedding of H on S_h with $r+1$ regions in which each of v_i and v_j belongs to the boundaries of (the same) two distinct regions.*

Proof

By hypothesis, there exists a 2-cell embedding of the graph G of order n and size m on S_h with r regions such that v_i is on the boundary of region R_i and v_j is on the boundary of region R_j. By Theorem 7.14, an n-tuple $(\pi_1, \pi_2, \ldots, \pi_n)$ of cyclic permutations corresponds to this embedding, namely for $t = 1, 2, \ldots, n$, π_t: $V(t) \to V(t)$ is a cyclic permutation of the subscripts of the vertices of $N(v_t)$ in counterclockwise order about v_t.

Let D denote the symmetric digraph obtained from G by replacing each edge by a symmetric pair of arcs and let $\pi: E(D) \to E(D)$ be defined by $\pi(v_\alpha, v_\beta) = (v_\beta, v_{\pi_\beta(\alpha)})$. Since the given embedding has r regions, π has r orbits. Denote each region and its corresponding orbit by the same symbol; in particular, R_i and R_j are orbits of π.

Suppose that $R_i \neq R_j$. We can therefore represent orbits R_i and R_j as

$$R_i: v_i - v_k - \cdots - v_{k'} - v_i$$

and

$$R_j: v_j - v_\ell - \cdots - v_{\ell'} - v_j.$$

It therefore follows that

$$\pi_i(k') = k \quad \text{and} \quad \pi_j(\ell') = \ell.$$

We now consider the graph $H = G + v_i v_j$ and define

$$V'(t) = \{r \mid v_r v_t \in E(H)\}$$

for $t = 1, 2, \ldots, n$. Thus $V'(t) = V(t)$ for $t \neq i, j$, and $V'(i) = V(i) \cup \{j\}$ while $V'(j) = V(j) \cup \{i\}$. For the graph H, we define an n-tuple $(\pi'_1, \pi'_2, \ldots, \pi'_n)$ of cyclic permutations, where $\pi'_t \colon V'(t) \to V'(t)$ for $t = 1, 2, \ldots, n$ such that $\pi'_t = \pi_t$ for $t \neq i, j$. Furthermore,

$$\pi'_i(a) = \begin{cases} \pi_i(a) & \text{if} \quad a \neq k' \\ j & \text{if} \quad a = k' \\ k & \text{if} \quad a = j \end{cases}$$

and

$$\pi'_j(a) = \begin{cases} \pi_j(a) & \text{if} \quad a \neq \ell' \\ i & \text{if} \quad a = \ell' \\ \ell & \text{if} \quad a = i. \end{cases}$$

Let D' be the digraph obtained from H by replacing each edge of H by a symmetric pair of arcs. Define the permutation $\pi' \colon E(D') \to E(D')$ by $\pi'(v_\alpha, v_\beta) = (v_\beta, v_{\pi'_\beta(\alpha)})$. The orbits of π' then consist of all orbits of π different from R_i and R_j together with the orbit

$$R \colon v_i - v_j - v_\ell - \cdots - v_{\ell'} - v_j - v_i - v_k - \cdots - v_{k'} - v_i.$$

Thus, π' has $r - 1$ orbits and the corresponding 2-cell embedding of H has $r - 1$ regions. Moreover, v_i and v_j lie on the boundary of R. Since $n - m + r = 2 - 2h$, it follows that $n - (m + 1) + (r - 1) = 2 - 2(h + 1)$ and H is 2-cell embedded on S_{h+1}. This completes the proof of (i).

Suppose that $R_i = R_j$. We can represent the orbit $R_i \ (= R_j)$ as

$$R_i \colon v_i - v_k - \cdots - v_{\ell'} - v_j - v_{\ell'} - \cdots - v_{k'} - v_i.$$

(Note that v_i and v_j cannot be consecutive in R_i since $v_i v_j \notin E(G)$.) It follows that

$$\pi_i(k') = k \quad \text{and} \quad \pi_j(\ell') = \ell.$$

We again consider the graph $H = G + v_i v_j$ and once more define

$$V'(t) = \{s \mid v_s v_t \notin E(H)\}$$

for $t = 1, 2, \ldots, n$. We define an n-tuple $(\pi'_1, \pi'_2, \ldots, \pi'_n)$ of cyclic permutations, where $\pi'_t \colon V'(t) \to V'(t)$ for $t = 1, 2, \ldots, n$ such that $\pi'_t = \pi_t$ for $t \neq i, j$. Also,

$$\pi'_i(a) = \begin{cases} \pi_i(a) & \text{if} \quad a \neq k' \\ j & \text{if} \quad a = k' \\ k & \text{if} \quad a = j \end{cases}$$

and

$$\pi'_j(a) = \begin{cases} \pi_j(a) & \text{if} \quad a \neq \ell' \\ i & \text{if} \quad a = \ell' \\ \ell & \text{if} \quad a = i. \end{cases}$$

Again we denote by D' the digraph obtained from H by replacing each edge of H by a symmetric pair of arcs and define the permutation $\pi': E(D') \to E(D')$ by $\pi'(v_\alpha, v_\beta) = (v_\beta, v_{\pi'_\beta(\alpha)})$. The orbits of π' consist of all orbits of π different from R_i together with the orbits

$$R': v_i - v_j - v_\ell - \cdots - v_{k'} - v_i$$

and

$$R'': v_j - v_i - v_k - \cdots - v_{\ell'} - v_j.$$

Therefore, π' has $r+1$ orbits and the resulting 2-cell embedding of H has $r+1$ regions. Furthermore, each of v_i and v_j belongs to the boundaries of both R' and R''. Here $n - m + r = 2 - 2h$ implies that $n - (m+1) + (r+1) = 2 - 2h$, and H is 2-cell embedded on S_h, which verifies (ii). □

A consequence of Theorem 7.15 will prove to be useful.

Corollary 7.16

Let e and f be adjacent edges of a connected graph G. If there exists a 2-cell embedding of $G' = G - e - f$ with one region, then there exists a 2-cell embedding of G with one region.

Proof

Let $e = uv$ and $f = vw$, where then $u \neq w$. Let there be given a 2-cell embedding of G' with one region R. Thus all vertices of G' belong to the boundary of R, including u and v. By Theorem 7.15(ii), there exists a 2-cell embedding of $G' + e$ with two regions where u and v lie on the boundary of both regions. Therefore, v is on the boundary of one region and w is on the boundary of the other region in the 2-cell embedding of $G' + e$. Applying Theorem 7.15(i), we conclude that there exists a 2-cell embedding of $G' + e + f = G$ with one region. □

We now turn our attention for the remainder of the section to the following question: Given a (connected) graph G, on which surfaces S_k do there exist 2-cell embeddings of G? As a major step towards answering this question, we present the following 'interpolation theorem' of Duke [D9].

Theorem 7.17

If there exist 2-cell embeddings of a connected graph G on the surfaces S_p and S_q, where $p \leq q$, and k is any integer such that $p \leq k \leq q$, then there exists a 2-cell embedding of G on the surface S_k.

Proof

Observe that there exist 2-cell embeddings of K_1 only on the sphere; thus, we assume that G is nontrivial.

Assume that there exists a 2-cell embedding of G on some surface S_ℓ. Let $V(G) = \{v_1, v_2, \ldots, v_n\}$, $n \geq 2$. By Theorem 7.14, there exists an n-tuple $(\pi_1, \pi_2, \ldots, \pi_n)$ of cyclic permutations associated with this embedding such that for $i = 1, 2, \ldots, n$, $\pi_i \colon V(i) \to V(i)$ is a cyclic permutation of the subscripts of the vertices of $N(v_i)$ in counterclockwise order about v_i.

Let D be the symmetric digraph obtained from G by replacing each edge by a symmetric pair of arcs. Let $\pi \colon E(D) \to E(D)$ be the permutation defined by $\pi(v_i, v_j) = (v_j, v_{\pi_j(i)})$. Denote the number of orbits in π by r; that is, assume that there are r 2-cell regions in the given embedding of G on S_ℓ.

Assume there exists some vertex of G, say v_1, such that $\deg v_1 \geq 3$. Then $\pi_1 = (a\, b\, c \ldots)$, where a, b and c are distinct. Let v_x be any vertex adjacent with v_1 other then v_a and v_b, and suppose that $\pi_1(x) = y$. Thus

$$\pi_1 = (a\, b\, c \ldots x\, y \ldots),$$

where, possibly, $x = c$ or $y = a$. Let E_1 be the subset of $E(D)$ consisting of the three pairs of arcs

$$(v_a, v_1), (v_1, v_b); \quad (v_b, v_1), (v_1, v_c); \quad (v_x, v_1), (v_1, v_y). \tag{7.4}$$

Note that the six arcs listed in (7.4) are all distinct. By the definition of the permutation π, we have

$$\pi(v_a, v_1) = (v_1, v_b), \quad \pi(v_b, v_1) = (v_1, v_c), \quad \text{and} \quad \pi(v_x, v_1) = (v_1, v_y).$$

This implies that the arc (v_a, v_1) is followed by the arc (v_1, v_b) in some orbit of π, and that the edge $v_a v_1$ of G is followed by the edge $v_1 v_b$ as we proceed clockwise around the boundary of the corresponding region in the given embedding of G in S_ℓ. Also, (v_b, v_1) is followed by (v_1, v_c) in some orbit of π and (v_x, v_1) is followed by (v_1, v_y) in some orbit.

We now define a new permutation $\pi': E(D) \to E(D)$ with the aid of the n-tuple $(\pi'_1, \pi'_2, \ldots, \pi'_n)$, where for $i = 1, 2, \ldots, n$, $\pi'_i: V(i) \to V(i)$ is a cyclic permutation defined by

$$\pi'_i = \begin{cases} (a\,c \ldots x\,b\,y \ldots) & \text{if } i = 1 \\ \pi_i & \text{if } 2 \le i \le n. \end{cases}$$

We then define $\pi'(v_i, v_j) = (v_j, v_{\pi'_j(i)})$. By Theorem 7.14, the n-tuple $(\pi'_1, \pi'_2, \ldots, \pi'_n)$ determines a 2-cell embedding of G on some surface, where for $i = 1, 2, \ldots, n$, π'_i is a cyclic permutation of the subscripts of the vertices adjacent to v_i in counterclockwise order about v_i.

Three cases are now considered, depending on the possible distribution of the pairs (7.4) of arcs in E_1 among the orbits of π.

Case 1. Assume that all arcs of E_1 belong to a single orbit R of π. Suppose, first, that the orbit R has the form

$$R: v_1 - v_y - \cdots - v_b - v_1 - v_c - \cdots - v_a - v_1 - v_b - \cdots - v_x - v_1.$$

Here the orbits of π' are the orbits of π except that the orbit R is replaced by the three orbits

$$R'_1: v_1 - v_y - \cdots - v_b - v_1,$$
$$R'_2: v_1 - v_c - \cdots - v_a - v_1,$$
$$R'_3: v_1 - v_b - \cdots - v_x - v_1.$$

Hence, π' describes a 2-cell embedding of G with $r + 2$ regions on a surface S'. Necessarily, then, $S' = S_{\ell-1}$.

The other possible form that the orbit R may take is

$$R: v_1 - v_y - \cdots - v_a - v_1 - v_b - \cdots - v_b - v_1 - v_c - \cdots - v_x - v_1.$$

In this situation, the orbits of π' are the orbits of π, except for R, which is replaced by the orbit

$$R': v_1 - v_y - \cdots - v_a - v_1 - v_c - \cdots - v_x - v_1 - v_b - \cdots - v_b - v_1.$$

Hence, π' has r orbits.

Case 2. Assume that π has two orbits, say R_1 and R_2, with R_1 containing two of the pairs of arcs in E_1, and R_2 containing the remaining pair of arcs. In this case, the orbits of π' are those of π, except for R_1 and R_2, which are replaced by two orbits R'_1 and R'_2, where one of R'_1 and R'_2 contains two arcs of E_1 and the other contains the remaining four arcs of E_1. In this case, π' has r orbits.

Case 3. Assume that π has three orbits R_1, R_2 and R_3 such that (v_a, v_1) is followed by (v_1, v_b) in R_1, (v_b, v_1) is followed by (v_1, v_c) in R_2, and (v_x, v_1) is followed by (v_1, v_y) in R_3. In this case, the orbits of π' are the orbits of π, except for R_1, R_2 and R_3, which are replaced by a single orbit R' of the form

$$R': v_1 - v_y - \cdots - v_x - v_1 - v_b - \cdots - v_a - v_1 - v_c - \cdots - v_b - v_1.$$

In this case, π' has $r - 2$ orbits so that π' describes a 2-cell embedding of G on $S_{\ell+1}$.

Thus, we can now conclude that the shifting of a single term in π_1 (producing π_1') changes the genus of the resulting surface on which G is 2-cell embedded by at most 1. Having made this observation, we can now complete the proof.

Let $(\mu_1, \mu_2, \ldots, \mu_n)$ be the n-tuple of cyclic permutations associated with a 2-cell embedding of G on S_p and let $(v_1, v_2, \ldots, v_n)$ be the n-tuple of cyclic permutations associated with a 2-cell embedding of G on S_q. If $\deg v_i$ is 1 or 2 for each i, $1 \le i \le n$, then $\mu_i = v_i$ so that $p = q$ and the desired result follows. Hence, we may assume that for some i, $1 \le i \le n$, $\deg v_i \ge 3$. For each such i, μ_i can be transformed into v_i by a finite number of single term shifts, as described above. Each such single term shift describes an embedding of G on a surface whose genus differs by at most 1 from the genus of the surface on which G is embedded prior to the shift. Therefore, by performing sequences of single term shifts on those μ_i for which $\deg v_i \ge 3$, the n-tuple $(\mu_1, \mu_2, \ldots, \mu_m)$ can be transformed into $(v_1, v_2, \ldots, v_n)$. Since $p \le k \le q$, there must be at least one term $(\pi_1, \pi_2, \ldots, \pi_n)$ in the aforementioned sequence beginning with $(\mu_1, \mu_2, \ldots, \mu_m)$ and ending with $(v_1, v_2, \ldots, v_n)$ that describes a 2-cell embedding of G on S_k. □

EXERCISES 7.2

7.11 (a) For the 2-cell embedding of $K_{3,3}$ shown in Figure 7.1(a), determine the 6-tuple of cyclic permutations π_i associated with this embedding. Determine the orbits of the resulting permutation π.

(b) For the 2-cell embedding of K_7 on S_1 shown in Figure 7.2, determine the 7-tuple of cyclic permutations π_i associated with this embedding. Determine the orbits of the resulting permutation π.

7.12 Let $G = K_4 \times K_2$.

(a) Show that G is nonplanar.

(b) Show, in fact, that $\gamma(G) = 1$ by finding an 8-tuple of cyclic permutations that describes a 2-cell embedding of G on S_1. Determine the orbits of the resulting permutation π.

7.13 Let G be a graph with $V(G) = \{v_1, v_2, v_3, v_4, v_5, v_6\}$ and let $(\pi_1, \pi_2, \pi_3, \pi_4, \pi_5, \pi_6)$ describe a 2-cell embedding of G on the surface S_k, where

$$\pi_1 = (2\,5\,6\,3), \quad \pi_2 = (3\,6\,1\,4), \quad \pi_3 = (4\,1\,2\,5),$$
$$\pi_4 = (5\,2\,3\,6), \quad \pi_5 = (6\,3\,4\,1), \quad \pi_6 = (1\,4\,5\,2).$$

  (k) What is this familiar graph G?
  (l) What is k?
  (m) Is $k = \gamma(G)$?

7.14 How many of the 2-cell embeddings of K_4 are embeddings in the plane? On the torus? On the double torus?

7.15 (a) Describe an embedding of $K_{3,3}$ on S_2 by means of a 6-tuple of cyclic permutations.

  (b) Show that there exists no 2-cell embedding of $K_{3,3}$ on S_3.

7.3 THE MAXIMUM GENUS OF A GRAPH

If G is a connected graph with $\gamma(G) = p$, and q is the largest positive integer such that G is 2-cell embeddable on S_q, then it follows from Theorem 7.17 that G can be 2-cell embedded on S_k if and only if $p \le k \le q$. This suggests the following concept.

Let G be a connected graph. The *maximum genus* $\gamma_M(G)$ of G is the maximum among the genera of all surfaces on which G can be 2-cell embedded. At the outset, it may not even be clear that every graph has a maximum genus since, perhaps, some graphs may be 2-cell embeddable on infinitely many surfaces. However, there are no graphs that can be 2-cell embedded on infinitely many surfaces, for suppose that G is a nontrivial connected graph with $V(G) = \{v_1, v_2, \ldots, v_n\}$. By Theorem 7.14, there exists a one-to-one correspondence between the set of all 2-cell embeddings of G and the n-tuples $(\pi_1, \pi_2, \ldots, \pi_n)$, where for $i = 1, 2, \ldots, n$, π_i: $V(i) \rightarrow V(i)$ is a cyclic permutation. Since the number of such n-tuples is finite, and in fact is equal to

$$\prod_{i=1}^{n} (\deg v_i - 1)!,$$

it follows that there are only finitely many 2-cell embeddings of G and therefore that there exists a surface of maximum genus on which G can be 2-cell embedded. We can now state an immediate consequence of Theorem 7.17.

Corollary 7.18

A connected graph G has a 2-cell embedding on the surface S_k if and only if

$$\gamma(G) \le k \le \gamma_M(G).$$

We now present an upper bound for the maximum genus of any connected graph. This bound employs a new but very useful concept.

The *Betti number* $\mathcal{B}(G)$ of graph G of order n and size m having k components is defined as

$$\mathcal{B}(G) = m - n + k.$$

Thus, if G is connected, then

$$\mathcal{B}(G) = m - n + 1.$$

The following result is due to Nordhaus, Stewart and White [NSW1].

Theorem 7.19

If G is a connected graph, then

$$\gamma_M(G) \le \left\lfloor \frac{\mathcal{B}(G)}{2} \right\rfloor.$$

Furthermore, equality holds if and only if there exists a 2-cell embedding of G on the surface of genus $\gamma_M(G)$ with exactly one or two regions according to whether $\mathcal{B}(G)$ is even or odd, respectively.

Proof

Let G be a connected graph of order n and size m that is 2-cell embedded on the surface of genus $\gamma_M(G)$, producing r (2-cell) regions. By Theorem 7.1,

$$n - m + r = 2 - 2\,\gamma_M(G).$$

Thus,

$$\mathcal{B}(G) = m - n + 1 = 2\,\gamma_M(G) + r - 1,$$

so that

$$\gamma_M(G) = \frac{\mathcal{B}(G) + 1 - r}{2} \le \frac{\mathcal{B}(G)}{2},$$

producing the desired bound.

Moreover, we have

$$\gamma_M(G) = \frac{\mathcal{B}(G) + 1 - r}{2} = \left\lfloor \frac{\mathcal{B}(G)}{2} \right\rfloor,$$

if and only if $r = 1$ (which can only occur when $\mathcal{B}(G)$ is even) or $r = 2$ (which is only possible when $\mathcal{B}(G)$ is odd).

A (connected) graph G is called *upper embeddable* if the maximum genus of G attains the upper bound given in Theorem 7.19; that is, if $\gamma_M(G) = \lfloor \mathcal{B}(G)/2 \rfloor$. The graph G is said to be *upper embeddable on a surface S* if $S = S_{\gamma_M(G)}$. We can now state an immediate consequence of Theorem 7.19.

Corollary 7.20

Let G be a connected graph with even (odd) Betti number. Then G is upper embeddable on a surface S if and only if there exists a 2-cell embedding of G on S with one region (two regions).

In order to present a characterization of upper embeddable graphs, it is necessary to introduce a new concept.

A spanning tree T of a connected graph G is a *splitting tree* of G if at most one component of $G - E(T)$ has odd size. It follows therefore that if $G - E(T)$ is connected, then T is a splitting tree. For the graph G of Figure 7.11, the tree T_1 is a splitting tree. On the other hand, T_2 is not a splitting tree of G.

The following observation that relates splitting trees and Betti numbers is elementary, but useful.

Theorem 7.21

Let T be a splitting tree of a graph G of order n and size m. Then every component of $G - E(T)$ has even size if and only if $\mathcal{B}(G)$ is even.

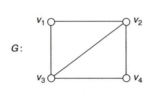

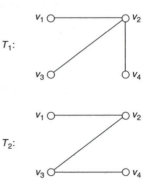

Figure 7.11 Splitting trees of graphs.

Proof

Suppose that every component of $G - E(T)$ has even size. Then $G - E(T)$ has even size. Since every tree of order n has size $n - 1$, the graph $G - E(T)$ has size $m - (n - 1) = m - n + 1$. Therefore, $\mathcal{B}(G) = m - n + 1$ is even.

Conversely, suppose that $\mathcal{B}(G)$ is even. The graph $G - E(T)$ has size $m - n + 1 = \mathcal{B}(G)$. Since T is a splitting tree of G, at most one component of $G - E(T)$ has odd size. Since the sum of the sizes of the components of $G - E(T)$ is even, it is impossible for exactly one such component to have odd size, producing the desired result. □

We now state a characterization of upper embeddable graphs, which was discovered independently by Jungerman [J3] and Xuong [X1].

Theorem 7.22

A graph G is upper embeddable if and only if G has a splitting tree.

Returning to the graph G of Figure 7.11, we now see that G is upper embeddable since G contains T_1 as a splitting tree. On the other hand, neither the graph G_1 nor the graph G_2 of Figure 7.12 has a single splitting tree, so, by Theorem 7.22, neither of these graphs is upper embeddable.

We mentioned earlier that no formula is known for the genus of an arbitrary graph. However, such is not the case with maximum genus. With the aid of Theorem 7.22, Xuong [X1] developed a formula for the maximum genus of any connected graph.

For a graph H we denote by $\xi_0(H)$ the number of components of H of odd size. For a connected graph G, we define the number $\xi(G)$ as follows:

$$\xi(G) = \min \xi_0(G - E(T)),$$

where the minimum is taken over all spanning trees T of G.

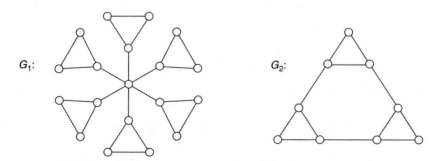

Figure 7.12 Graphs that are not upper embeddable.

Theorem 7.23

The maximum genus of a connected graph G is given by

$$\gamma_M(G) = \tfrac{1}{2}(\beta(G) - \xi(G)).$$

Returning to the graph G_1 of Figure 7.12, we see that $\mathcal{B}(G_1) = 6$ and that $\xi_0(G_1 - E(T)) = 6$ for every spanning tree T. Therefore, $\xi(G_1) = 6$ so that

$$\gamma_M(G_1) = \tfrac{1}{2}(\mathcal{B}(G_1) - \xi(G_1)) = 0$$

and G_1 is 2-cell embeddable only on the sphere.

With the aid of Theorem 7.23 (or Theorem 7.22), it is possible to show that a wide variety of graphs are upper embeddable. The following result is due to Kronk, Ringeisen and White [KRW1].

Corollary 7.24

Every complete k-partite graph, $k \geq 2$, is upper embeddable.

From Corollary 7.24, it follows at once that every complete graph is upper embeddable, a result due to Nordhaus, Stewart and White [NSW1]. We present a proof using Theorem 7.22.

Corollary 7.25

The maximum genus of K_n is given by

$$\gamma_M(K_n) = \left\lfloor \frac{(n-1)(n-2)}{4} \right\rfloor.$$

Proof

If T is a spanning path of K_n, then $K_n - E(T)$ contains at most one nontrivial component. Therefore, T is a splitting tree of K_n, and, by Theorem 7.22, K_n is upper embeddable. Since $\mathcal{B}(K_n) = (n-1)(n-2)/2$, the result follows. □

A formula for the maximum genus of complete bipartite graphs was discovered by Ringeisen [R4].

Corollary 7.26

The maximum genus of $K_{s,t}$ is given by

$$\gamma_M(K_{s,t}) = \left\lfloor \frac{(s-1)(t-1)}{2} \right\rfloor.$$

Zaks [Z1] discovered a formula for the maximum genus of the n-cube.

Corollary 7.27

The maximum genus of Q_n, $n \geq 2$, is given by

$$\gamma_M(Q_n) = (n-2)2^{n-2}.$$

Finally, we also note that it is possible to embed graphs on nonorientable surfaces such as the Möbius strip, projective plane and Klein bottle. As might be expected, every planar graph (as well as some nonplaner graphs) can be embedded on such surfaces. Figure 7.13 shows K_5 embedded on the Möbius strip. These topics shall not be the subject of further discussion, however.

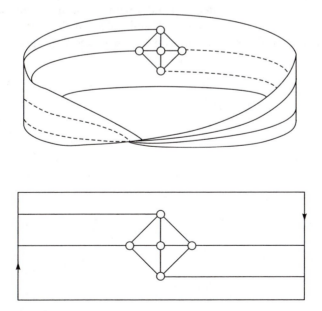

Figure 7.13   An embedding of K_5 on the Möbius strip.

EXERCISES 7.3

7.16 Describe an embedding of K_5 on $S_{\gamma_M(K_5)}$ by means of a 5-tuple of cyclic permutations.

7.17 Determine the maximum genus of the graph G_2 of Figure 7.12.

7.18 Determine the maximum genus of the Petersen graph.

7.19 (a) Let G be a connected graph with blocks $B_1, B_2, \ldots, B_k$. Prove that

$$\gamma_M(G) \geq \sum_{i=1}^{k} \gamma_M(B_i).$$

    (b) Show that the inequality in (a) may be strict.

7.20 Prove Theorem 7.22 as a corollary to Theorem 7.23.

7.21 Prove Corollary 7.24.

7.22 Prove Corollary 7.26.

7.23 Prove Corollary 7.27.

7.24 Prove or disprove: For every positive integer k, there exists a connected graph G_k such that $\lfloor \mathcal{B}(G)/2 \rfloor - \gamma_M(G_k) = k$.

7.25 Prove or disprove: If H is a connected spanning subgraph of an upper embeddable graph G, then H is upper embeddable.

7.26 For $G = C_s \times C_t$ $(s, t \geq 3)$, determine $\gamma(G)$ and $\gamma_M(G)$.

7.27 Prove that if each vertex of a connected graph G lies on at most one cycle, then G is only 2-cell embeddable on the sphere.

7.28 Prove, for positive integers p and q with $p \leq q$, that there exists a graph G of genus p that can be 2-cell embedded on S_q.

7.29 Prove that if G is upper embeddable, then $G \times K_2$ is upper embeddable.

8

Graph colorings

The graph-theoretic parameter that has received the most attention over the years is the chromatic number. Its prominence in graph theory is undoubtedly due to its involvement with the Four Color Problem, which is discussed in this chapter. The main goal of this chapter, however, is to describe the many ways in which a graph can be colored and to present results on these topics.

8.1 VERTEX COLORINGS

A *coloring* of a graph G is an assignment of colors (which are actually considered as elements of some set) to the vertices of G, one color to each vertex, so that adjacent vertices are assigned different colors. A coloring in which k colors are used is a *k-coloring*. A graph G is *k-colorable* if there exists an *s*-coloring of G for some $s \leq k$. It is obvious that if G has order n, then G can be n-colored, so that G is n-colorable.

The minimum integer k for which a graph G is k-colorable is called the *vertex chromatic number*, or simply the *chromatic number* of G, and is denoted by $\chi(G)$. If G is a graph for which $\chi(G) = k$, then G is *k-chromatic*. Certainly, if $H \subseteq G$, then $\chi(H) \leq \chi(G)$.

In a given coloring of a graph G, a set consisting of all those vertices assigned the same color is referred to as a *color class*. The chromatic number of G may be defined alternatively as the minimum number of independent subsets into which $V(G)$ can be partitioned. Each such independent set is then a color class in the $\chi(G)$-coloring of G so defined.

For some graphs, the chromatic number is quite easy to determine. For example,

$$\chi(C_{2k}) = 2, \quad \chi(C_{2k+1}) = 3, \quad \chi(K_n) = n,$$

and, in general,

$$\chi(K(n_1, n_2, \ldots, n_k)) = k.$$

G:

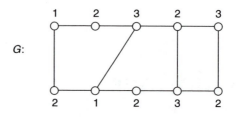

Figure 8.1 A 3-chromatic graph.

The graph G of Figure 8.1 is 3-colorable; a 3-coloring of G is indicated with the colors denoted by the integers 1, 2, 3. Therefore, G is k-colorable for $k \geq 3$ and $\chi(G) \leq 3$. Since C_5 is a subgraph of G and $\chi(C_5) = 3$, it follows that $\chi(G) \geq 3$. These two inequalities imply that $\chi(G) = 3$, that is, G is 3-chromatic.

If G is a k-partite graph, then $\chi(G) \leq k$ since the partite sets of G can be color classes in a k-coloring of G. Conversely, every graph G with $\chi(G) \leq k$ is necessarily k-partite. Similarly, $\chi(G) = k$ if and only if G is k-partite but G is *not* ℓ-partite for $\ell < k$. Consequently, the 1-chromatic graphs are precisely the empty graphs and the 2-chromatic graphs are the nonempty bipartite graphs. However, for no value of k greater than 2 is such an applicable characterization known.

We need only be concerned with determining the chromatic numbers of nonseparable graphs since the chromatic number of a disconnected graph is the maximum of the chromatic numbers of its components and the chromatic number of a connected graph with cut-vertices is the maximum of the chromatic numbers of its blocks.

Although the chromatic number is one of the most studied parameters in graph theory, no formula exists for the chromatic number of an arbitrary graph. Thus, for the most part, one must be content with supplying bounds for the chromatic number of graphs. In order to present such bounds, we now discuss graphs that are critical or minimal with respect to chromatic number.

For an integer $k \geq 2$, we say that a graph G is *critically k-chromatic* if $\chi(G) = k$ and $\chi(G - v) = k - 1$ for all $v \in V(G)$; G is *minimally k-chromatic* if $\chi(G) = k$ and $\chi(G - e) = k - 1$ for all $e \in E(G)$. There are several results dealing with critically k-chromatic graphs and minimally k-chromatic graphs, many of which are due to Dirac [D3]. We shall consider here only one of the more elementary, yet very useful, of these.

Every critically k-chromatic graph is nonseparable, while every minimally k-chromatic graph without isolated vertices is nonseparable. Furthermore, every minimally k-chromatic graph (without isolated vertices) is critically k-chromatic. The converse is not true in general, however; for example, the graph of Figure 8.2 is critically 4-chromatic but not minimally 4-chromatic. For $k = 2$ and $k = 3$, the converse is true. In fact, K_2 is the only critically 2-chromatic graph as well as the only minimally 2-chromatic graph without isolated vertices; while the odd cycles are the only critically 3-chromatic graphs and the only minimally 3-chromatic

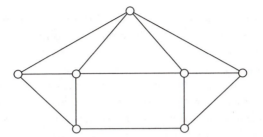

Figure 8.2 A critically 4-chromatic graph that is not minimally 4-chromatic.

graphs having no isolated vertices. For $k \geq 4$, neither the critically k-chromatic graphs nor the minimally k-chromatic graphs have been characterized. Although it is quite difficult, in general, to determine whether a given k-chromatic graph G is critical or minimal, G contains both critically k-chromatic subgraphs and minimally k-chromatic subgraphs. A k-chromatic subgraph of G of minimum order is critically k-chromatic, while a k-chromatic subgraph of G of minimum size is minimally k-chromatic.

The first theorem of this chapter concerns the structure of critically (and minimally) k-chromatic graphs.

Theorem 8.1

Every critically k-chromatic graph, $k \geq 2$, is $(k-1)$-edge-connected.

Proof

Let G be critically k-chromatic, $k \geq 2$. If $k = 2$ or $k = 3$, then $G = K_2$ or G is an odd cycle, respectively; therefore, G is 1-edge-connected or 2-edge-connected.

Assume that $k \geq 4$ and suppose, to the contrary, that G is not $(k-1)$-edge-connected. Hence by Theorem 3.20, there exists a partition of $V(G)$ into subsets V_1 and V_2 such that the set E' of edges joining V_1 and V_2 contains fewer than $k-1$ elements. Since G is critically k-chromatic, the subgraphs $G_1 = \langle V_1 \rangle$ and $G_2 = \langle V_2 \rangle$ are $(k-1)$-colorable. Let each of G_1 and G_2 be colored with at most $k-1$ colors, using the same set of $k-1$ colors. If each edge in E' is incident with vertices of different colors, then G is $(k-1)$-colorable. This contradicts the fact that $\chi(G) = k$. Hence we may assume that there are edges of E' incident with vertices assigned the same color. We show that the colors assigned to the elements of V_1 may be permuted so that each edge in E' joins vertices assigned different colors. Again this will imply that $\chi(G) \leq k-1$, produce a contradiction, and complete the proof.

In the coloring of G_1, let $U_1, U_2, \ldots, U_t$ be those color classes of G_1 such that for each i, $1 \leq i \leq t \leq k - 2$, there is at least one edge joining U_i and G_2. For $i = 1$, $2, \ldots, t$, assume that there are k_i edges joining U_i and G_2. Hence, for each i, $1 \leq i \leq t$, it follows that $k_i > 0$ and $\sum_{i=1}^{t} k_i \leq k - 2$.

If each u_1 in U_1 is adjacent only with vertices assigned colors different from that assigned to u_1, then the assignment of colors to the vertices of G is not altered. On the other hand, if some vertex u_1 of U_1 is adjacent with a vertex of G_2 that is assigned the same color as that of u_1, then in G_1 we may permute the $k - 1$ colors so that in the new assignment of colors to the vertices of G, no vertex of U_1 is adjacent to a vertex of G_2 having the same color. This is possible since the vertices of U_1 may be assigned any one of at least $k - 1 - k_1$ colors and $k - 1 - k_1 > 0$.

If, in this new assignment of colors to the vertices of G, each vertex u_2 of U_2 is adjacent only with vertices assigned colors different from that assigned to u_2, then no (additional) permutation of colors of G_1 occurs. However, if some vertex u_2 of U_2 is adjacent with a vertex of G_2 that is assigned the same color as that of u_2, then in G_1 we may permute the $k - 1$ colors, leaving the color assigned to U_1 fixed, so that no vertex of $U_1 \cup U_2$ is adjacent to a vertex of G_2 having the same color. This can be done since the vertices of U_2 can be assigned any of $(k - 1) - (k_2 + 1)$ colors, and $(k - 1) - (k_2 + 1) \geq (k - 1) - (k_1 + k_2) > 0$. Continuing this process, we arrive at a $(k - 1)$-coloring of G, producing the desired contradiction. □

Since every connected, minimally k-chromatic graph is critically k-chromatic, the preceding result has an immediate consequence.

Corollary 8.2

If G is a connected, minimally k-chromatic graph, $k \geq 2$, then G is $(k - 1)$-edge-connected.

Theorem 8.1 and Corollary 8.2 imply that $\kappa_1(G) \geq k - 1$ for every critically k-chromatic graph G or connected, minimally k-chromatic graph G. The next corollary now follows directly from Theorem 3.18.

Corollary 8.3

If G is critically k-chromatic or connected and minimally k-chromatic, then $\delta(G) \geq k - 1$.

We are now prepared to present bounds for the chromatic number of a graph. We give here several upper bounds, beginning with the best known and most applicable. The theorem is due to Brooks [B11] but the proof here is due to Lovász [L4].

Theorem 8.4

If G is a connected graph that is neither an odd cycle nor a complete graph, then

$$\chi(G) \leq \Delta(G).$$

Proof

Let G be a connected graph that is neither an odd cycle nor a complete graph, and suppose that $\chi(G) = k$, where, necessarily, $k \geq 2$. Let H be a critically k-chromatic subgraph of G. Then H is nonseparable and $\Delta(H) \leq \Delta(G)$.

Suppose that $H = K_k$ or that H is an odd cycle. Then $G \neq H$. Since G is connected, $\Delta(G) > \Delta(H)$. If $H = K_k$, then $\Delta(H) = k - 1$ and $\Delta(G) \geq k$; so

$$\chi(G) = k \leq \Delta(G).$$

If H is an odd cycle, then

$$\Delta(G) \geq 3 = k = \chi(G).$$

Hence, we may assume that H is critically k-chromatic and is neither an odd cycle nor a complete graph; this implies that $k \geq 4$.

Let H have order n. Since $\chi(H) = k \geq 4$ and H is not complete, it follows that $n \geq 5$. We now consider two cases, depending on the connectivity of H.

Case 1. Suppose that H is 3-connected. Let x and y be vertices of H such that $d_H(x, y) = 2$, and suppose that x, w, y is a path in H. The graph $H - x - y$ is connected. Let $x_1 = w, x_2, \ldots, x_{n-2}$ be the vertices of $H - x - y$, listed so that each vertex x_i $(2 \leq i \leq n - 2)$ is adjacent to at least one vertex preceding it. By letting $x_{n-1} = x$ and $x_n = y$, we have the sequence

$$x_1 = w, x_2, \ldots, x_{n-2}, x_{n-1} = x, x_n = y.$$

Assign the color 1 to the vertices x_{n-1} and x_n. We successively color $x_{n-2}, x_{n-3}, \ldots, x_2$ with one of the colors $1, 2, \ldots, \Delta(H)$ that was not used in coloring adjacent vertices following it in the sequence. Such a color is available since each x_i $(2 \leq i \leq n - 2)$ is adjacent to at most $\Delta(H) - 1$ vertices following it in the sequence. Since $x_1 = w$ is adjacent to two vertices colored 1 (namely, x_{n-1} and x_n), a color is available for x_1. Therefore,

$$\chi(G) = \chi(H) \leq \Delta(H) \leq \Delta(G).$$

Case 2. Suppose that $\kappa(H) = 2$. Since H is critically k-chromatic, it follows by Corollary 8.3 that H is $(k - 1)$-edge-connected. By Theorem 3.18 and the fact that

$k \geq 4$, we have $\delta(H) \geq \kappa_1(H) \geq k - 1 \geq 3$. Since H is not complete, $\delta(H) \leq n - 2$. Hence there exists a vertex u in H such that $3 \leq \deg_H u \leq n - 2$. If $\kappa (H - u) = 2$, then let v be a vertex with $d_H(u, v) = 2$. We may let $x = u$ and $y = v$, and proceed as in Case 1. On the other hand, if $\kappa(H - u) = 1$, then we consider two end-blocks B_1 and B_2 containing cut-vertices w_1 and w_2, respectively, of $H - u$. Since H is 2-connected, there exist vertices u_1 in $B_1 - w_1$ and u_2 in $B_2 - w_2$ that are adjacent to u. Let $x = u_1$ and $y = u_2$, and proceed as in Case 1.

This completes the proof. □

We may now state an upper bound for the chromatic number of an arbitrary graph.

Corollary 8.5

For every graph G, $\chi(G) \leq 1 + \Delta(G)$.

The bound for the chromatic number given in Theorem 8.4 is not particularly good for certain classes of graphs. For example, the bound provided for the star graphs $K_{1,n-1}$ differs from its chromatic number by $n - 2$. We shall see in Section 8.3 that 4 serves as an upper bound for the chromatic number of all planar graphs; however, Theorem 8.4 gives no bound for the entire class. Thus, there are several important classes of graphs G for which the bound $\chi(G) \leq \Delta(G)$ is poor indeed. A better bound in many cases is given by an inequality observed by Szekeres and Wilf [SW1]. The reader will notice the similarity if this result and Theorem 3.11.

Theorem 8.6

For every graph G,

$$\chi(G) \leq 1 + \max \delta(G')$$

where the maximum number is taken over all induced subgraphs G' of G.

Proof

The result follows immediately for empty graphs, so we assume G is a graph with $\chi(G) = k \geq 2$. Let H be an induced k-critical subgraph of G. Since H is an induced subgraph of G,

$$\delta(H) \leq \max \delta(G') \tag{8.1}$$

where the maximum is taken over all induced subgraphs G' of G.

By Corollary 8.3, $\delta(H) \geq k-1$, so by (8.1),

$$\max \delta(G') \geq k-1 = \chi(G) - 1$$

where the maximum is taken over all induced subgraphs G' of G, giving the result. $\qquad\square$

Theorem 8.6 gives an upper bound of 2 for the chromatic numbers of the graphs $K_{1,n}$, which is exact. Since every planar graph has minimum degree at most 5 (by Corollary 6.5) and since every subgraph of a planar graph is planar, a bound of 6 is provided for the chromatic number of planar graphs by Theorem 8.6. In each of these two cases, a marked improvement is shown over the result offered by Theorem 8.4. If G is a regular graph of degree r, then both Theorems 8.4 and 8.6 give $r+1$ as an upper bound for $\chi(G)$; however, this bound is poor for many r-regular graphs, such as $K_{r,r}$.

There are other upper bounds that have been obtained for chromatic numbers. We consider two of these. The following upper bound in terms of a longest path is due to Gallai [G2].

Theorem 8.7

For every graph G,

$$\chi(G) \leq 1 + \ell(G)$$

where $\ell(G)$ denotes the length of a longest path in G.

Proof

The result is obvious if G is empty, so we may assume that $\chi(G) = k \geq 2$. Let H be a critically k-chromatic subgraph of G, so that by Corollary 8.3, $\delta(H) \geq k-1$. By Theorem 3.8, H (and therefore G) contains a path of length $k-1$. Hence $\ell(G) \geq k-1 = \chi(G) - 1$, producing the desired result. $\qquad\square$

We now direct our attention briefly to lower bounds for the chromatic number. The *clique number* $\omega(G)$ of a graph G is the maximum order among the complete subgraphs of G. Clearly, $\omega(G) = \beta(\overline{G})$ for every graph G. If $K_k \subseteq G$ for some k, then $\chi(G) \geq \chi(K_k) = k$. It follows that $\chi(G) \geq \omega(G)$. Although this lower bound for $\chi(G)$ is not particularly good in general, $\chi(G)$ actually equals $\omega(G)$ for some special but important classes of graphs. For example, if G is bipartite, then either G is empty and $\chi(G) = 1 = \omega(G)$, or else $\chi(G) = 2 = \omega(G)$. In order to give another

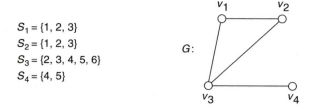

$S_1 = \{1, 2, 3\}$
$S_2 = \{1, 2, 3\}$
$S_3 = \{2, 3, 4, 5, 6\}$
$S_4 = \{4, 5\}$

Figure 8.3 An intersection graph.

example where equality holds, we introduce the notions of intersection graphs and interval graphs.

Let $\mathcal{F}$ be a finite family of not necessarily distinct nonempty sets. The *intersection graph* of $\mathcal{F}$ is obtained by representing each set in $\mathcal{F}$ by a vertex and then adding an edge between two vertices whose corresponding sets have a nonempty intersection. A graph G is called an *intersection graph* if it is the intersection graph of some family $\mathcal{F}$. A family $\mathcal{F} = \{S_1, S_2, S_3, S_4\}$ and its intersection graph are shown in Figure 8.3. Here, vertex v_i corresponds to set S_i.

When $\mathcal{F}$ is allowed to be an arbitrary family of sets, the class of graphs obtained as intersection graphs is simply *all* graphs (see Marczewski [M3]). By restricting the sets in $\mathcal{F}$, many interesting classes of graphs are obtained. For example, the intersection graph of a family of closed intervals of real numbers is called an *interval graph*. The graph G of Figure 8.3 is seen to be an interval graph by considering the intervals $I_1 = [1, 3], I_2 = [1, 3], I_3 = [2, 6]$ and $I_4 = [4, 5]$. However, not all graphs are interval graphs. For example, C_4 is not an interval graph.

If G is an interval graph, then the vertices of G correspond to closed intervals; say v_i corresponds to $I_i = [\ell_i, r_i]$. We show that G has an $\omega(G)$-coloring. Assume that the vertices of G have been labeled so that $\ell_i \leq \ell_j$ if $i \leq j$. Color $v_1, v_2, \ldots, v_n$ in order, assigning to v_i the smallest color (positive integer) j that has not been assigned to a neighbor of v_i. (Such a coloring is called a *greedy* coloring.) For example, using this procedure on the graph G of Figure 8.3, we obtain the 3-coloring of G given in Figure 8.4. Note that $\omega(G) = 3$.

In general, suppose that we have given a greedy coloring to an interval graph G and that vertex v_i has been assigned color k. This implies that v_i is adjacent to $v_{i_1}, v_{i_2}, \ldots, v_{i_{k-1}} (i_1 < i_2 < \cdots < i_{k-1} < i)$ colored $1, 2, \ldots, k-1$, in some order. We show that these vertices, together with v_i, induce a complete graph. The left endpoints of the corresponding intervals satisfy

$$\ell_{i_1} \leq \ell_{i_2} \leq \cdots \leq \ell_{i_{k-1}} \leq \ell_i.$$

If $\ell_{i_1} = \ell_{i_2} = \cdots = \ell_{i_{k-1}} = \ell_i$, then certainly $\langle \{v_{i_1}, v_{i_2}, \ldots, v_{i_{k-1}}, v_i\} \rangle$ is complete. Otherwise there is an integer t with $1 \leq t \leq k-1$ for which $\ell_{i_j} < \ell_i$

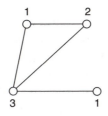

Figure 8.4 A greedy 3-coloring.

for $1 \leq j \leq t$ and $\ell_{i_j} = \ell_i$ for each j with $t+1 \leq j \leq k-1$. Clearly, $v_{i_{t+1}}, v_{i_{t+2}}, \ldots, v_{i_{k-1}}, v_i$ induce a complete graph. Now, since $\ell_{i_j} < \ell_i$ for $1 \leq j \leq t$ and v_{i_j} and v_i are adjacent, it follows that $r_{i_j} \geq \ell_i$. Thus $v_{i_1}, v_{i_2}, \ldots, v_{i_{k-1}}, v_i$ induce a complete graph. We conclude that the greedy coloring so produced is an $\omega(G)$-coloring.

As we have seen, if G is a bipartite graph or an interval graph, then $\chi(G) = \omega(G)$. Furthermore, since every induced subgraph of a bipartite (interval) graph is also a bipartite (interval) graph, we see that $\chi(H) = \omega(H)$ for every induced subgraph H of a bipartite or interval graph. A graph G is called *perfect* if $\chi(H) = \omega(H)$ for each induced subgraph H of G. Thus bipartite graphs and interval graphs are examples of perfect graphs.

Interval graphs form a subset of a larger class of perfect graphs called chordal graphs. A graph G is called *chordal* if every cycle of G of length greater than 3 has a chord, that is, an edge joining two non-consecutive vertices of the cycle. In the literature, chordal graphs have also been called *triangulated, rigid circuit* and *perfect elimination* graphs.

Since every induced subgraph of a chordal graph is also chordal, to show that chordal graphs are perfect we need only show that $\chi(G) = \omega(G)$ for every chordal graph G. This follows from the following characterization of chordal graphs due to Hajnal and Surányi [HS1] and Dirac [D6].

Theorem 8.8

A nonempty graph G is a chordal graph if and only if either G is complete or G can be obtained from two chordal graphs G_1 and G_2 (having orders less than that of G) by identifying two complete subgraphs of the same order in G_1 and G_2.

Proof

If G can be obtained as described, then clearly G is chordal. Conversely, since every complete graph is chordal, let G be a noncomplete chordal graph and let S be any minimal vertex-cut of G. Let A be the vertex set of one component of $G - S$ and let $B = V(G) - S - A$. Define the (chordal) subgraphs G_1 and G_2 of G by $G_1 = \langle A \cup S \rangle$ and $G_2 = \langle B \cup S \rangle$. Then G can be obtained from G_1 and G_2 by identifying the vertices of S. We show that $\langle S \rangle$ is complete. This is certainly true if $|S| = 1$, so we may assume that $|S| \geq 2$.

Since S is minimal, each $x \in S$ is adjacent to some vertex of each component of $G - S$. Therefore, for each pair $x, y \in S$, there exist paths $x, a_1, a_2, \ldots, a_r, y$ and $x, b_1, b_2, \ldots, b_t, y$, where each $a_i \in A$ and $b_i \in B$, such that these paths are chosen to be of minimum length. Thus, $C: x, a_1, a_2, \ldots, a_r, y, b_t, b_{t-1}, \ldots, b_1, x$ is a cycle of G of length at least 4, implying that C has a chord. However, $a_i b_j \notin E(G)$ since S is a vertex-cut and $a_i a_j \notin E(G)$ and $b_i b_j \notin E(G)$ by the minimality of r and t. Thus $xy \in E(G)$. $\qquad\square$

Corollary 8.9

If G is a chordal graph, then $\chi(G) = \omega(G)$.

Proof

If G is empty, then $\chi(G) = 1 = \omega(G)$. Thus, we assume G is nonempty.

We proceed by induction on the order n of G. If $n = 1$, then $G = K_1$ and $\chi(G) = \omega(G) = 1$. Assume that the clique and chromatic numbers are equal for chordal graphs of order less than n and let G be a chordal graph of order $n \geq 2$.

If G is complete, then $\chi(G) = \omega(G)$. If G is not complete, then G can be obtained from two chordal graphs G_1 and G_2 of order less than that of G by identifying two complete subgraphs of the same order in G_1 and G_2. Let S be the set of vertices of the identified complete graphs. Since there are no edges between $V(G_1) - S$ and $V(G_2) - S$, it follows that

$$\omega(G) = \max\{\omega(G_1), \omega(G_2)\}.$$

Clearly, $\chi(G) \geq \max\{\chi(G_1), \chi(G_2)\}$. However, since a $\chi(G_i)$-coloring of G_i assigns distinct colors to the vertices of S ($i = 1, 2$), it follows that we can make the colorings agree on S to obtain a $\max\{\chi(G_1), \chi(G_2)\}$-coloring of G. Thus,

$$\chi(G) = \max\{\chi(G_1), \chi(G_2)\}.$$

By the inductive hypothesis, $\chi(G_1) = \omega(G_1)$ and $\chi(G_2) = \omega(G_2)$. Thus, $\chi(G) = \omega(G)$. $\qquad\square$

Corollary 8.10

Every chordal graph is perfect.

Perfect graphs were introduced by Berge [B5], who conjectured that a graph G is perfect if and only if $\overline{G}$ is perfect. This conjecture (sometimes referred to as the *Perfect Graph Conjecture*) was verified by Lovász [L3].

Theorem 8.11

A graph G is perfect if and only if $\overline{G}$ is perfect.

Since $\chi(C_{2k+1}) = 3 \neq \omega(C_{2k+1}), k \geq 2$, it follows that if an induced subgraph of a graph G is an odd cycle of length at least 5, then G is not perfect. Similarly, if $\overline{G}$ contains an induced odd cycle of length at least 5, then $\overline{G}$ and, by Theorem 8.14, G are not perfect. Berge conjectured ([B7, p. 361]) that every graph that is not perfect contains either an induced odd cycle of length at least 5 or its complement contains such a cycle. This conjecture (referred to as the *Strong Perfect Graph Conjecture*) was verified by Chudnovsky, Robertson, Seymour and Thomas [CRST1].

Theorem 8.12

A graph G is perfect if and only if no induced subgraph of G or $\overline{G}$ is an odd cycle of length at least 5.

As we have noted, $\chi(G) \geq \omega(G)$ for every graph G. Hence if G contains triangles, then $\chi(G) \geq 3$. However, there exist graphs G that are triangle-free such that $\chi(G) \geq 3$. For example, the odd cycles C_{2k+1}, with $k \geq 2$, have chromatic number 3 and are, of course, triangle-free. The graph of Figure 8.5, called the *Grötzsch graph*, is 4-chromatic and triangle-free and is, in fact, the smallest such graph (in terms of order).

It may be surprising that there exist triangle-free graphs with arbitrarily large chromatic number. This fact has been established by a number of mathematicians, including Descartes [D2], Kelly and Kelly [KK1] and Zykov [Z3]. The following construction is due to Mycielski [M11], however.

Theorem 8.13

For every positive integer k, there exists a k-chromatic triangle-free graph.

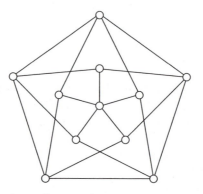

Figure 8.5 The Grötzsch graph: a 4-chromatic triangle-free graph.

Proof

The proof is by induction on k. If k is 1, 2 or 3, then the graphs K_1, K_2 and C_5, respectively, have the required properties. Assume that H is a triangle-free graph with $\chi(H) = k$, where $k \geq 3$. We show that there exists a triangle-free graph with chromatic number $k + 1$. Let $V(H) = \{v_1, v_2, \ldots, v_n\}$. We construct a graph G from H by adding $n + 1$ new vertices $u, u_1, u_2, \ldots, u_n$. The vertex u is joined to each vertex u_i and, in addition, u_i is joined to each neighbor of v_i.

To see that G is triangle-free, first observe that u belongs to no triangle. Since no two vertices u_i are adjacent, any triangle would consist of a vertex u_i and vertices v_j and $v_\ell, i \neq j, \ell$, but by the construction, this would imply that $\langle\{v_i, v_j, v_\ell\}\rangle$ is a triangle in H, which is impossible.

Let a k-coloring of H be given. Now assign to u_i the same color assigned to v_i and assign a $(k + 1)$st color to u. This produces a $(k + 1)$-coloring of G. Hence $\chi(G) \leq k + 1$. Suppose that $\chi(G) \leq k$, and let there be given a k-coloring of G, with colors $1, 2, \ldots, k$, say. Necessarily the vertex u is colored differently from each u_i. Suppose that u is assigned color k. Since $\chi(H) = k$, the color k is assigned to some vertices of H. Recolor each v_i that is colored k with the color assigned to u_i. This produces a $(k - 1)$-coloring of H and a contradiction. Thus, $\chi(G) = k + 1$, and the proof is complete. ☐

This result has been extended significantly by Erdös [E3] and Lovász [L2]. We postpone the proof of Theorem 8.14 until Chapter 13 (Theorem 13.5). The *girth* of a graph is the length of its shortest cycle.

Theorem 8.14

For every two integers $k \geq 2$ and $\ell \geq 3$ there exists a k-chromatic graph whose girth exceeds ℓ.

According to Theorem 8.13, a k-chromatic graph may contain no triangles and, therefore, no large complete subgraphs. In particular, a k-chromatic graph need not contain K_k. There are two well-known conjectures related to this observation.

For $k \leq 3$, it is trivial to see that every k-chromatic graph contains K_k or a subdivision of K_k. Dirac [D3] showed that this is also true for $k = 4$ and Hajos [H2] conjectured that for each positive integer k, every k-chromatic graph contains K_k or a subdivision of K_k. This conjecture, however, was shown to be false for $k \geq 7$. A weaker conjecture was proposed by Hadwiger [H1].

Hadwiger's Conjecture

If G is a k-chromatic graph, then K_k is a subcontraction of G.

This conjecture was verified for $k = 4$ by Dirac [D3]. The proofs for $k = 5$ and $k = 6$ depend on results of Wagner [W2] and of Robertson, Seymour and Thomas [RST1] and on the proof of the famous Four Color Theorem which we will encounter in Section 8.3. For $k \geq 7$, this conjecture remains open.

Our next result, due to Nordhaus and Gaddum [NG1], is the best known result on the chromatic numbers of a graph and its complement. The proof is based on one by H. V. Kronk.

Theorem 8.15

If G is a graph of order n, then

(i) $2\sqrt{n} \leq \chi(G) + \chi(\overline{G}) \leq n + 1$,
(ii) $n \leq \chi(G) \cdot \chi(\overline{G}) \leq ((n+1)/2)^2$.

Proof

Let a $\chi(G)$-coloring for G and a $\chi(\overline{G})$-coloring for $\overline{G}$ be given. Using these colorings, we obtain a coloring of K_n. Assign a vertex v of K_n the color (c_1, c_2), where c_1 is the color assigned to v in G and c_2 is the color assigned to v in $\overline{G}$. Since every two vertices of K_n are adjacent either in G or in $\overline{G}$, they are assigned different colors in that subgraph of K_n. Thus, this is a coloring of K_n using at most $\chi(G) \cdot \chi(\overline{G})$ colors, so

$$n = \chi(K_n) \leq \chi(G) \cdot \chi(\overline{G}).$$

This establishes the lower bound in (ii). Since the arithmetic mean of two positive numbers is always at least as large as their geometric mean, we have

$$\sqrt{n} \leq \sqrt{\chi(G) \cdot \chi(\overline{G})} \leq \frac{\chi(G) + \chi(\overline{G})}{2}.$$

This verifies the lower bound of (i).

To verify the upper bound in (i), we make use of Theorem 8.6. Suppose that $k = \max \delta(H)$, where the maximum is taken over all induced subgraphs H of G. Hence every induced subgraph of G has minimum degree at most k and, by Theorem 8.6, it follows that $\chi(G) \leq 1 + k$. Next we show that every induced subgraph of $\overline{G}$ has minimum degree at most $n - k - 1$. Assume, to the contrary, that there is an induced subgraph H of G such that $\delta(\overline{H}) \geq n - k$. Thus the vertices of H have degree at most $k - 1$ in G.

Let F be an induced subgraph of G with $\delta(F) = k$. Thus no vertex of F belongs to H. Since the order of F is at least $k + 1$, the order of H is at most $n - k - 1$, contradicting the fact that $\delta(\overline{H}) \geq n - k$. We may therefore conclude that every induced subgraph of $\overline{G}$ has minimum degree at most $n - k - 1$ and, by Theorem 8.6, that $\chi(\overline{G}) \leq 1 + (n - k - 1) = n - k$. Thus,

$$\chi(G) + \chi(\overline{G}) \leq (1 + k) + (n - k) = n + 1,$$

completing the proof of the upper bound in (i). The upper bound in (ii) now follows by consideration of geometric and arithmetic means. □

We close this section with some examples of variations of graph coloring. We observed that the chromatic number of a graph G can be defined as the minimum number of independent sets into which $V(G)$ can be partitioned. More generally, let $\mathcal{P}$ denote a family of graphs. The $\mathcal{P}$ *chromatic number* $\chi_{\mathcal{P}}(G)$ is the minimum number of sets into which $V(G)$ can be partitioned so that each set induces a graph that belongs to $\mathcal{P}$. This parameter is well-defined whenever $K_1 \in \mathcal{P}$. If $\mathcal{P}$ is the family of empty graphs, then $\chi_{\mathcal{P}}(G) = \chi(G)$. If $\mathcal{P}$ is the family of forests, then $\chi_{\mathcal{P}}(G) = a(G)$, the vertex-arboricity of G. If $\mathcal{P}$ is the family of empty graphs and complete graphs, then $\chi_{\mathcal{P}}(G)$ is the *cochromatic number* of G, first introduced in [LS1]. Alternatively, the cochromatic number of G is the minimum number of subsets in a partition of $V(G)$ so that each subset is independent in G or in $\overline{G}$. Certainly for each family $\mathcal{P}$ described above, $\chi_{\mathcal{P}}(G) \leq \chi(G)$ for every graph G in $\mathcal{P}$.

We now describe another coloring number. Suppose that for each vertex v for a graph G there is associated a list $L(v)$ of allowable colors for v. The *list chromatic number* $\chi_\ell(G)$ is the smallest positive integer k such that for each assignment of a list $L(v)$ of cardinality at least k to every vertex v of G, it is possible to color G so that every vertex is assigned a color from its list. Thus $\chi(G) \leq \chi_\ell(G)$ for every graph G. That this inequality may be strict is illustrated by the bipartite graph G of Figure 8.6 for which $\chi_\ell(G) = 3$. A 2-coloring of G cannot be given from the indicated lists. A graph G is *k-choosable* if $\chi_\ell(G) \leq k$.

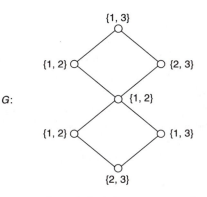

G:

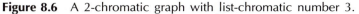

Figure 8.6 A 2-chromatic graph with list-chromatic number 3.

EXERCISES 8.1

8.1 Prove, for every graph G of order n, that $n/\beta(G) \le \chi(G) \le n + 1 - \beta(G)$.

8.2 Determine and prove a result analogous to Exercise 8.1 for vertex-arboricity.

8.3 Prove a result analogous to Theorem 8.4 for disconnected graphs.

8.4 What bound is given for $\chi(G)$ by Theorem 8.6 in the case that G is (a) a tree? (b) an outerplanar graph (Exercise 6.11)?

8.5 Let G be a k-chromatic graph, where $k \ge 2$, and let r be a positive integer such that $r \ge \Delta(G)$. Prove that there exists an r-regular k-chromatic graph H such that G is an induced subgraph of H.

8.6 Determine (and prove) a necessary and sufficient condition for a graph to have a 2-colorable line graph.

8.7 Let G be a connected, cubic graph of order $n > 4$ having girth 3. Determine $\chi(G)$.

8.8 Let $G_1, G_2, \ldots, G_k$ be pairwise disjoint graphs, and let $G = G_1 + G_2 + \cdots + G_k$. Prove that

$$\chi(G) = \sum_{i=1}^{k} \chi(G_i) \quad \text{and} \quad \omega(G) = \sum_{i=1}^{k} \omega(G_i).$$

8.9 (a) Show that every graph is an intersection graph.

(b) Let G be a nonempty graph. Show that a set $\mathcal{F}$ can be associated with G so that the intersection graph of $\mathcal{F}$ is the line graph of G.

8.10 Show that every induced subgraph of an interval graph is an interval graph.

8.11 Show that every interval graph is a chordal graph.

8.12 (a) Show that if G is a chordal graph, then for every proper complete subgraph H of G there is a vertex $v \in V(G) - V(H)$ for which the neighbors of v in G induce a complete subgraph in G.

(b) Show, without using Theorem 8.11, that the complement of a chordal graph is perfect.

8.13 For each integer $n \geq 7$, give an example of a graph G_n of order n such that no induced subgraph of G_n is an odd cycle of length at least 5 but G_n is *not* perfect.

8.14 Determine G if, in the proof of Theorem 8.13,

(a) $H = K_2$; (b) $H = C_5$.

8.15 Prove that for every two integers $k \geq 3$ and $\ell \geq 3$, with $k \geq \ell$, there exists a graph G such that $\chi(G) = k$ and $\omega(G) = \ell$.

8.16 Show that the conjectures of Hajos and Hadwiger are true for $k \leq 3$.

8.17 Show that all the bounds given in Theorem 8.15 are sharp.

8.18 Determine and prove a theorem analogous to Theorem 8.15 for vertex-arboricity.

8.19 Define a graph G to be *k-degenerate*, $k \geq 0$, if for every induced subgraph H of G, $\delta(H) \leq k$. Then the 0-degenerate graphs are the empty graphs, and by Exercise 3.2, the 1-degenerate graphs are precisely the forests. By Corollary 6.5, every planar graph is 5-degenerate. A k-degenerate graph is *maximal k-degenerate* if, for every two nonadjacent vertices u and v of G, the graph $G + uv$ is not k-degenerate.

For $k \geq 0$, let $\mathcal{P}_k$ denote the family of k-degenerate graphs. Then $\chi_{P_k}(G)$ is the minimum number of subsets into which $V(G)$ can be partitioned so that each subset induces a k-degenerate subgraph of G. A graph is said to be *ℓ-critical with respect to* χ_{P_k}, $\ell \geq 2$, if $\chi_{P_k}(G) = \ell$ and $\chi_{P_k}(G - v) = \ell - 1$ for every $v \in V(G)$.

(a) Prove that if G is a maximal k-degenerate graph of order n, where $n \geq k + 1$, then $\delta(G) = k$.

(b) Determine $\chi_{P_k}(K_n)$.

(c) Prove that if G is a graph that is ℓ-critical with respect to χ_{P_k}, then $\delta(G) \geq (k + 1)(\ell - 1)$.

8.21 (a) Give an example of a graph H for which the cochromatic number is 3.

(b) Give an example of a graph H that is not the union of (disjoint) complete graphs for which the cochromatic number equals $\chi(H)$.

(c) Give an example of a graph H that is not a union of complete graphs for which the cochromatic number is less than $\chi(H)$.

8.22 Show that $K_{3,3}$ has list-chromatic number 3.

8.2 EDGE COLORINGS

We now switch out attention from coloring the vertices of a graph to coloring its edges. An assignment of colors to the edges of a nonempty graph G so that adjacent edges are colored differently is an *edge coloring* of G (a *k-edge coloring* if k colors are used). The graph G is *k-edge colorable* if there exists an ℓ-edge coloring of G for some $\ell \leq k$. The minimum k for which a graph G is k-edge colourable is its *edge chromatic number* (or *chromatic index*) and is denoted by $\chi_1(G)$.

The determination of $\chi_1(G)$ can be transformed into a problem dealing with chromatic numbers; namely, from the definitions it is immediate that

$$\chi_1(G) = \chi(L(G)),$$

where $L(G)$ is the line graph of G. This observation appears to be of little value in computing edge chromatic numbers, however, since chromatic numbers are extremely difficult to evaluate in general.

It is obvious that $\Delta(G)$ is lower bound for $\chi_1(G)$. In what must be considered the fundamental result on edge colorings, Vizing [V3] proved that $\chi_1(G)$ equals $\Delta(G)$ or $1 + \Delta(G)$.

Theorem 8.16 (Vizing)

If G is a nonempty graph, then

$$\chi_1(G) \leq 1 + \Delta(G).$$

Proof

Suppose the theorem is not true. Then among the graphs for which the theorem is false, let G be one of minimum size. Hence G is not $(1 + \Delta)$-edge colorable, where $\Delta = \Delta(G)$; however, if $e = uv$ is an edge of G, then $G - e$ is $(1 + \Delta(G - e))$-edge colorable. Since $\Delta(G - e) \leq \Delta(G)$, we have that $G - e$ is $(1 + \Delta)$-edge colorable.

Let there be given a $(1 + \Delta)$-edge coloring of $G - e$; that is, every edge of G except e is assigned one of $1 + \Delta$ colors so that adjacent edges are colored differently. For each edge $e' = uv'$ of G that is incident with u, we define its *dual color* as any one of the $1 + \Delta$ colors that is not used to color edges incident with v'. Since no vertex of G has degree exceeding Δ, there is at least one color available for the dual color. It may occur that distinct edges have the same dual color.

Let $e = e_0$ have dual color α_1. (The color α_1 is not the color of any edge of G incident with v.) There must be some edge e_1 incident with u that has been assigned the color α_1; for if not, then the edge e could be colored α_1, thereby producing a $(1 + \Delta)$-edge coloring of G. Let α_2 be the dual color of e_1. If there is an edge incident with u that has been assigned the color α_2, then we denote it by e_2 and call its dual color α_3. In this manner, we construct a sequence $e_0, e_1, \ldots, e_k$, $k \geq 1$, containing a maximum number of distinct edges. The final edge e_k of this sequence is therefore colored α_k and has dual color α_{k+1}.

If there is no edge of G incident with u that is assigned the color α_{k+1}, then we may assign each of the edges $e_0, e_1, \ldots, e_k$ with its dual color and obtain a $(1 + \Delta)$-edge coloring of G. This, of course, is impossible. Hence we may assume that there exists an edge e_{k+1} of G incident with u that is colored α_{k+1}. Since e_0, $e_1, \ldots, e_k$ is maximum as to the number of distinct edges, we must have $e_{k+1} = e_j$ for some j, $1 \leq j \leq k$, or equivalently, $\alpha_{k+1} = \alpha_j$. Now certainly $\alpha_{k+1} \neq \alpha_k$ since the color assigned to e_k cannot be the same as its dual color; thus, $1 \leq j < k$. It is convenient to let $j = t + 1$, where then $0 \leq t < k - 1$. Hence $\alpha_{k+1} = \alpha_{t+1}$, so that e_k and e_t have the same dual color.

We now make some observations that will be important in the remainder of the proof. Since the edge e cannot be assigned any of the $1 + \Delta$ colors without producing two adjacent edges having the same color, it follows that for each color α among the $1 + \Delta$ colors, there is an edge of G adjacent with e that is colored α. This implies that there must be colors assigned to edges incident with v that are not assigned to any edge incident with u. Let β be one such color. Furthermore, let $e_i = uv_i$, $i = 0, 1, \ldots, k$, where then $v_0 = v$. The color β must be assigned to some edge incident with v_i for each $i = 1, 2, \ldots, k$; for suppose there is a vertex v_m, $1 \leq m \leq k$, such that no edge incident with v_m is colored β. Then we may change the color of e_m to β and color each e_i, $0 \leq i < m$, with its dual color to obtain a $(1 + \Delta)$-edge coloring of G.

We define two paths P and Q as follows. Let P be a path with initial vertex v_k of maximum length whose edges are alternately colored β and α_{k+1}, while Q is a path with initial vertex v_t having maximum length whose edges are alternately colored β and $\alpha_{t+1} = \alpha_{k+1}$. Suppose P terminates at w and Q at w'. We consider four cases according to certain possibilities for w and w'.

Case 1. Assume $w = v_m$ for some m, $0 \leq m \leq k - 1$. In this case, the initial and terminal edges of P are colored β. Also, no edge incident with v_m is assigned the color α_{k+1}. We note also that unless $v_m = v_t$, the vertex v_t is not on P. Interchange the colors β and α_{k+1} of the edges of P. Upon doing this, we have no edge incident with v_m that is assigned the color β and, moreover, the dual color of each e_i, $i < m$, is not altered. If $m = 0$, assign e the color β. If $m > 0$, then, as described earlier, we may change the color of e_m to β and color each e_i, $0 \leq i < m$, with its dual color. This implies that G is $(1 + \Delta)$-edge colorable, which is contradictory.

Case 2. Assume $w' = v_m$ for some m, $0 \leq m \leq k$, $m \neq t$. Here also, the initial and terminal edges of Q are assigned the color β, and no edge incident with v_m is colored α_{k+1}. Also, Q does not contain v_k unless $v_m = v_k$. Interchange the colors β

and α_{k+1} of the edges of Q. If $m < t$, then we proceed as in Case 1. If $m > t$, change the color of e to β if $t = 0$, while if $m > t > 0$, change the color of e_t to β and color each e_i, $0 \leq i < t$, with its dual color. Once again this implies that G is $(1 + \Delta)$-edge colorable, which is impossible.

Case 3. Assume $w \neq v_m$, $0 \leq m \leq k - 1$, and $w \neq u$; or $w' \neq v_m$ for any $m \neq t$ and $w' \neq u$. We consider w only, the conclusion being identical for w'. Observe that by interchanging the colors β and α_{k+1} of P, the color β is assigned to no edge incident with v_k and the dual color of each e_i, $0 \leq i < k$, remains the same. This situation, as we have seen, yields a contradiction.

Thus, only one other case remains.

Case 4. Assume $w = u$ and $w' = u$. Since u is incident with no edge colored β, the initial edge of both paths P and Q is colored β while each terminal edge is assigned α_{k+1}. If P and Q are edge-disjoint, then u is incident with two distinct edges colored α_{k+1} which cannot occur. Thus P and Q have an edge in common. But then there is a vertex incident with three edges belonging to P or Q. At least two of these edges are colored either β or α_{k+1}. This is a contradiction. □

With the aid of Theorem 8.16, the set of all nonempty graphs can be divided naturally into two classes. A nonempty graph G is said to be of *class one* if $\chi_1(G) = \Delta(G)$ and of *class two* if $\chi_1(G) = 1 + \Delta(G)$. The main problem, then, is to determine whether a given graph is of class one or of class two.

The set of all edges of a graph G receiving the same color in an edge coloring of G is called an *edge color class*. By Vizing's theorem, the edge chromatic number of an r-regular graph G is either r or $r + 1$. For example, C_n $(n \geq 3)$ is of class one if n is even and of class two if n is odd. If $\chi_1(G) = r$ for an r-regular graph, then necessarily each color class in a $\chi_1(G)$-edge coloring of G induces a spanning, 1-regular subgraph of G. Thus, as we shall see in Chapter 9, K_n is of class one if n is even and of class two if n is odd. More generally, every regular graph of odd order is of class two. It is not true, however, that every regular graph of even order is of class one; the Petersen graph, for example, is of class two.

Although it is probably not obvious, there are considerably more class one graphs than class two graphs, relatively speaking. Indeed, Erdös and Wilson [EW1] have proved that the probability that a graph of order n is of class one approaches 1 as n approaches infinity. However, the problem of determining which graphs belong to which class is unsolved.

The following result, due to Beineke and Wilson [BW1], gives a sufficient condition for a graph to be of class two. An *independent set of edges* in a graph G is a set of edges, each two of which are independent (nonadjacent). The *edge independence number* $\beta_1(G)$ of G is the maximum cardinality among the independent sets of edges of G.

Theorem 8.17

Let G be a graph of size m. If

$$m > \Delta(G) \cdot \beta_1(G),$$

then G is of class two.

Proof

Assume that G is of class one. Then $\chi_1(G) = \Delta(G)$. Let a $\chi_1(G)$-edge coloring of G be given. Each edge color class of G has at most $\beta_1(G)$ edges. Therefore, $m \leq \Delta(G) \cdot \beta_1(G)$. □

Since $\beta_1(G) \leq \lfloor \frac{1}{2}n \rfloor$ for every graph G of order n, we have an immediate consequence of the preceding result. A graph G of order n and size m is called *overfull* if $m > \Delta(G) \cdot \lfloor \frac{1}{2}n \rfloor$.

Corollary 8.18

Every overfull graph is of class two.

It should be emphasized that Theorem 8.17 and its corollary provide strictly sufficient conditions for a graph to be of class two. There exist graphs with relatively few edges that are of class two. Of course, the odd cycles are of class two, but, then, they are regular of odd order. The Petersen graph is of class two. A cubic graph of class two whose girth is at least 5 is called a *snark*. Thus the Petersen graph is a snark. Isaacs [I1] has shown that there exist infinitely many snarks. For example, the graph of Figure 8.7 is called the *double-star snark*.

Hilton [H12] and Chetwynd and Hilton [CH2] conjectured that a graph G of order n with $\Delta(G) > \frac{1}{3}n$ is of class two if and only if G contains an overfull subgraph H with $\Delta(G) = \Delta(H)$. Certainly if G contains an overfull subgraph H with $\Delta(G) = \Delta(H)$, then G is of class two. The converse has been established for several classes of graphs. For example, if G is a complete k-partite graph for some k, then G is of class two only if G contains an overfull subgraph H with $\Delta(G) = \Delta(H)$.

When discussing vertex colorings, we found it useful to consider graphs that are critical with respect to chromatic number. Now that we are investigating edge colorings, it proves valuable to consider certain minimal graphs.

A graph G with at least two edges is *minimal with respect to edge chromatic number* (or simply *minimal* if the parameter is clear from context) if $\chi_1(G - e) = \chi_1(G) - 1$ for every edge e of G. Since isolated vertices have no effect on edge

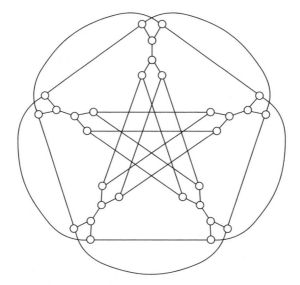

Figure 8.7 The double-star snark.

colorings, it is natural to rule out isolated vertices when considering such minimal graphs. Also, since the edge chromatic number of a disconnected graph G having only nontrivial components is the maximum of the edge chromatic numbers of the components of G, every minimal graph without isolated vertices is connected. Therefore, the added hypothesis that a minimal graph G is connected is equivalent to the assumption that G has no isolated vertices.

Two of the most useful results dealing with these minimal graphs are also results of Vizing [V4].

Theorem 8.19

Let G be a connected graph of class two that is minimal with respect to edge chromatic number. Then every vertex of G is adjacent to at least two vertices of degree $\Delta(G)$. In particular, G contains at least three vertices of degree $\Delta(G)$.

Theorem 8.20

Let G be a connected graph of class two that is minimal with respect to edge chromatic number. If u and v are adjacent vertices with deg $u = k$, then v is adjacent to at least $\Delta(G) - k + 1$ vertices of degree $\Delta(G)$.

We next examine to which class a graph belongs if it is minimal with respect to edge chromatic number.

Theorem 8.21

Let G be a connected graph with $\Delta(G) = d \geq 2$. Then G is minimal with respect to edge chromatic number if and only if either:

(i) *G is of class one and $G = K_{1,d}$ or*
(ii) *G is of class two and $G - e$ is of class one for every edge e of G.*

Proof

Assume first that $G = K_{1,d}$. Then $\chi_1(G) = \Delta(G) \geq 2$ while $\chi_1(G - e) = \Delta(G) - 1$ for every edge e of G. Next, suppose that G is of class two and that $G - e$ is of class one for every edge e of G. Then, for an arbitrary edge e of G, we have

$$\chi_1(G - e) = \Delta(G - e) < 1 + \Delta(G) = \chi_1(G).$$

Conversely, assume that $\chi_1(G - e) < \chi_1(G)$ for every edge e of G. If G is of class one, then

$$\Delta(G) \leq \Delta(G - e) + 1 \leq \chi_1(G - e) + 1 = \chi_1(G) = \Delta(G).$$

Therefore, $\Delta(G - e) = \Delta(G) - 1$ for every edge e of G, which implies that $G = K_{1,d}$.

If G is of class two, then

$$\chi_1(G - e) + 1 = \chi_1(G) = \Delta(G) + 1$$

so that $\chi_1(G - e) = \Delta(G)$ for every edge e of G. Suppose that G contains an edge e_1 such that $G - e_1$ is of class two. Then $\chi_1(G - e_1) = \Delta(G - e_1) + 1$. Hence, $\Delta(G - e_1) < \Delta(G)$, implying that G has at most two vertices of degree $\Delta(G)$. This, however, contradicts Theorem 8.19 and completes the proof. ☐

In the next section we shall be discussing various colorings of planar graphs, primarily vertex colorings and region colorings. We briefly consider edge colorings of planar graphs here. In this context, our chief problem remains to determine which planar graphs are of class one and which are of class two. It is easy to find planar graphs G of class one for which $\Delta(G) = d$ for each $d \geq 2$ since all star graphs are planar and of class one. There exist planar graphs G of class two with $\Delta(G) = d$ for $d = 2, 3, 4, 5$. For $d = 2$, the graph K_3 has the desired properties. For $d = 3, 4, 5$, the graphs of Figure 8.8 satisfy the required conditions.

It is not known whether there exist planar graphs of class two having maximum degree 6 or 7; however, Vizing [V4] proved that if G is planar and $\Delta(G) \geq 8$, then G must be of class one. We prove the following, somewhat weaker, result.

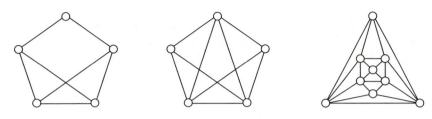

Figure 8.8 Planar graphs of class two.

Theorem 8.22

If G is a planar graph with $\Delta(G) \geq 10$, *then G is of class one.*

Proof

Suppose that the theorem is not true. Then among the graphs for which the theorem is false, let G be a connected graph of minimum size. Thus, G is planar, $\Delta(G) = d \geq 10$, and $\chi_1(G) = d + 1$. Furthermore, G is minimal with respect to edge chromatic number. By Corollary 6.5, G contains vertices of degree 5 or less. Let S denote the set of all such vertices. Define $H = G - S$. Since H is planar, H contains a vertex w such that $\deg_H w \leq 5$. Because $\deg_G w > 5$, the vertex w is adjacent to one or more vertices of S. Let $v \in S$ such that $wv \in E(G)$, and let $\deg_G v = k \leq 5$. Then, by Theorem 8.20, w is adjacent to at least $d - k + 1$ vertices of degree d, but $d - k + 1 \geq 6$ so that w is adjacent to at least six vertices of degree d. Since $d \geq 10$, w is adjacent to at least six vertices of H, contradicting the fact that $\deg_H w \leq 5$. □

Seymour [S2] conjectured that a planar graph is of class two if and only if G contains an overfull subgraph H with $\Delta(G) = \Delta(H)$. If true, this conjecture would imply that every planar graph G with $\Delta(G) \geq 6$ is of class one.

There is a coloring that assigns colors to both the vertices and the edges of a graph. A *total coloring* of a graph G is an assignment of colors to the elements (vertices and edges) of G so that adjacent elements and incident elements of G are colored differently. A *k-total coloring* is a total coloring that uses k colors. The minimum k for which a graph G admits a k-total coloring is called the *total chromatic number* of G and is denoted by $\chi_2(G)$. Certainly, $\chi_2(G) \geq 1 + \Delta(G)$. The total chromatic number was introduced by Behzad [B1] who made the following conjecture.

Total Coloring Conjecture

For every graph G,

$$\chi_2(G) \leq 2 + \Delta(G).$$

Strong evidence for the truth of the Total Coloring Conjecture was provided by McDiarmid and Reed [MR1] who showed that the probability that a graph G of order n satisfies $\chi_2(G) \leq 2 + \Delta(G)$ approaches 1 as n approaches infinity.

EXERCISES 8.2

8.22 Show that every nonempty regular graph of odd order is of class two.

8.23 Let H be a nonempty regular graph of odd order, and let G be a graph obtained from H by deleting $\frac{1}{2}(\Delta(H) - 1)$ or fewer edges. Show that G is of class two.

8.24 Prove or disprove: If G_1 and G_2 are class one graphs and H is a graph with $G_1 \subseteq H \subseteq G_2$, then H is of class one.

8.25 Show that the Petersen graph is of class two.

8.26 Prove that every hamiltonian cubic graph is of class one.

8.27 Show that each graph in Figure 8.8 is of class two.

8.28 Determine the class of each of the five regular polyhedra.

8.29 Determine the class of $K_{r,r}$.

8.30 Show that a cubic graph with a bridge has edge chromatic number 4.

8.31 (a) Prove that $\chi_2(G) \geq 1 + \Delta(G)$ for every graph G.

(b) Verify the Total Coloring Conjecture for graphs G with $\Delta(G) \leq 2$.

(c) Determine $\chi(G)$, $\chi_1(G)$ and $\chi_2(G)$ for the graph G of order 5 and size 7 in Figure 8.8.

8.3. MAP COLORINGS AND FLOWS

It has been said that the mapmakers of many centuries past were aware of the 'fact' that any map on the plane (or sphere) could be colored with four or fewer colors so that no two adjacent countries were colored alike. Two countries are considered to be *adjacent* if they share a common boundary line (not simply a single point). As was pointed out by May [M4], however, there has been no indication in ancient atlases, books on cartography, or books on the history of mapmaking that people were familiar with this so-called fact. Indeed, it is probable that the *Four Color Problem*, that is, the problem of determining whether the countries of any map on the plane (or sphere) can be colored with four or fewer colors such that adjacent countries are colored differently, originated and grew in the minds of mathematicians.

What, then, is the origin of the Four Color Problem? The first written reference to the problem appears to be in a letter, dated October 23, 1852, by Augustus De Morgan, mathematics professor at University College, London, to

Sir William Rowan Hamilton (after whom 'hamiltonian graphs' are named) of Trinity College, Dublin. The letter by De Morgan reads in part:

> A student of mine asked me today to give him a reason for a fact which I did not know was a fact — and do not yet. He says that if a figure be anyhow divided and the compartments differently coloured so that figures with any portion of common boundary line are differently coloured — four colours may be wanted, but no more. . . . Query cannot a necessity for five or more be invented. . . But it is tricky work. . . . what do you say? And has it, if true, been noticed? My pupil says he guessed it in colouring a map of England. The more I think of it, the more evident it seems. If you retort with some very simple case which makes me out a stupid animal, I think I must do as the Sphynx did. . . .

The student referred to by De Morgan was Frederick Guthrie. By 1880 the problem had become quite well-known. During that year, Frederick Guthrie published a note in which he stated that the originator of the question asked of De Morgan was his brother, Francis Guthrie. We quote from Frederick Guthrie's note [G7]:

> Some thirty years ago, when I was attending Professor De Morgan's class, my brother, Francis Guthrie, who had recently ceased to attend them (and who is now professor of mathematics at the South African University, Cape Town), showed me the fact that the greatest necessary number of colors to be used in colouring a map so as to avoid identity of colour in lineally contiguous districts is four. I should not be justified, after this lapse of time, in trying to give his proof. . . .
>
> With my brother's permission I submitted the theorem to Professor De Morgan, who expressed himself very pleased with it; accepted it as new; and, as I am informed by those who subsequently attended his classes, was in the habit of acknowledging whence he got his information.
>
> If I remember rightly, the proof which my brother gave did not seem altogether satisfactory to himself; but I must refer to him those interested in the subject.

On the basis of this note, we seem to be justified in proclaiming that the Four Color Problem was the creation of one Francis Guthrie.

Returning to the letter of De Morgan to Hamilton, we note the very prompt reply of disinterest by Hamilton to De Morgan on October 26, 1852:

> I am not likely to attempt your 'quaternion of colours' very soon.

Before proceeding further with this brief historical encounter with the Four Color Problem, we pause in order to give a more precise mathematical statement of the problem.

A plane graph G is said to be *k-region colorable* if the regions of G can be colored with k or fewer colors so that adjacent regions are colored differently. The *Four Color Problem* is thus the problem of settling the following conjecture.

The Four Color Conjecture

Every map (plane graph) is 4-region colorable.

In dealing with the Four Color Conjecture, one need not consider all plane graphs, as we shall now see. The *region chromatic number* $\chi^*(G)$ of a plane graph G is the minimum k for which G is k-region colorable. Since $\chi^*(G)$ is the maximum region chromatic number among its blocks, the Four Color Problem can be restated as determining whether every plane nonseparable graph is 4-region colorable.

In graph theory the Four Color Problem is more often stated in terms of coloring the vertices of a graph; that is, coloring the graph. In this form, the Four Color Conjecture is stated as follows.

The Four Color Conjecture

Every planar graph is 4-colorable.

It is in terms of this second statement that the Four Color Problem will be primarily considered. We now verify that these two formulations of the Four Color Conjecture are indeed equivalent. Before doing this, however, we require the concept of the dual of a plane graph.

For a given connected plane graph G we construct a multigraph G_d as follows. A vertex is placed in each region of G, and these vertices constitute the vertex set of G_d. Two distinct vertices of G_d are then joined by an edge for each edge common to the boundaries of the two corresponding regions of G. In addition, a loop is added at a vertex v of G_d for each bridge of G that belongs to the boundary of the corresponding region. Each edge of G_d is drawn so that it crosses its associated edge of G but no other edge of G or G_d (which is always possible); hence, G_d is planar. The multigraph G_d is referred to as the *dual* of G. In addition to being planar, G_d has the property that it has the same size as G and can be drawn so that each region of G_d contains a single vertex of G; indeed, $(G_d)_d = G$. If each set of parallel edges of G_d joining the same two vertices is replaced by a single edge and all loops are deleted, the result is a graph, referred to as the *underlying graph* $\tilde{G}_d$ of G_d. These concepts are illustrated in Figure 8.9, with the vertices of G_d represented by solid circles.

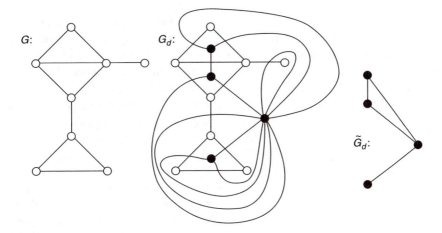

Figure 8.9 The dual (and its underlying graph) of a plane graph.

Theorem 8.23

Every planar graph is 4-colorable if and only if every plane graph is 4-region colorable.

Proof

Without loss of generality, we may assume that the graphs under consideration are connected.

Suppose that every planar graph is 4-colorable. Let G be an arbitrary connected plane graph, and consider $\widetilde{G}_d$, the underlying graph of its dual G_d. Two regions of G are adjacent if and only if the corresponding vertices of $\widetilde{G}_d$ are adjacent. Since $\widetilde{G}_d$ is planar, it follows, by hypothesis, that $\widetilde{G}_d$ is 4-colorable; thus, G is 4-region colorable.

For the converse, assume that every plane graph is 4-region colorable, and let G be an arbitrary connected plane graph. As we have noted, the dual G_d of G can be embedded in the plane so that each region of G_d contains exactly one vertex of G. If G_d is not a graph, then it can be converted into a graph G' by inserting two vertices into each loop of G_d and by placing a vertex in all but one edge in each set of parallel edges joining the same two vertices. Two vertices of G are adjacent if and only if the corresponding regions of G' are adjacent. Since G' is 4-region colorable, G is 4-colorable. □

With these concepts at hand, we now return to our historical account of the Four Color Problem. We indicated that this problem was evidently invented in 1852 by Francis Guthrie. The growing awareness of the problem was quite probably aided by De Morgan, who often spoke of it to other mathematicians. The first known published reference to the Four Color Problem is attributed to De Morgan in an anonymous article in the April 14, 1860 issue of the journal *Athenaeum*. By the 1860s the problem was becoming more widely known. The Four Color Problem received added attention when on June 13, 1878, Arthur Cayley asked, during a meeting of the London Mathematical Society, whether the problem had been solved. Soon afterwards, Cayley [C2] published a paper in which he presented his views on why the problem appeared to be so difficult. From his discussion, one might very well infer the existence of planar graphs with an arbitrarily large chromatic number.

One of the most important events related to the Four Color Problem occurred on July 17, 1879, when the magazine *Nature* carried an announcement that the Four Color Conjecture had been verified by Alfred Bray Kempe. His proof of the conjecture appeared in a paper [K2] published in 1879 and was also described in a paper [K3] published in 1880. For approximately ten years, the Four Color Conjecture was considered to be settled. Then in 1890, Percy John Heawood [H8] discovered an error in Kempe's proof. However, using Kempe's technique, Heawood was able to prove that every planar graph is 5-colorable. This result was referred to, quite naturally, as the Five Color Theorem.

Theorem 8.24

Every planar graph is 5-colorable.

Proof

The proof is by induction on the order n of the graph. For $n \leq 5$, the result is obvious. Assume that all planar graphs with $n-1$ vertices, $n > 5$, are 5-colorable, and let G be a plane graph of order n. By Corollary 6.5, G contains a vertex v of degree 5 or less. By deleting v from G, we obtain the plane graph $G - v$. Since $G - v$ has order $n - 1$, it is 5-colorable by the inductive hypothesis. Let there be given a 5-coloring of $G - v$, denoting the colors by 1, 2, 3, 4 and 5. If some color is not used in coloring the vertices adjacent with v, then v may be assigned that color, producing a 5-coloring of G itself. Otherwise, deg $v = 5$ and all five colors are used for the vertices adjacent with v.

Without loss of generality, we assume that v_1, v_2, v_3, v_4, v_5 are the five vertices adjacent with and arranged cyclically about v and that v_i is assigned the color i, $1 \leq i \leq 5$. Now consider any two colors assigned to nonconsecutive vertices v_i, say 1 and 3, and let H be the subgraph of $G - v$ induced by all those vertices colored 1 or 3. If v_1 and v_3 belong to different components of H, then by interchanging the colors assigned to vertices in the component of H containing v_1, for example, a 5-coloring of $G - v$ is produced in which no vertex adjacent with v is assigned the color 1. Thus if we color v with 1, a 5-coloring of G results.

Suppose then that v_1 and v_3 belong to the same component of H. Consequently, there exists a v_1–v_3 path P, all of whose vertices are colored 1 or 3. The path P, together with the path v_3, v, v_1, produces a cycle C in G that encloses v_2, or v_4 and v_5. Hence there exists no v_2–v_4 path in G, all of whose vertices are colored 2 or 4. Denote by F the subgraph of G induced by all those vertices colored 2 or 4. Interchanging the colors of the vertices in the component of F containing v_2, we arrive at a 5-coloring of $G - v$ in which no vertex adjacent with v is assigned the color 2. If we color v with 2, a 5-coloring of G results. □

In the 86 years that followed the appearance of Heawood's paper, numerous attempts were made to unlock the mystery of the Four Color Problem. Then on June 21, 1976, Kenneth Appel and Wolfgang Haken announced that they, with the aid of John Koch, had verified the Four Color Conjecture.

Appel and Haken's proof [AHK1] was logically quite simple; in fact, many of the essential ideas were the same as those used (unsuccessfully) by Kempe and, then, by Heawood. However, their proof was combinatorially complicated by the extremely large number of necessary case distinctions, and nearly 1200 hours of computer time were required to perform extensive computations. A simpler, and more easily checked, proof of the Four Color Theorem was obtained much later by Robertson, Sanders, Seymour and Thomas [RSST1]. Even their proof, however, required extensive computer calculations.

Theorem 8.25

Every planar graph is 4-colorable.

Although the Four Color Theorem has been established, other approaches have been suggested that might eliminate the heavy dependence on the computer. Here, the idea is to solve an equivalent problem. In Section 8.1, for example, we saw the conjecture of Hadwiger that K_k is a subcontraction of every k-chromatic graph. For $k = 5$, this conjecture is equivalent to the Four Color Theorem. We sketch the proof below. The details can be found in Wagner [W2].

Theorem 8.26

Every planar graph is 4-colorable if and only if K_5 is a subcontraction of every 5-chromatic graph.

Proof

Suppose that K_5 is a subcontraction of every 5-chromatic graph. Let G be a planar graph. Then, by Theorem 6.16, neither K_5 nor $K_{3,3}$ is a subcontraction of G. It follows that $\chi(G) \leq 4$.

For the converse, assume that every planar graph is 4-colorable, but that there are 5-chromatic graphs for which K_5 is not a subcontraction. Let G be a counterexample of minimum order. It is straightforward to show that G is 4-connected. But then G is a 4-connected graph for which K_5 is not a subcontraction, implying that G is planar (Wagner [W2]). Since $\chi(G) = 5$, this produces a contradiction to the Four Color Theorem. □

Another approach, due to Tait [T1], involves coloring the edges of bridgeless cubic planar graphs. In establishing this result it is convenient to make use of the group $\mathbb{Z}_2 \times \mathbb{Z}_2$ denoting its elements by $(0, 0)$, $(0, 1)$, $(1, 0)$ and $(1, 1)$.

Theorem 8.27

Every planar graph is 4-colorable if and only if every bridgeless cubic planar graph is 3-edge colorable.

Proof

By Theorem 8.23, it suffices to show that every plane graph is 4-region colorable if and only if every bridgeless cubic planar graph is 3-edge colorable.

Assume that every plane graph is 4-region colorable and let G be a bridgeless cubic plane graph. Let the regions of G be colored with the elements of $\mathbb{Z}_2 \times \mathbb{Z}_2$. Since G contains no bridges, each edge of G belongs to the boundary of two (adjacent) regions. Define the color of an edge to be the sum of the colors of those two regions bounded, in part, by the edge. Since every element of $\mathbb{Z}_2 \times \mathbb{Z}_2$ is self-inverse, no edge of G is assigned the color $(0, 0)$. However, since $\mathbb{Z}_2 \times \mathbb{Z}_2$ is a group, it follows that the three edges incident with a vertex are assigned the colors $(0, 1)$, $(1, 0)$ and $(1, 1)$. Hence G is 3-edge colorable.

Conversely, assume that every bridgeless cubic planar graph is 3-edge colorable. We show that every plane graph is 4-region colorable. Certainly, every plane graph is 4-region colorable if and only if every bridgeless plane graph is 4-region colorable. Furthermore, suppose that every cubic bridgeless plane graph is 4-region colorable. Let H be a bridgeless plane graph. We now construct a cubic plane block H' from H as follows. If H contains a vertex v of degree 2 that is incident with edges e and f, we subdivide e and f by introducing vertices v_1 and v_2 into e and f, respectively, remove v, and then identify v_1 and v_2, respectively, with the vertices of

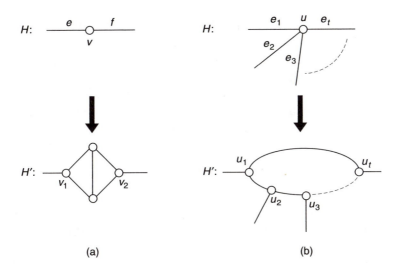

Figure 8.10 Constructing a cubic graph H' from a graph H.

degree 2 in a copy of the graph $K_{1,1,2}$ (Figure 8.10[a]). If H contains a vertex u of degree $t \geq 4$, incident with the consecutive edges $e_1, e_2, \ldots, e_t$, then we subdivide each e_i by inserting a vertex u_i in each e_i, $i = 1, 2, \ldots, t$, removing the vertex u, and identifying each u_i with the corresponding vertex of the t-cycle $u_1, u_2, \ldots, u_t, u_1$ (Figure 8.10[b]). By hypothesis, $\chi^*(H') \leq 4$, for the resulting cubic plane block H'; hence there exists a k-region coloring, $k \leq 4$, of H'. However, by identifying all vertices of the graph $K_{1,1,2}$ for each vertex of degree 2 and by identifying the vertices of the t-cycle for each vertex of degree $t \geq 4$, the graph H is reproduced and a k-region coloring of H is induced. Hence H is 4-region colorable.

Thus the proof will be complete once we have shown that if every bridgeless cubic planar graph is 3-edge colorable, then every bridgeless cubic plane graph is 4-region colorable.

Let G be a bridgeless cubic plane graph. Then G is 3-edge colorable. Let the edges of G be colored with the nonzero elements of $\mathbb{Z}_2 \times \mathbb{Z}_2$. Let R be some region of G and assign the color $(0, 0)$ to it. Let S be some other region of G. We now assign a color (an element of $\mathbb{Z}_2 \times \mathbb{Z}_2$) according to the following rule. Let A be a continuous curve joining a point of region R with a point of region S such that A passes through no vertex of G. We now define the color of S to be the sum of the colors of those edges crossed by A, where the color of an edge e is counted as many times as e is crossed. In order to show that the color of S is well-defined, we verify that the color assigned to S is independent of the curve A; however, this will be accomplished once it has been shown that if C is any simple closed curve not passing through vertices of G, then the sum of the colors of the edges crossed by C is $(0, 0)$. Let C be such a curve. If no vertex of G lies interior to C, then each edge crossed by C is crossed an even number of times; and since each element of

$\mathbb{Z}_2 \times \mathbb{Z}_2$ is self-inverse, it follows that the sum of the colors of the edges crossed by C is $(0,0)$. If C encloses vertices, then, without loss of generality, we may assume that any edge crossed by C is crossed exactly once. We proceed as follows. Let $e_1, e_2, \ldots, e_s$ be those edges crossed by or lying interior to C, and suppose that the first r of these edges are crossed by C. Observe that the sum of the colors of the three edges incident with any vertex is $(0,0)$; hence, if we were to total these sums for all vertices lying interior to C, we, of course, arrive at $(0,0)$ also. However, this sum also equals

$$c(e_1) + c(e_2) + \cdots + c(e_r) + 2[c(e_{r+1}) + c(e_{r+2}) + \cdots + c(e_s)],$$

where $c(e_i)$ indicates the color of the edge e_i. Therefore, $c(e_1) + c(e_2) + \cdots + c(e_r) = (0, 0)$, that is, the sum of the color of the edges crossed by C is $(0,0)$.

It now remains to show that this procedure yields a 4-region coloring of G. However, if R_1 and R_2 are two adjacent regions, sharing the edge e in their boundaries, then the colors assigned to R_1 and R_2 differ by $c(e) \neq (0, 0)$. This completes the proof. □

It follows from Theorems 8.25 and 8.27 that every bridgeless cubic planar graph is 3-edge colorable.

Corollary 8.28

Every bridgeless cubic planar graph is 3-edge colorable.

Several conjectured extensions of the Four Color Theorem remain open. For example, Hadwiger's Conjecture, equivalent to the Four Color Theorem for $k = 5$, is unsettled for $k \geq 7$. In another direction, Corollary 8.28 states that every bridgeless cubic planar graph is 3-edge colorable. The condition that the cubic graph is bridgeless is certainly necessary here since *no* cubic graph with a bridge is 3-edge colorable (Exercise 8.30). Furthermore, since the Petersen graph is bridgeless and has edge-chromatic number 4, we cannot drop the requirement of planarity in Corollary 8.28. Tutte [T12], however, made the following conjecture.

Tutte's First Conjecture

If G is a bridgeless cubic graph with $\chi_1(G) = 4$, then the Petersen graph is a subcontraction of G.

Since the Petersen graph is a subcontraction of no planar graph, Tutte's First Conjecture, if true, together with Theorem 8.27, implies the Four Color Theorem.

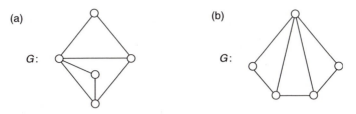

Figure 8.11 A nearly maximal planar graph G and an embedding of G.

Before turning to Tutte's Second Conjecture we consider a rather surprising result about list coloring planar graphs. Recall that $\chi_\ell(G)$ denotes the list chromatic number of a graph G and that $\chi_G \leq \chi_\ell(G)$ for every graph G. Our next result shows that not only is every planar graph 5-colorable but is, in fact, 5-list colorable. The proof of this result does not use the fact that every planar graph has a vertex of degree at most 5 and thus does not depend on Euler's Identity.

Our goal is to show that $\chi_\ell(G) \leq 5$ for every planar graph G. Certainly, it suffices to verify this result for connected planar graphs. Also, if the result can be established for maximal planar graphs, the more general result will follow. Maximal planar graphs (of order at least 3) are those graphs that can be embedded in the plane such that the boundary of every region is a triangle.

Define a graph G to be *nearly maximal planar* if there exists an embedding of G such that the boundary of every region of G is a cycle, at most one of which is not a triangle. If G is a nearly maximal planar graph, we may assume that G is embedded in the plane such that the boundary of every interior region is a triangle, while the boundary of the exterior region is a cycle of length 3 or more.

The graph G of Figure 8.11(a) is nearly maximal planar since there is a planar embedding of G (see Figure 8.11[b]) in which the boundary of every interior region is a triangle while the boundary of the exterior region is a cycle. Our next result is due to Thomassen [T4].

Theorem 8.29

Let G be a nearly maximal planar graph, and let C be the cycle that is the boundary of the exterior region of G. Suppose that the color lists $L(v), v \in V(G)$, satisfy the conditions:

(1) *Two consecutive vertices x and y of C are colored with distinct colors α and β, respectively;*
(2) *$|L(v)| \geq 3$ for all other vertices v of C;*
(3) *$|L(v)| \geq 5$ for all vertices v of G that are not on C.*

Then the coloring of x and y can be extended to a proper coloring of G by choosing colors from the respective lists of the vertices of G distinct from x and y and so $\chi_\ell(G) \le 5$.

Proof

We proceed by induction on the order of G. It follows trivially if the order of G is 3. Assume that the result holds for all nearly maximal planar graphs H for which $3 \le |V(H)| < n$ for an integer $n \ge 4$, and let G be a nearly maximal planar graph of order n. We consider two cases.

Case 1. Assume the cycle C has a chord, that is, G has an edge uv, where u and v are non-consecutive vertices on C. Hence there is a nearly maximal planar subgraph G_1 of G the boundary of whose exterior region is a cycle C', obtained from the edge uv and the u–v path on C containing x and y and the edge uv. This graph satifies the induction hypothesis and therefore has a 5-list coloring. Regarding u and v as precolored, we see that the inductive hypothesis is also satisfied for the nearly maximal planar graph G_2 the boundary of whose exterior region is a cycle C'' obtained from the edge uv and the u–v path on C that does not contain x and y and the edge uv. Hence G_2 can be 5-list colored with the available colors and thus the same is true of G.

Case 2. Assume C has no chord. Let v_0 be the vertex other than y on C that is adjacent to x, and let $x, v_1, v_2, \ldots, v_t, w$ be the neighbors of v_0, where w is on C. Since G is nearly maximal planar, we have the situation shown in Figure 8.12.

Consider the nearly maximal planar graph $G' = G - v_0$. Since $|L(v_0)| \ge 3$, there exist two colors ε and v in $L(v_0)$ different from α. Now we replace every color list $L(v_i)$ by $L(v_i) - \{\varepsilon, v\}$ keeping the original color sets for all other vertices in G'. Then G' satisfies the induction hypothesis and is thus 5-list coloring by induction. Choosing ε or v for v_0 we can extend the list coloring of G' to G, and the proof is complete. □

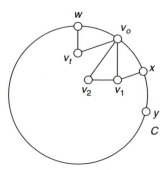

Figure 8.12 The case where the boundary cycle C has no chord.

Tutte's First Conjecture about the edge-chromatic number of bridgeless cubic graphs is a special case of yet another conjecture of Tutte concerning nowhere-zero flows.

Let D be an oriented graph and let $k \geq 2$ be an integer. A *nowhere-zero k-flow* in D is a function ϕ defined on $E(D)$ so that

(i) $\phi(e) \in \{\pm 1, \pm 2, \ldots, \pm(k-1)\}$ for each arc e of D, and

(ii) for each vertex v of D,

$$\sum_{(v,u) \in E(D)} \phi(v, u) = \sum_{(u,v) \in E(D)} \phi(u, v).$$

If an orientation D of a graph G has a nowhere-zero k-flow ϕ and D' is the orientation of G obtained by replacing some arc (u, v) with the arc (v, u), then D' has the nowhere-zero k-flow $\phi_{D'}$ defined by

$$\phi_{D'}(e) = \begin{cases} \phi(e) & \text{if } e \neq (u, v) \\ -\phi(e) & \text{if } e = (u, v). \end{cases}$$

Thus if some orientation of G has a nowhere-zero k-flow, then so does *every* orientation of G. Consequently, when we say that a graph G has a nowhere-zero k-flow, we mean, in fact, that every orientation of G has a nowhere-zero k-flow.

Nowhere-zero flows in planar graphs are of particular interest because of their relationship to region colorings. Let G be a bridgeless plane graph, with the edges of G oriented arbitrarily, and let c be a k-region coloring of G. Thus for each region R of G, the region R is colored $c(R)$, where $c(R) \in \{1, 2, \ldots, k\}$. For each oriented edge $e = (u, v)$ of G, define $\phi(e)$ to be $c(R_1) - c(R_2)$, where R_1 is the region to the right of $e = (u, v)$ as we travel along e from u to v and R_2 is the region to the left of e. An example of this construction is given in Figure 8.13.

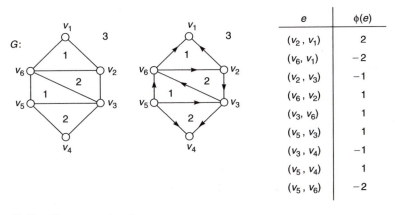

e	$\phi(e)$
(v_2, v_1)	2
(v_6, v_1)	-2
(v_2, v_3)	-1
(v_6, v_2)	1
(v_3, v_6)	1
(v_5, v_3)	1
(v_3, v_4)	-1
(v_5, v_4)	1
(v_5, v_6)	-2

Figure 8.13 Flow construction.

It is straightforward to verify that the integer-valued function ϕ defined above on the bridgeless plane graph G is a nowhere-zero k-flow. Thus if a bridgeless plane graph G is k-region colorable, then G has a nowhere-zero k-flow. Conversely, Tutte [T9] showed that if a bridgeless plane graph G has a nowhere-zero k-flow, then G is k-region colorable. Consequently, another equivalent form of the Four Color Theorem can be given.

Theorem 8.30

Every planar graph is 4-colorable if and only if every bridgeless planar graph has a nowhere-zero 4-flow.

Applying Theorems 8.25 and 8.30, we obtain Corollary 8.31, the 'flow analogue' of Corollary 8.28.

Corollary 8.31

Every bridgeless planar graph has a nowhere-zero 4-flow.

As with Corollary 8.28, the condition 'bridgeless' is necessary in Corollary 8.31 since no graph with a bridge has a nowhere-zero k-flow for any $k \geq 2$ (Exercise 8.34). Perhaps not surprisingly, the Petersen graph has no nowhere-zero 4-flow and so the planarity condition in Corollary 8.31 cannot be deleted.

Theorem 8.32

The Petersen graph has no nowhere-zero 4-flow.

Proof

Suppose, to the contrary, that the Petersen graph P has a nowhere-zero 4-flow ϕ corresponding to some orientation P' of P. If $\phi(u, v) < 0$ for some arc (u, v), then we can reverse the direction of the arc (u, v) to produce (v, u) and assign the value $\phi(v, u) = -\phi(u, v)$, obtaining another nowhere-zero 4-flow in P'. By repeating this procedure, we obtain a nowhere-zero 4-flow ϕ in P' with $0 < \phi(e) < 4$ for every arc e of P'. Since P is cubic, it follows that for each vertex v of P' the three arcs incident with v have flow values 1, 1, 2 or 1, 2, 3.

Now, those arcs e with $\phi(e) = 2$ produce a 1-regular spanning subgraph of P. Thus, the corresponding edges can be assigned a single color in an edge coloring of P. Therefore, those arcs e for which $\phi(e) = 1$ or $\phi(e) = 3$ produce a 2-regular spanning subgraph H of P.

A vertex v of H is said to be *of type* I if there is an arc e of P' with $\phi(e) = 1$ that is incident to v; while v is *of type* II if there is an arc e of P' with $\phi(e) = 1$ that is incident from v. Consequently, every vertex of H is of type I or of type II, but not both. Moreover, in any cycle of H the vertices alternate between type I and type II. Thus every component of H is an even cycle. It follows that H is 2-edge colorable, so P is 3-edge colorable, which produces a contradiction. ☐

The technique used in the proof of Theorem 8.32 can be used to show that if a bridgeless cubic graph G has a nowhere-zero 4-flow, then G is 3-edge colorable. In fact, the converse is also true.

Theorem 8.33

Let G be a bridgeless cubic graph. Then G has a nowhere-zero 4-flow if and only if G is 3-edge colorable.

In view of Corollary 8.31 and Theorem 8.32, Tutte [T12] proposed a second conjecture.

Tutte's Second Conjecture

If G is a bridgeless graph with no nowhere-zero 4-flow, then the Petersen graph is a subcontraction of G.

For cubic graphs, the two Tutte conjectures are equivalent. Tutte also conjectured that *every* bridgeless graph has a nowhere-zero 5-flow. Although this conjecture was made in 1954, it was not even known until 1975 whether 5 could be replaced by some larger number. Jaeger [J2] showed that every bridgeless graph has a nowhere-zero 8-flow, which was improved by Seymour [S4] who showed that every bridgeless graph has a nowhere-zero 6-flow. Tutte's 5-flow conjecture remains open.

The Four Color Theorem deals with the maximum chromatic number among all graphs that can be embedded in the plane. The *chromatic number of a surface* (where, as always, a surface is a compact orientable 2-manifold) S_k of genus k, denoted $\chi(S_k)$, is the maximum chromatic number among all graphs that can be embedded on S_k. The surface S_0 is the sphere and the Four Color Theorem states that $\chi(S_0) = 4$. Heawood [H8] showed that $\chi(S_1) = 7$; that is, the chromatic number of the torus is 7. Moreover, Heawood was under the impression that he had proved

$$\chi(S_k) = \left\lfloor \frac{7 + \sqrt{1 + 48k}}{2} \right\rfloor$$

for all $k > 0$. However, Heffter [H9] pointed out that Heawood had only established the upper bound:

$$\chi(S_k) \leq \left\lfloor \frac{7 + \sqrt{1 + 48k}}{2} \right\rfloor. \tag{8.2}$$

The statement that $\chi(S_k) = \left\lfloor (7 + \sqrt{1 + 48k})/2 \right\rfloor$ for all $k > 0$ eventually became known as the *Heawood Map Coloring Conjecture*. In 1968, Ringel and Youngs [RY1] completed a remarkable proof of the conjecture, which involved a number of people over a period of many decades. This result is now known as the *Heawood Map Coloring Theorem*. The proof we present assumes inequality (8.2).

Theorem 8.34

For every positive integer k,

$$\chi(S_k) = \left\lfloor \frac{7 + \sqrt{1 + 48k}}{2} \right\rfloor.$$

Proof

Because of inequality (8.2), it remains only to verify that

$$\chi(S_k) \geq \left\lfloor \frac{7 + \sqrt{1 + 48k}}{2} \right\rfloor$$

for all $k > 0$. Define

$$n = \left\lfloor \frac{7 + \sqrt{1 + 48k}}{2} \right\rfloor,$$

so that $n \leq \left(7 + \sqrt{1 + 4k}\right)/2$. From this, it follows that $k \geq (n - 3)(n - 4)/12$. Therefore,

$$k \geq \left\lceil \frac{(n - 3)(n - 4)}{12} \right\rceil. \tag{8.3}$$

Since the right-hand expression of (8.3) equals the genus of K_n (by Theorem 7.10), $\gamma(K_n) \leq k$ so that

$$\chi(S_{\gamma(K_n)}) \leq \chi(S_k).$$

Clearly K_n is embeddable on $S_{\gamma(K_n)}$; consequently, $\chi(S_{\gamma(K_n)}) \geq n$, implying that $\chi(S_k) \geq n$. □

As a consequence of the Four Color Theorem, Theorem 8.34 also holds for $k = 0$.

EXERCISES 8.3

8.32 Use a proof similar to that of Theorem 8.24 to show that $a(G) \leq 3$ for every planar graph G.

8.33 Show that if a bridgeless plane graph is k-region colorable, then G has a nowhere-zero k-flow.

8.34 Show that if G has a nowhere-zero k-flow, $k \geq 2$, then G is bridgeless.

8.35 Give an example of a graph G for which $\gamma(G) = 2$ and $\chi(G) = \chi(S_2)$. Verify that your example has these properties.

8.36 Use the result given in Theorem 8.6 to establish an upper bound for the chromatic numbers of graphs embeddable on the torus. Discuss the sharpness of your bound.

9

Matchings, factors and decompositions

We now consider special subgraphs that a graph may contain or into which a graph may be decomposed. In particular, we emphasize isomorphic decompositions. This leads us to a consideration of graph labelings.

9.1 MATCHINGS AND INDEPENDENCE IN GRAPHS

Recall that two edges in a graph G are independent if they are not adjacent in G. A set of pairwise independent edges of G is called a *matching* in G, while a matching of maximum cardinality is a *maximum matching* in G. Thus the number of edges in a maximum matching of G is the edge independence number $\beta_1(G)$ of G. In the graph G of Figure 9.1, the set $M_1 = \{e_1, e_4\}$ is a matching that is not a maximum matching, while $M_2 = \{e_1, e_3, e_5\}$ and $M_3 = \{e_1, e_3, e_6\}$ are maximum matchings in G.

If M is a matching in a graph G with the property that every vertex of G is incident with an edge of M, then M is a *perfect matching* in G. Clearly, if G has a perfect matching M, then G has even order and $\langle M \rangle$ is a 1-regular spanning subgraph of G. Thus, the graph G of Figure 9.1 cannot have a perfect matching.

If M is a specified matching in a graph G, then every vertex of G is incident with at most one edge of M. A vertex that is incident with no edges of M is called an $\overline{M}$-vertex. The following theorem will prove to be useful.

Theorem 9.1

Let M_1 and M_2 be matchings in a graph G. Then each component of the spanning subgraph H of G with $E(H) = (M_1 - M_2) \cup (M_2 - M_1)$ is one of the following types:

(i) *an isolated vertex,*

(ii) *an even cycle whose edges are alternately in M_1 and in M_2,*

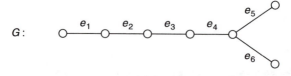

Figure 9.1 Matchings and maximum matchings.

(iii) *a nontrivial path whose edges are alternately in M_1 and in M_2 and such that each end-vertex of the path is either an $\overline{M}_1$-vertex or an $\overline{M}_2$-vertex but not both.*

Proof

First we note that $\Delta(H) \leq 2$, for if H contains a vertex v such that $\deg_H v \geq 3$, then v is incident with at least two edges in the same matching. Since $\Delta(H) \leq 2$, every component of H is a path (possibly trivial) or a cycle. Since no two edges in a matching are adjacent, the edges of each cycle and path in H are alternately in M_1 and in M_2. Thus each cycle in H is even.

Suppose that $e = uv$ is an edge of H and u is the end-vertex of a path P that is a component of H. The proof will be complete once we have shown that u is an $\overline{M}_1$-vertex or an $\overline{M}_2$-vertex but not both. Since $e \in E(H)$, it follows that $e \in M_1 - M_2$ or $e \in M_2 - M_1$. If $e \in M_1 - M_2$, then u is not an $\overline{M}_1$-vertex. We show that u is an $\overline{M}_2$-*vertex*. If this is not the case, then there is an edge f in M_2 (thus $f \neq e$) such that f is incident to u. Since e and f are adjacent, $f \notin M_1$. Thus, $f \in M_2 - M_1 \subseteq E(H)$. Thus, however, is impossible since u is the end-vertex of P. Therefore, u is an $\overline{M}_2$-*vertex*; similarly, if $e \in M_2 - M_1$, then u is an $\overline{M}_1$-*vertex*. $\qquad\square$

In order to present a characterization of maximum matchings, we introduce two new terms. Let M be a matching in a graph G. An *M-alternating path* of G is a path whose edges are alternately in M and not in M. An *M-augmenting path* is an M-alternating path both of whose end-vertices are $\overline{M}$-vertices. The following characterization of maximum matchings is due to Berge [B4].

Theorem 9.2

A matching M in a graph G is a maximum matching if and only if there exists no M-augmenting path in G.

Proof

Assume that M is a maximum matching in G and that there exists an M-augmenting path P of G. Necessarily, P has odd length. Let M' denote the edges of P belonging to M, and let $M'' = E(P) - M'$. Since $|M''| = |M'| + 1$, the set $(M - M') \cup M''$ is a matching having cardinality exceeding that of M, producing a contradiction.

Conversely, let M_1 be a matching in a graph G, and suppose that there exists no M_1-augmenting path in G. We verify that M_1 is a maximum matching. Let M_2 be a maximum matching in G. By the first part of the proof, there exists no M_2-augmenting path in G. Let H be the spanning subgraph of G with $E(H) = (M_1 - M_2) \cup (M_2 - M_1)$. Suppose that H_1 is a component of H that is neither an isolated vertex nor an even cycle. Then it follows from Theorem 9.1 that H_1 is a path of even length whose edges are alternately in M_1 and in M_2, for otherwise, there would exist a path in G that is either M_1-augmenting or M_2-augmenting, which is impossible. It now follows by Theorem 9.1 that $|M_1 - M_2| = |M_2 - M_1|$, which, in turn, implies that $|M_1| = |M_2|$. Hence, M_1 is a maximum matching. $\qquad\square$

According to Theorem 9.2, if a matching M is given, it is possible to decide whether M is a maximum matching by determining whether G has an M-augmenting path.

In applications, maximum matchings in bipartite graphs have proved to be most useful. The next result, namely Theorem 9.3, attributed to König [K9] and Hall [H5], is of interest in its own right.

In a graph G, a nonempty subset U_1 of $V(G)$ is said to be *matched* to a subset U_2 of $V(G)$ disjoint from U_1 if there exists a matching M in G such that each edge of M is incident with a vertex of U_1 and a vertex of U_2, and every vertex of U_1 is incident with an edge of M, as is every vertex of U_2. If $M \subseteq M^*$, where M^* is also a matching in G, we also say that U_1 is *matched under* M^* to U_2.

Let U be a nonempty set of vertices of a graph G and let its *neighborhood* $N(U)$ denote the set of all vertices of G adjacent with at least one element of U. Then the set U is said to be *nondeficient* if $|N(S)| \geq |S|$ for every nonempty subset S of U.

Theorem 9.3

Let G be a bipartite graph with partite sets V_1 and V_2. The set V_1 can be matched to a subset of V_2 if and only if V_1 is nondeficient.

Proof

Suppose the V_1 can be matched to a subset of V_2 under a matching M^*. Then every nonempty subset S of V_1 can be matched under M^* to some subset of V_2, implying that $|N(S)| \geq |S|$; so V_1 is nondeficient.

To verify the converse, let G be a bipartite graph for which V_1 is nondeficient and suppose that V_1 cannot be matched to a subset of V_2. Let M be a maximum matching in G. By assumption, there is a vertex v in V_1 that is an $\overline{M}$-vertex. Let S be the set of all vertices of G that are connected to v by an M-alternating path. Since M is a maximum matching, an application of Theorem 9.2 yields v as the only $\overline{M}$-vertex in S.

Let $W_1 = S \cap V_1$ and let $W_2 = S \cap V_2$. Using the definition of the set S, together with the fact that no vertex of $S - \{v\}$ is an $\overline{M}$-vertex, we conclude that $W_1 - \{v\}$ is matched under M to W_2. Therefore, $|W_2| = |W_1| - 1$ and $W_2 \subseteq N(W_1)$. Furthermore, for every $w \in N(W_1)$, the graph G contains an M-alternating v–w path so that $N(W_1) \subseteq W_2$. Thus, $N(W_1) = W_2$ and

$$|N(W_1)| = |W_2| = |W_1| - 1 < |W_1|.$$

This, however, contradicts the fact that V_1 is nondeficient. □

We are now in a position to present a well-known theorem due to Hall [H5]. A collection $S_1, S_2, \ldots, S_k, k \geq 1$, of finite nonempty sets is said to have a *system of distinct representatives* or a *transversal* if there exists a set $\{s_1, s_2, \ldots, s_k\}$ of distinct elements such that $s_i \in S_i$ for $1 \leq i \leq k$.

Theorem 9.4

A collection $S_1, S_2, \ldots, S_k, k \geq 1$, of finite sets has a system of distinct representatives if and only if the union of any j of these sets contains at least j elements, for each j such that $1 \leq j \leq k$.

Proof

From the collection $S_1, S_2, \ldots, S_k, k \geq 1$, of finite sets we construct a bipartite graph G with partite sets V_1 and V_2 in the following manner. Let V_1 be the set $\{v_1, v_2, \ldots, v_k\}$ of distinct vertices, where v_i corresponds to the set S_i, and let V_2 be a set of vertices disjoint from V_1 such that $|V_2| = \left|\cup_{i=1}^{k} S_i\right|$, where there is a one-to-one correspondence between the elements of V_2 and those of $\cup_{i=1}^{k} S_i$. The construction of G is completed by joining a vertex v of V_1 with a vertex w of V_2 if and only if v corresponds to a set S_i and w corresponds to an element of S_i. From the manner in which G is defined, it follows that V_1 is nondeficient if and only if the union of any j of the sets S_i contains at least j elements. Now obviously, the sets S_i have a system of distinct representatives if and only if V_1 can be matched to a subset of V_2. Theorem 9.3 now produces the desired result. □

The preceding discussion is directly related to a well-known combinatorial problem called the *Marriage Problem*: Given a set of boys and a set of girls where each girl knows some of the boys, under what conditions can all girls get married, each to a boy she knows? In this context, Theorem 9.4 may be reformulated to produce what is often referred to as *Hall's Marriage Theorem*: If there are k girls, then the Marriage Problem has a solution if and only if every subset of j girls $(1 \leq j \leq k)$ collectively know at least j boys.

We have already noted that if M is a perfect matching in a graph G, then $\langle M \rangle$ is a 1-regular spanning subgraph of G. Any spanning subgraph of a graph G is referred to as a *factor* of G. A k-regular factor is called a *k-factor*. Thus F is a 1-factor of a graph G if and only if $E(F)$ is a perfect matching in G. The determination of whether a given graph contains a 1-factor is a problem that has received much attention in the literature. Of course, if a graph G has a 1-factor, then G has even order. A characterization of graphs that contain 1-factors has been obtained by Tutte [T10]. The following proof of Tutte's theorem is due to Anderson [A1]. An *odd component* of a graph is a component of odd order.

Theorem 9.5

A nontrivial graph G has a 1-factor if and only if for every proper subset S of $V(G)$, the number of odd components of $G - S$ does not exceed $|S|$.

Proof

Let F be a 1-factor of G. Assume, to the contrary, that there exists a proper subset W of $V(G)$ such that the number of odd components of $G - W$ exceeds $|W|$. For each odd component H of $G - W$, there is necessarily an edge of F joining a vertex of H with a vertex of W. This implies, however, that at least one vertex of W is incident with at least two edges of F, which is impossible. This establishes the necessity.

Next we consider the sufficiency. For a subset S of $V(G)$, denote the number of odd components of $G - S$ by $k_0(G - S)$. Hence, the hypothesis of G may now be restated as $k_0(G - S) \leq |S|$ for every proper subset S of $V(G)$. In particular, $k_0(G - \emptyset) \leq |\emptyset| = 0$, implying that G has only even components and therefore has even order n. Furthermore, we note that for each proper subset S of $V(G)$, the numbers $k_0(G - S)$ and $|S|$ are of the same parity since n is even.

We proceed by induction on even positive integers n. If G is a graph of order $n = 2$ such that $k_0(G - S) \leq |S|$ for every proper subset S of $V(G)$, then $G = K_2$ and G has a 1-factor.

Assume for all graphs H of even order less than n (where $n \geq 4$ is an even integer) that if $k_0(H - W) \leq |W|$ for every proper subset W of $V(H)$, then H has a 1-factor. Let G be a graph of order n and assume that $k_0(G - S) \leq |S|$ for each proper subset S of $V(G)$. We consider two cases.

Case 1. Suppose that $k_0(G-S) < |S|$ for all subsets S of $V(G)$ with $2 \leq |S| < n$. Since $k_0(G-S)$ and $|S|$ are of the same parity, $k_0(G-S) \leq |S| - 2$ for all subsets S of $V(G)$ with $2 \leq |S| < n$. Let $e = uv$ be an edge of G and consider $G - u - v$. Let T be a proper subset of $V(G - u - v)$. It follows that $k_0(G - u - v - T) \leq |T|$, for suppose, to the contrary, that $k_0(G - u - v - T) > |T|$. Then

$$k_0(G - u - v - T) > |T| = |T \cup \{u, v\}| - 2,$$

so that $k_0(G - (T \cup \{u, v\})) \geq |T \cup \{u, v\}|$, contradicting our supposition. Thus, by the inductive hypothesis, $G - u - v$ has a 1-factor and, hence, so does G.

Case 2. Suppose that there exists a subset R of $V(G)$ such that $k_0(G-R) = |R|$, where $2 \leq |R| < n$. Among all such sets R, let S be one of maximum cardinality, where $k_0(G-S) = |S| = k$. Further, let $G_1, G_2, \ldots, G_k$ denote the odd components of $G - S$. These are the only components of $G - S$, for if G_0 were an even component of $G - S$ and $u_0 \in V(G_0)$, then $k_0(G - (S \cup \{u_0\})) \geq k + 1 = |S \cup \{u_0\}|$ implying necessarily that $k_0(G - (S \cup \{u_0\})) = |S \cup \{u_0\}|$, which contradicts the maximum property of S.

For $i = 1, 2, \ldots, k$, let S_i denote the set of those vertices of S adjacent to one or more vertices of G_i. Each set S_i is nonempty; otherwise some G_i would be an odd component of G. The union of any j of the sets $S_1, S_2, \ldots, S_k$ contains at least j vertices for each j with $1 \leq j \leq k$; for otherwise, there exists j $(1 \leq j \leq k)$ such that the union T of some j sets contains less than j vertices. This would imply, however, that $k_0(G - T) > |T|$, which is impossible. Thus, we may employ Theorem 9.4 to produce a system of distinct representatives for $S_1, S_2, \ldots, S_k$. This implies that S contains vertices $v_1, v_2, \ldots, v_k$, and each G_i contains a vertex u_i $(1 \leq i \leq k)$ such that $u_i v_i \in E(G)$ for $i = 1, 2, \ldots, k$.

Let W be a proper subset of $V(G_i - u_i)$, $1 \leq i \leq k$. We show that $k_0(G_i - u_i - W) \leq |W|$, for suppose that $k_0(G_i - u_i - W) > |W|$. Since $G_i - u_i$ has even order, $k_0(G_i - u_i - W)$ and $|W|$ are of the same parity and so $k_0(G_i - u_i - W) \geq |W| + 2$. Thus,

$$\begin{aligned} k_0(G - (S \cup W \cup \{u_i\})) &= k_0(G_i - u_i - W) + k_0(G - S) - 1 \\ &\geq |S| + |W| + 1 \\ &= |S \cup W \cup \{u_i\}|. \end{aligned}$$

This, however, contradicts the maximum property of S. Therefore, $k_0(G_i - u_i - W) \leq |W|$ as claimed, implying by the inductive hypothesis that, for $i = 1, 2, \ldots, k$, the subgraph $G_i - u_i$ has a 1-factor. This 1-factor, together with the existence of the edges $u_i v_i$ $(1 \leq i \leq k)$, produces a 1-factor in G. □

By Tutte's theorem, it follows, of course, that if G is a graph of order $n \geq 2$ such that for each proper subset S of $V(G)$, the number of odd components of $G - S$ does not exceed $|S|$, then $\beta_1(G) = n/2$. Using a similar proof technique to that used in the proof of Theorem 9.5, Berge [B4] obtained the following extension of Tutte's theorem.

Theorem 9.6

Let G be a graph of order n. If k is the smallest nonnegative integer such that for each proper subset S of V(G), the number of odd components of G − S does not exceed |S| + 2k, then

$$\beta_1(G) = \left\lfloor \frac{n - 2k}{2} \right\rfloor.$$

Again, by Tutte's theorem, if G is a graph such that for every proper subset S of $V(G)$, the number of odd components of $G − S$ does not exceed $|S|$, then G has a 1-factor. Of course, if G is a graph of even order such that for every vertex-cut S of $V(G)$, the number of components (odd *or* even) of $G − S$ does not exceed $|S|$, then G has a 1-factor. Hence every 1-tough graph of even order contains a 1-factor. Enomoto, Jackson, Katernis and Saito [EJKS1] extended this result.

Theorem 9.7

If G is a k-tough graph of order $n \geq k + 1$, where kn is even, then G has a k-factor.

By definition, every 1-regular graph contains a 1-factor. A 2-regular graph G contains a 1-factor if and only if every component of G is an even cycle. This brings us to the 3-regular or cubic graphs. Petersen [P2] provided a sufficient condition for a cubic graph to contain a 1-factor.

Theorem 9.8

Every bridgeless cubic graph contains a 1-factor.

Proof

Let G be a bridgeless cubic graph and assume that $V(G)$ has a proper subset S such that the number of odd components of $G − S$ exceeds $|S|$. Let $j = |S|$ and let $G_1, G_2, \ldots, G_k$ $(k > j)$ be the odd components of $G − S$. There must be at least one edge joining a vertex of G_i to a vertex of S, for each $i = 1, 2, \ldots, k$; for otherwise, G_i is a cubic graph of odd order. On the other hand, since G contains no bridges, there cannot be exactly one such edge; that is, there are at least two edges joining G_i and S, for each $i = 1, 2, \ldots, k$.

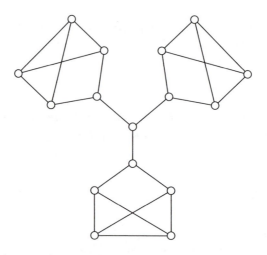

Figure 9.2 A cubic graph containing no 1-factors.

Suppose that for some $i = 1, 2, \ldots, k$, there are exactly two edges joining G_i and S. Then there is an odd number of odd vertices in the component G_i of $G - S$, which cannot happen. Hence, for each $i = 1, 2, \ldots, k$, there are at least three edges joining G_i and S. Therefore, the total number of edges joining $\cup_{i=1}^k V(G_i)$ and S is at least $3k$. However, since each of the j vertices of S has degree 3, the number of edges joining $\cup_{i=1}^k V(G_i)$ and S is at most $3j$. Therefore, $3j \geq 3k$, which is a contradiction since $3k > 3j$. Hence, no such set S exists. By Theorem 9.5, then, we conclude that G has a 1-factor. $\qquad\square$

Indeed, Petersen [P2] showed that if a cubic graph G contains at most two bridges, then G has a 1-factor (Exercise 9.5). On the other hand, this result cannot be extended further since the cubic graph of Figure 9.2 has three bridges but contains no 1-factor. In fact, it is the unique cubic graph of minimum order with this property.

From Petersen's work, we know that if G is a cubic graph of order n containing at most two bridges, then $\beta_1(G) = n/2$. The following result [CKLS1] provides a sharp bound on the edge independence number of a cubic graph in terms of the number of bridges it possesses.

Theorem 9.9

If G is a connected cubic graph of order n containing fewer than $3(k + 1)$ bridges, then

$$\beta_1(G) \geq \frac{n - 2k}{2}.$$

Errera [E6] showed, however, that if all the bridges of a cubic graph, regardless of how many there might be, lie on a single path, then G has a 1-factor.

Theorem 9.10

If all the bridges of a connected cubic graph G lie on a single path of G, then G has a 1-factor.

The three bridges of the cubic graph G of Figure 9.2 do not lie on a single path of G, of course, and G does not have a 1-factor. It is a direct consequence of Errera's theorem that if the bridges of a cubic graph G of order n lie on a single path, then $\beta_1(G) = n/2$. If the bridges of a cubic graph G do not lie on a single path, then it may very well occur that $\beta_1(G) < n/2$. But more (see [CKOR1]) can be said about this.

Theorem 9.11

If G is a connected cubic graph of order n all of whose bridges lie on r edge-disjoint paths of G, then

$$\beta_1(G) \geq \frac{n}{2} - \left\lfloor \frac{2r}{3} \right\rfloor.$$

Recall that an independent set of vertices in a graph G is one whose elements are pairwise independent (nonadjacent) and that the vertex independence number $\beta(G)$ of G is the maximum cardinality among the independent sets of vertices in G. For example, if $s \leq t$, then $\beta(K_{s,t}) = t$ and $\beta_1(K_{s,t}) = s$.

A vertex and an edge are said to *cover* each other in a graph G if they are incident in G. A *vertex cover* in G is a set of vertices that covers all the edges of G. An *edge cover* in a graph G without isolated vertices is a set of edges that covers all vertices of G.

The minimum cardinality of a vertex cover in a graph G is called the *vertex covering number* of G and is denoted by $\alpha(G)$. As expected, the *edge covering number* $\alpha_1(G)$ of a graph G (without isolated vertices) is the minimum cardinality of an edge cover in G. For $s \leq t$, we have $\alpha(K_{s,t}) = s$ and $\alpha_1(K_{s,t}) = t$. As another illustration of these four parameters, we note that for $n \geq 2$, $\beta(K_n) = 1$, $\beta_1(K_n) = \lfloor n/2 \rfloor$, $\alpha(K_n) = n - 1$ and $\alpha_1(K_n) = \lceil n/2 \rceil$. Observe that for the two graphs G of order n considered above, namely $K_{s,t}$ with $n = s + t$, and K_n, we have

$$\alpha(G) + \beta(G) = \alpha_1(G) + \beta_1(G) = n.$$

These two examples serve to illustrate the next theorem, due to Gallai [G1].

Theorem 9.12

If G is a graph of order n having no isolated vertices, then

$$\alpha(G) + \beta(G) = n \qquad\qquad (9.1)$$

and

$$\alpha_1(G) + \beta_1(G) = n \qquad\qquad (9.2)$$

Proof

We begin with (9.1). Let U be an independent set of vertices of G with $|U| = \beta(G)$. Clearly, the set $V(G) - U$ is a vertex cover in G. Therefore, $\alpha(G) \leq n - \beta(G)$. If, however, W is a set of $\alpha(G)$ vertices that covers all edges of G, then $V(G) - W$ is independents; thus $\beta(G) \geq n - \alpha(G)$. This proves (9.1).

To verify (9.2), let E_1 be an independent set of edges of G with $|E_1| = \beta_1(G)$. Obviously, E_1 covers $2\beta_1(G)$ vertices of G. For each vertex of G not covered by E_1, select an incident edge and define E_2 to be the union of this set of edges and E_1. Necessarily, E_2 is an edge cover in G so that $|E_2| \geq \alpha_1(G)$. Also we note that $|E_1| + |E_2| = n$; hence $\alpha_1(G) + \beta_1(G) \leq n$. Now suppose that E' is an edge cover in G with $|E'| = \alpha_1(G)$. The minimality of E' implies that each component of $\langle E' \rangle$ is a tree. Select from each component of $\langle E' \rangle$ one edge, denoting the resulting set of edges by E''. We observe that $|E''| \leq \beta_1(G)$ and that $|E'| + |E''| = n$. These two facts imply that $\alpha_1(G) + \beta_1(G) \geq n$, completing the proof of (9.2) and the theorem. $\qquad\square$

If C is a vertex cover in a graph G and E is an independent set of edges, then for each edge e of E there is a vertex v_e in C that is incident with e. Furthermore, if $e, f \in E$, then $v_e \neq v_f$. Thus for any independent set E of edges and any vertex cover C in G, we have $|C| \geq |E|$. This, of course, implies that $\alpha(G) \geq \beta_1(G)$. In general, equality does not hold here. If however, G is bipartite, then we do have $\alpha(G) = \beta_1(G)$, as was shown by König [K9].

Theorem 9.13

If G is a bipartite graph, then

$$\alpha(G) = \beta_1(G).$$

Proof

Since $\alpha(G) \geq \beta_1(G)$, it suffices to show that $\alpha(G) \leq \beta_1(G)$. Let V_1 and V_2 be the partite sets of G and let M be a maximum matching in G. Then $\beta_1(G) = |M|$. Denote by U the set of all $\overline{M}$-vertices in V_1. (If $U = \emptyset$, then the proof is, of course, complete.) Observe that $|M| = |V_1| - |U|$. Let S be the set of all vertices of G that are connected to some vertex in U by an M-alternating path. Define $W_1 = S \cap V_1$ and $W_2 = S \cap V_2$.

As in the proof of Theorem 9.3, we have that $W_1 - U$ is matched to W_2 and that $N(W_1) = W_2$. Since $W_1 - U$ is matched to W_2, it follows that $|W_1| - |W_2| = |U|$.

Observe that $C = (V_1 - W_1) \cup W_2$ is a vertex cover in G; for otherwise, there is an edge vw in G such that $v \in W_1$ and $w \notin W_2$. Furthermore,

$$|C| = |V_1| - |W_1| + |W_2| = |V_1| - |U| = |M|.$$

Therefore, $\alpha(G) \leq |C| = |M| = \beta_1(G)$ and the proof is complete. $\qquad\square$

Next we present upper and lower bounds for the edge independence number, due to Weinstein [W3].

Theorem 9.14

Let G be a graph of order n without isolated vertices. Then

$$\left\lceil \frac{n}{1 + \Delta(G)} \right\rceil \leq \beta_1(G) \leq \left\lfloor \frac{n}{2} \right\rfloor.$$

Furthermore, these bounds are sharp.

Proof

It suffices to prove the theorem for connected graphs. The upper bound for $\beta_1(G)$ is immediate and clearly sharp.

In order to verify the lower bound, we employ induction on the size m of a connected graph. If $m = 1$ or $m = 2$, then the lower bound follows. Assume that the lower bound holds for all connected graphs of positive size not exceeding k, where $k \geq 2$, and let G be a connected graph of order n having size $k + 1$. If G has a cycle edge e, then

$$\beta_1(G) \geq \beta_1(G - e) \geq \frac{n}{1 + \Delta(G - e)} \geq \frac{n}{1 + \Delta(G)}.$$

Otherwise, G is a tree. If $G = K_{1,n-1}$, then $\beta_1(G) = n/(1 + \Delta(G)) = 1$ (which also shows the sharpness of the lower bound). If $G \neq K_{1,n-1}$, then G contains an edge e such that $G - e$ has two nontrivial components G_1 and G_2. Let n_i denote the order of G_i, $i = 1, 2$. Applying the inductive hypothesis to G_1 and G_2, we obtain

$$\beta_1(G) \geq \beta_1(G_1) + \beta_1(G_2) \geq \frac{n_1}{1 + \Delta(G_1)} + \frac{n_2}{1 + \Delta(G_2)}$$

$$\geq \frac{n_1}{1 + \Delta(G)} + \frac{n_2}{1 + \Delta(G)} = \frac{n}{1 + \Delta(G)}. \qquad \square$$

Combining Theorems 9.12 and 9.14, we have our next result.

Corollary 9.15

Let G be a graph of order n without isolated vertices. Then

$$\left\lceil \frac{n}{2} \right\rceil \leq \alpha_1(G) \leq \left\lfloor \frac{n \cdot \Delta(G)}{1 + \Delta(G)} \right\rfloor.$$

Furthermore, these bounds are sharp.

It is easy to see for a graph G of order n without isolated vertices that $1 \leq \beta(G) \leq n - 1$ and that these bounds are sharp. This implies that $1 \leq \alpha(G) \leq n - 1$ are sharp bounds for $\alpha(G)$.

A set S of vertices or edges in a graph G is said to be *maximal with respect to a property P* if S has property P but no proper superset of S has property P; while S is *minimal with respect to P* if S has property P but no proper subset of S has property P. Although isolated instances of these concepts have been discussed earlier, we will be encountering such ideas in a more systematic manner in this chapter and the next. In particular, for certain properties P, we will be interested in the maximum and/or minimum cardinality of a maximal or minimal set with property P.

For example, a maximal independent set of vertices of maximum cardinality in a graph G is called a *maximum independent set of vertices*. The number of vertices in a maximum independent set has been called the independence number of G, which we have denoted by $\beta(G)$. We define the *lower independence number $i(G)$* of G as the minimum cardinality of a maximal independent set of vertices of G. Thus, for $K_{s,t}$ where $s < t$, there are only two maximal independent sets of vertices, namely, the partite sets of $K_{s,t}$. Hence, $\beta(K_{s,t}) = t$, while $i(K_{s,t}) = s$.

Likewise, a maximal matching or a maximal independent set of edges of maximum cardinality in a graph G is a maximum matching. The number of edges in a maximum matching is the edge independence number $\beta_1(G)$ of G. The minimum cardinality of a maximal matching is the *lower edge independence number* of G and is denoted by $i_1(G)$. For example, for the path P_6, $\beta_1(P_6) = 3$ and $i_1(P_6) = 2$. Two results [JRS1] dealing with maximal independent sets of edges are stated next (Exercises 9.10 and 9.11).

Theorem 9.16

For every nonempty graph G,

$$i_1(G) \leq \beta_1(G) \leq 2i_1(G).$$

Theorem 9.17

Let G be a nonempty graph. If k is an integer such that $i_1(G) \leq k \leq \beta_1(G)$, then G contains a maximal matching with k edges.

EXERCISES 9.1

9.1 Show that every tree has at most one perfect matching.

9.2 Determine the maximum size of a graph of order n having a maximum matching of k edges if (a) $n = 2k$, (b) $n = 2k + 2$.

9.3 Use Menger's theorem to prove Theorem 9.3.

9.4 Let G be a bipartite graph with partite sets V_1 and V_2, where $|V_1| \leq |V_2|$. The *deficiency* def (U) of a set $U \subseteq V_1$ is defined as $\max\{|S| - |N(S)|\}$, where the maximum is taken over all nonempty subsets S of U. Show that $\beta_1(G) = \min\{|V_1|, |V_1| - \text{def}(V_1)\}$.

9.5 Prove that every cubic graph with at most two bridges contains a 1-factor.

9.6 (a) Let G be an odd graph and let $\{V_1, V_2\}$ be a partition of $V(G)$, where E' is the set of edges joining V_1 and V_2. Prove that $|V_1|$ and $|E'|$ are of the same parity.

(b) Prove that every $(2k+1)$-regular, $2k$-edge-connected graph, $k \geq 1$, contains a 1-factor.

9.7 Prove that if G is an r-regular, $(r-2)$-edge-connected graph $(r \geq 3)$ of even order containing at most $r-1$ distinct edge-cuts of cardinality $r-2$, then G has a 1-factor.

9.8 Show that a graph G is bipartite if and only if $\beta(H) \geq \frac{1}{2}|V(H)|$ for every subgraph H of G.

9.9 Let G be a graph and let $U \subseteq V(G)$. Use Tutte's theorem to prove that G has a matching that covers U if and only if for every proper subset S of $V(G)$, the number of odd components of $G - S$ containing only vertices of U does not exceed $|S|$.

9.10 Prove Theorem 9.16.

9.11 Prove Theorem 9.17.

9.12 Show that Theorems 9.16 and 9.17 have no analogues to maximal independent sets of vertices.

9.13 Characterize those nonempty graphs with the property that every pair of distinct maximal independent sets of vertices is disjoint.

9.14 The *matching graph* $M(G)$ of a nonempty graph G has the maximum matchings of G as its vertices, and two vertices M_1 and M_2 of $M(G)$ are adjacent if M_1 and M_2 differ in only one edge. Show that each cycle $C_n, n = 3, 4, 5, 6$, is the matching graph of some graph.

9.15 Prove or disprove: A graph G without isolated vertices has a perfect matching if and only if $\alpha_1(G) = \beta_1(G)$.

9.16 Show that if G is a bipartite graph without isolated vertices, then $\alpha_1(G) = \beta(G)$.

9.2 FACTORIZATIONS AND DECOMPOSITIONS

A graph G is said to be *factorable* into the factors $G_1, G_2, \ldots, G_t$ if these factors are pairwise edge-disjoint and $\cup_{i=1}^{t} E(G_i) = E(G)$. If G is factored into $G_1, G_2, \ldots, G_t$, then we represent this by $G = G_1 \oplus G_2 \oplus \cdots \oplus G_t$, which is called a *factorization* of G.

If there exists a factorization of a graph G such that each factor is a k-factor (for a fixed k), then G is *k-factorable*. If G is a k-factorable graph, then necessarily G is r-regular for some integer r that is a multiple of k.

If a graph G is factorable into $G_1, G_2, \ldots, G_t$, where each $G_i = H$ for some graph H, then we say that G is *H-factorable* and that G has an *isomorphic factorization* into the factor H. Certainly, if a graph G is H-factorable, then the size of H divides the size of G. A graph G of order $n = 2k$ is 1-factorable if and only if G is kK_2-factorable.

The problem in this area that has received the most attention is the determination of which graphs are 1-factorable. Of course, only regular graphs of even order can be 1-factorable. Trivially, every 1-regular graph is 1-factorable. Since a 2-regular graph contains a 1-factor if and only if every component is an even cycle, it is precisely these 2-regular graphs that are 1-factorable. The situation for r-regular graphs, $r \geq 3$, in general, or even only 3-regular graphs in particular, is considerably more complicated. By Petersen's theorem, every bridgeless cubic graph contains a 1-factor. Consequently, every bridgeless cubic graph can be factored into a 1-factor and a 2-factor. Not every bridgeless cubic graph is 1-factorable, however. Indeed, as Petersen himself observed [P3], the Petersen graph (Figure 9.3) is not 1-factorable; for otherwise, it would have edge chromatic number 3, which is not the case (Exercise 8.25).

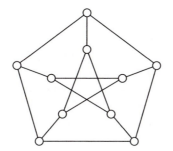

Figure 9.3 The Petersen graph: a bridgeless cubic graph that is not 1-factorable.

We now describe two classes of 1-factorable graphs. The first of these is due to König [K8].

Theorem 9.18

Every regular bipartite graph of degree $r \geq 1$ is 1-factorable.

Proof

We proceed by induction on r, the result being obvious for $r = 1$. Assume, then, that every regular bipartite graph of degree $r - 1, r \geq 2$, is 1-factorable, and let G be a regular bipartite graph of degree r, where V_1 and V_2 are the partite sets of G.

We now show that V_1 is nondeficient. Let S be a nonempty subset of V_1. The number of edges of G incident with the vertices of S is $r|S|$. These edges are, of course, also incident with the vertices of $N(S)$. Since G is r-regular, the number of edges joining S and $N(S)$ cannot exceed $r|N(S)|$. Hence, $r|N(S)| \geq r|S|$ so that $|N(S)| \geq |S|$. Therefore, V_1 is nondeficient, implying by Theorem 9.3 that V_1 can be matched to a subset of V_2. Since G is regular of positive degree, $|V_1| = |V_2|$; thus, G has a 1-factor F. The removal of the edges of F from G results in a bipartite graph G' that is regular of degree $r - 1$. By the inductive hypothesis, G' is 1-factorable, implying that G is 1-factorable as well. □

The following result is part of mathematical folklore.

Theorem 9.19

The complete graph K_{2k} is 1-factorable.

Proof

The result is obvious for $k=1$, so we assume that $k \geq 2$. Denote the vertex set of K_{2k} by $\{v_0, v_1, \ldots, v_{2k-1}\}$ and arrange the vertices $v_1, v_2, \ldots, v_{2k-1}$ in a regular $(2k-1)$-gon, placing v_0 in the center. Now join every two vertices by a straight line segment, producing K_{2k}. For $i = 1, 2, \ldots, 2k-1$, define the 1-factor F_i to consist of the edge $v_0 v_i$ together with all those edges perpendicular to $v_0 v_i$. Then $K_{2k} = F_1 \oplus F_2 \oplus \cdots \oplus F_{2k-1}$, so K_{2k} is 1-factorable. $\qquad\square$

The construction described in the proof of Theorem 9.19 is illustrated in Figure 9.4 for the graph K_6.

We return briefly to the 1-factorization of K_{2k} described in the proof of Theorem 9.19. Recall that the 1-factor F_1 consists of the edge $v_0 v_1$ and all edges perpendicular to $v_0 v_1$, namely, $v_2 v_{2k-1}, v_3 v_{2k-2}, \ldots, v_k v_{k+1}$. If the k edges of F_1 are rotated clockwise through an angle of $2\pi/(2k-1)$ radians, then the 1-factor F_2 is obtained. In general, if the edges of F_1 are rotated clockwise through an angle of $2\pi j/(2k-1)$ radians, where $0 \leq j \leq 2k-2$, then the 1-factor F_{j+1} is produced. A factorization of a graph obtained in this manner is referred to as a *cyclic factorization*.

Such a factorization can be viewed in another way. Let K_{2k} be drawn as described in the proof of Theorem 9.19. We now label each edge of K_{2k} with one of the integers $0, 1, \ldots, k-1$. Indeed, we will assign $2k-1$ edges of K_{2k} the label i for $i = 0, 1, \ldots, k-1$. Every edge of the type $v_0 v_i$ $(1 \leq i \leq 2k-1)$ is labeled 0. Now let C denote the cycle $v_1, v_2, \ldots, v_{2k-1}, v_1$. For $1 \leq s < t \leq 2k-1$, the edge

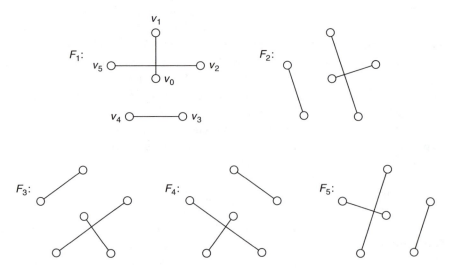

Figure 9.4 A 1-factorization of K_6.

$v_s v_t$ is assigned the distance label $d_C(v_s, v_t)$. Observe that $1 \le d_C(v_s, v_t) \le k-1$. Thus, the $2k-1$ edges of C are labeled 1; in general, then, $2k-1$ edges of K_{2k} are labeled the integer i for $0 \le i \le k-1$. Observe, further, that F_1 contains k edges, one of which is labeled i for $0 \le i \le k-1$. Moreover, when an edge of F_1 labeled i ($0 \le i \le k-1$) is rotated clockwise through an angle of $2\pi j/(2k-1)$ radians, $0 \le j \le 2k-2$, an edge of F_{j+1} also labeled i is obtained. Hence a 1-factorization of K_{2k} is produced.

We now turn to 2-factorable graphs. Of course, for a graph to be 2-factorable, it is necessary that it be $2k$-regular for some integer $k \ge 1$. Petersen [P2] showed that this obvious necessary condition is sufficient as well.

Theorem 9.20

A graph G is 2-factorable if and only if G is $2k$-regular for some integer $k \ge 1$.

Proof

We have already noted that if G is a 2-factorable graph, then G is regular of positive even degree. Conversely, suppose that G is $2k$-regular for some integer $k \ge 1$. Assume, without loss of generality, that G is connected. Hence, G is eulerian and so contains an eulerian circuit C.

Let $V(G) = \{v_1, v_2, \ldots, v_n\}$. We define a bipartite graph H with partite sets $U = \{u_1, u_2, \ldots, u_n\}$ and $W = \{w_1, w_2, \ldots, w_n\}$, where

$$E(H) = \{u_i w_j \mid v_j \text{ immediately follows } v_i \text{ on } C\}.$$

The graph H is k-regular and so, by Theorem 9.18, is 1-factorable. Hence, $H = F_1 \oplus F_2 \oplus \cdots \oplus F_k$ is a 1-factorization of H.

Corresponding to each 1-factor F_ℓ ($1 \le \ell \le k$) of H is a permutation α_ℓ on the set $\{1, 2, \ldots, n\}$, defined by $\alpha_\ell(i) = j$ if $u_i w_j \in E(F_\ell)$. Let α_ℓ be expressed as a product of disjoint permutation cycles. There is no permutation cycle of length 1 in this product; for if (i) were a permutation cycle, then this would imply that $\alpha_\ell(i) = i$. However, this further implies that $u_i w_i \in E(F_\ell)$ and that $v_i v_i \in E(C)$, which is impossible. Also there is no permutation cycle of length 2 in this product; for if $(i\ j)$ were a permutation cycle, then $\alpha_\ell(i) = j$ and $\alpha_\ell(j) = i$. This would indicate that $u_i w_j, u_j w_i \in E(F_\ell)$ and that v_j both immediately follows and precedes v_i on C, contradicting the fact that no edge is repeated on a circuit. Thus, the length of every permutation cycle in α_ℓ is at least 3.

Each permutation cycle in α_ℓ therefore gives rise to a cycle in G, and the product of disjoint permutation cycles in α_ℓ produces a collection of mutually disjoint cycles in G containing all vertices of G, that is, a 2-factor in G. Since the

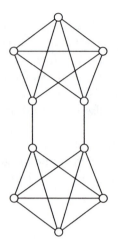

Figure 9.5 A 2-factorable graph that is not hamiltonian factorable.

1-factors F_ℓ in H are mutually edge-disjoint, the resulting 2-factors in G are mutually edge-disjoint. Hence, G is 2-factorable. □

By Theorem 9.20, then, there exists a factorization of every regular graph G of positive even degree in which every factor is a union of cycles. We next consider the problem of whether there exists a factorization of G such that every factor is a single cycle. A *hamiltonian factorization* of a graph G is a factorization of G such that every factor is a hamiltonian cycle of G. Certainly, if a graph G has a hamiltonian factorization, then G is a 2-connected regular graph of positive even degree. The converse of this statement is not true, however, as the graph of Figure 9.5 shows.

For complete graphs, 2-factorable and hamiltonian factorable are equivalent concepts.

Theorem 9.21

For every positive integer k, the graph K_{2k+1} is hamiltonian factorable.

Proof

Since the result is clear for $k=1$, we may assume that $k\geq 2$. Let $V(K_{2k+1})=\{v_0,v_1,\ldots,v_{2k}\}$. Arrange the vertices $v_1,v_2,\ldots,v_{2k}$ in a regular $2k$-gon and place v_0 in some convenient position. Join every two vertices by a straight line segment, thereby producing K_{2k+1}. We define the edge set of F_1 to consist of v_0v_1, v_0v_{k+1}, all edges parallel to v_1v_2 and all edges parallel to $v_{2k}v_2$ (see F_1 in Figure 9.6 for the

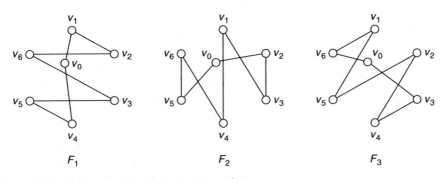

Figure 9.6 A hamiltonian factorization of K_7.

case $k = 3$). In general, for $i = 1, 2, \ldots, k$, we define the edge set of the factor F_i to consist of $v_0 v_i$, $v_0 v_{k+i}$, all edges parallel to $v_i v_{i+1}$ and all edges parallel to $v_{i-1} v_{i+1}$, where the subscripts are expressed modulo $2k$. Then $K_{2k+1} = F_1 \oplus F_2 \oplus \cdots \oplus F_k$, where F_i is the hamiltonian cycle

$$v_0, v_i, v_{i+1}, v_{i-1}, v_{i+2}, v_{i-2}, \ldots v_{k+i-1}, v_{k+i+1}, v_{k+i}, v_0. \qquad \square$$

This result is illustrated in Figure 9.6 for K_7.

The factorization described in the proof of Theorem 9.21 is again a cyclic factorization. If we place the vertex v_0 in the center of the regular $2k$-gon and rotate the edges of the hamiltonian cycle F_1 clockwise through an angle of $2\pi/2k = \pi/k$ radians, then the hamiltonian cycle F_2 is produced. Indeed, if we rotate the edges of F_1 clockwise through an angle of $\pi j/k$ radians for any integer j with $1 \le j \le k-1$, then the hamiltonian cycle F_{j+1} is produced and the desired hamiltonian factorization of K_{2k+1} is obtained (Exercise 9.21).

Another factorization result now follows readily from Theorem 9.21.

Corollary 9.22

The complete graph K_{2k} can be factored into k hamiltonian paths.

Theorems 9.19 and 9.21 have an interesting consequence of a different nature. If G is an r-regular graph of order n, then, of course, $0 \le r \le n-1$. On the other hand, if r and n are odd positive integers (with $0 \le r \le n-1$), then there can be no r-regular graph of order n. With this lone exception, every other type of regular graph is possible.

Corollary 9.23

Let r and n be integers with $0 \le r \le n-1$. Then there exists an r-regular graph of order n if and only if r and n are not both odd.

Proof

It suffices to show that there exists an r-regular graph of order n if at least one of r and n is even (and $0 \leq r \leq n-1$). Suppose first that n is even. Then $n = 2k$ for some positive integer k. By Theorem 9.19, K_{2k} can be factored into 1-factors $F_1, F_2, \ldots, F_{2k-1}$. The union of r of these 1-factors produces an r-regular graph of order n.

Next, suppose that n is odd. Then r is necessarily even. The graph K_1 is 0-regular of order 1, so we may assume that $n = 2k + 1 \geq 3$. By Theorem 9.21, K_{2k+1} can be factored into hamiltonian cycles $F_1, F_2, \ldots, F_k$. The union of $r/2$ of these hamiltonian cycles gives an r-regular graph of order n. □

Using the construction employed in the proof of Theorem 9.21, we can obtain another factorization result.

Theorem 9.24

The graph K_{2k} can be factored into $k-1$ hamiltonian cycles and a 1-factor.

Very similar to the concept of factorization is decomposition. A *decomposition* of a graph G is a collection $\{H_i\}$ of nonempty subgraphs such that $H_i = \langle E_i \rangle$ for some (nonempty) subset E_i of $E(G)$, where $\{E_i\}$ is a partition of $E(G)$. Thus no subgraph H_i in a decomposition of G contains isolated vertices. If $\{H_i\}$ is a decomposition of G, then we write $G = H_1 \oplus H_2 \oplus \cdots \oplus H_t$, as we do with factorizations, and say G is decomposed into the subgraphs $H_1, H_2, \ldots, H_t$, where then $|\{H_i\}| = t$. Indeed, if $G = H_1 \oplus H_2 \oplus \cdots \oplus H_t$ is a decomposition of a graph G of order n and we define $F_i = H_i \cup [n - |V(H_i)|]K_1$ for $1 \leq i \leq t$, then $F_1 \oplus F_2 \oplus \cdots \oplus F_t$ is a factorization of G. On the other hand, every factorization of a nonempty graph G also gives rise to a decomposition of G. Suppose that $G = F_1 \oplus F_2 \oplus \cdots \oplus F_s$ is a factorization of a nonempty graph G, so written that $F_1, F_2, \ldots, F_t$, are nonempty ($t \leq s$). Let $H_i = \langle E(F_i) \rangle$ for $i = 1, 2, \ldots, t$. Then $H_1 \oplus H_2 \oplus \cdots \oplus H_t$ is a decomposition of G.

If $\{H_i\}$ is a decomposition of a graph G such that $H_i = H$ for some graph H for each i, then G is said to be *H-decomposable*. If G is an *H*-decomposable graph, then we also write $H \mid G$ and say that H *divides* G. Also H is said to be a *divisor* of G, and G is a *multiple* of H. For $G = K_{2,2,2}$ (the graph of the octahedron) and for the graph H shown in Figure 9.7, we have that G is *H*-decomposable. An *H*-decomposition of G is also shown in Figure 9.7.

The decomposition shown in Figure 9.7 is a cyclic decomposition. In general, a *cyclic decomposition* of a graph G into k copies of a subgraph H is obtained by (a) drawing G in an appropriate manner, (b) selecting a suitable subgraph H_1 of G that is isomorphic to H, and (c) rotating the vertices and edges of H_1 through an appropriate angle $k-1$ times to produce the k copies of H in the decomposition.

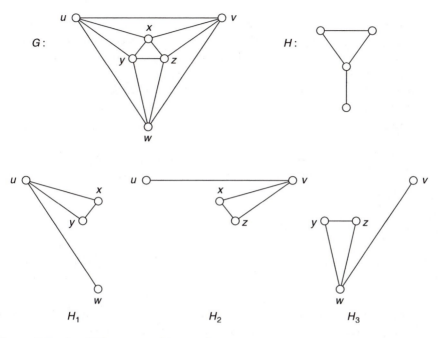

G: H:

H_1 H_2 H_3

Figure 9.7 An *H*-decomposable graph.

If G is an H-decomposable graph for some graph H, then certainly H is a subgraph of G and the size of H divides the size of G. Although this last condition is necessary, it is not sufficient. For example, the graph $K_{1,4}$ is a subgraph of the graph G of Figure 9.7 and the size 4 of $K_{1,4}$ divides the size 12 of G, but G is not $K_{1,4}$-decomposable (Exercise 9.26).

The basic problem in this context is, for a graph G and a subgraph H of G whose size divides that of G whether, the graph G is H-decomposable. First, we consider H-decomposable graphs for graphs H of small size. Of course, every (nonempty) graph is K_2-decomposable. Every component of a P_3-decomposable graph must have even size. In fact, this condition is sufficient for a graph to be P_3-decomposable (see [CPS1]).

Theorem 9.25

A nontrivial connected graph G is P_3-decomposable if and only if G has even size.

Proof

We have already noted that if G is P_3-decomposable, then G has even size. For the converse, assume that G has even size. Suppose, first, that G is eulerian, where the edges of G are encountered in the order $e_1, e_2, \ldots, e_m$. Then each of the sets $\{e_1, e_2\}$, $\{e_3, e_4\}, \ldots, \{e_{m-1}, e_m\}$ induce a copy of P_3; so G is P_3-decomposable. Otherwise, G has $2k$ odd vertices for some $k \geq 1$. By Theorem 4.3, $E(G)$ can be partitioned into subsets $E_1, E_2, \ldots, E_k$, where for each i, $\langle E_i \rangle$ is an open trail T_i of even length connecting odd vertices of G. Then, as with the eulerian circuit above, the edges of each trail T_i can be paired off so that each pair of edges induces a copy of P_3. Thus G is P_3-decomposable. $\square$

The only other graph of size 2 without isolated vertices is $2K_2$. The class of $2K_2$-decomposable graphs was discovered by Y. Caro (unpublished) and Ruiz [R12].

Theorem 9.26

A nontrivial graph G is $2K_2$-decomposable if and only if G has even size m, $\Delta(G) \leq \frac{1}{2}m$ and $G \neq K_3 \cup K_2$.

Most of the interest in K_3-decompositions has involved complete graphs. A K_3-decomposition of a complete graph is called a *Steiner triple system*. Kirkman [K5] characterized Steiner triple systems.

Theorem 9.27

The complete graph K_n is K_3-decomposable if and only if n is odd and $3 | \binom{n}{2}$.

For K_n to be K_{p+1}-decomposable, the conditions $p \mid (n-1)$ and $\binom{p+1}{2} \mid \binom{n}{2}$ are certainly necessary. These conditions are not sufficient in general, however. For $n = p^2 + p + 1$, Ryser [R13] showed that K_n is K_{p+1}-decomposable if and only if there exists a projective plane of order p; and in order for a projective plane of order p to exist, p must satisfy the Bruck–Ryser conditions [BR2] that $p \equiv 0 \pmod 4$ or $p \equiv 1 \pmod 4$, and $p = x^2 + y^2$ for some integers x and y. The smallest value of p for which the existence of a projective plane of order p is unknown is $p = 10$.

Whenever K_n is K_{p+1}-decomposable, we have an example of a combinatorial structure referred to as a *balanced incomplete block design*. Thus graph decompositions may be viewed as generalized block designs.

There is another important interpretation of a special type of decomposition. In particular, the minimum number of 1-regular subgraphs into which a nonempty graph G can be decomposed is the edge chromatic number $\chi_1(G)$ of G. By Theorem 8.16, the edge chromatic number of an r-regular graph G ($r \geq 1$) is r or $r + 1$. If $\chi_1(G) = r$, then each edge color class in a $\chi_1(G)$-edge coloring of G induces a 1-factor of G. Thus, an r-regular graph has edge chromatic number r if and only if it is 1-factorable.

The vast majority of factorization and decomposition results deal with factoring or decomposing complete graphs into a specific graph or graphs. R. M. Wilson [W7] proved that for every graph H without isolated vertices, there exist infinitely many positive integers n such that K_n is H-decomposable.

Theorem 9.28

For every graph H without isolated vertices and having size m, there exists a positive integer N such that if (i) $n \ge N$, (ii) $m|\binom{n}{2}$ *and* (iii) $d|(n-1)$, *where*

$$d = \gcd\{\deg v \mid v \in V(H)\},$$

Then K_n is H-decomposable.

As an immediate consequence of Theorem 9.28, there exist regular H-decomposable graphs for every graph H without isolated vertices. This result also appears in Fink [F2], where specific H-decomposable (not necessarily complete) graphs are described. We give a proof of this result, but prior to doing this, it is convenient to introduce some additional terminology, which will be explored further in Section 9.3.

A graph G of size m is called *graceful* if it is possible to label the vertices of G with distinct elements from the set $\{0, 1, \ldots, m\}$ in such a way that the induced edge labeling, which prescribes the integer $|i - j|$ to the edge joining vertices labeled i and j, assigns the labels $1, 2, \ldots, m$ to the edges of G. Such a labeling is called a *graceful labeling*. Thus, a graceful graph is a graph that admits a graceful labeling.

The graphs $K_3, K_4, K_4 - e$ and C_4 are graceful as is illustrated in Figure 9.8. Here the vertex labels are placed within the vertices and the induced edge labels are placed near the relevant edges. Not every graph is graceful, however. For example, the graphs K_5, C_5 and $K_1 + 2K_2$ are not graceful.

The *gracefulness* $\operatorname{grac}(G)$ of a graph G with $V(G) = \{v_1, v_2, \ldots, v_n\}$ and without isolated vertices is the smallest positive integer k for which it is possible to label the vertices of G with distinct elements from the set $\{0, 1, \ldots, k\}$ in such a way that

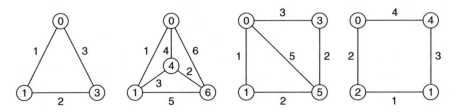

Figure 9.8 Graceful graphs.

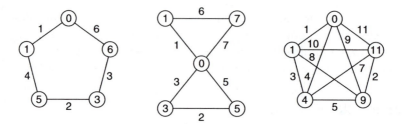

Figure 9.9 The three connected graphs of order 5 that are not graceful.

distinct edges receive distinct labels. Such vertex labelings always exist, one of which is to label v_i by 2^{i-1}. Hence for every graph G of order n and size m without isolated vertices, $m \leq \text{grac}(G) \leq 2^{n-1}$. If G is a graph of size m with $\text{grac}(G) = m$, then G is graceful. Thus the gracefulness of a graph G is a measure of how close G is to being graceful. By definition, it is possible to label the vertices of a graph G with distinct elements of the set $\{0, 1, \ldots, \text{grac}(G)\}$ so that the edges of G receive distinct labels. Of course, some vertex of G must be labeled $\text{grac}(G)$, but it is not known whether an edge of G must then be labeled $\text{grac}(G)$.

All connected graphs of order at most 4 are graceful. There are exactly three connected nongraceful graphs of order 5. In each case the gracefulness is one more than the size. These three graphs with an appropriate labeling are shown in Figure 9.9.

We now return to our discussion of decompositions and give a constructive proof that for every graph H without isolated vertices, there exists a regular H-decomposable graph. The proof is due to Fink and Ruiz [FR1] and was inspired by a proof technique of Rosa [R10].

Theorem 9.29

For every graph H without isolated vertices, there exists a regular H-decomposable graph.

Proof

Let H be a graph of order n and size m without isolated vertices and suppose that $\text{grac}(H) = k$. Hence there exists a labeling $\phi: V(H) \to \{0, 1, \ldots, k\}$ of vertices of H so that distinct edges of H are labeled differently and $\max\{\phi(x)|x \in V(H)\} = k$.

We now construct a regular H-decomposable graph G of order $p = 2k + 1$. Let $V(G) = \{v_0, v_1, \ldots, v_{p-1}\}$ and arrange these vertices cyclically in clockwise order about a regular p-gon. Next we define a graph H_1 by

$$V(H_1) = \{v_{\phi(x)} | x \in V(H)\}$$

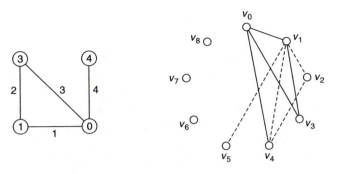

Figure 9.10 Construction of a regular H-decomposable graph.

and

$$E(H_1) = \{v_{\phi(x)}v_{\phi(y)}|xy \in V(H)\}.$$

For $i = 2, 3, \ldots, p$, define H_i by cyclically rotating H_1 through a clockwise angle of $2\pi(i-1)/p$ radians. Therefore, for $1 \le i \le p$,

$$V(H_i) = \{v_{\phi(x)+i-1}|x \in V(H)\}$$

and

$$E(H_i) = \{v_{\phi(x)+i-1}v_{\phi(y)+i-1}|xy \in V(H)\}.$$

The definition of G is completed by defining $E(G) = \cup_{i=1}^{p}E(H_i)$. (See Figure 9.10 for a given graph H, a possible labeling of the vertices of H, the induced edge labels of H, the vertices of G, and the subgraphs H_1 and H_2 of G, where the edges of H_2 are drawn with dashed lines.)

The graph G is therefore decomposable into the graphs $H_1, H_2, \ldots, H_p$, each of which is isomorphic to H, and G is $2m$-regular. □

If, in the proof of Theorem 9.29, the graph H is graceful, then $\mathrm{grac}(H) = m$ and G is a $2m$-regular graph of order $2m+1$, that is, $G = K_{2m+1}$ and, consequently, K_{2m+1} is H-decomposable. Indeed, then, for any graceful graph H of size m, the complete graph K_{2m+1} is H-decomposable; in fact, there is a cyclic decomposition of K_{2m+1} into H. This observation is due to Rosa [R10]. Because of its importance, we give a direct proof due to Rosa of this result.

Theorem 9.30

If H is a graceful graph of size m, then K_{2m+1} is H-decomposable. Indeed, K_{2m+1} can be cyclically decomposed into copies of H.

Proof

Since H is graceful, there is a graceful labeling of H, that is, the vertices of H can be labeled from a subset of $\{0, 1, \ldots, m\}$ so that the induced edge labels are $1, 2, \ldots, m$. Let $V(K_{2m+1}) = \{v_0, v_1, \ldots, v_{2m}\}$ where the vertices of K_{2m+1} are arranged cyclically in a regular $(2m+1)$-gon, denoting the resulting $(2m+1)$-cycle by C. A vertex labeled i $(0 \leq i \leq m)$ in H is placed at v_i in K_{2m+1} and this is done for each vertex of H. Every edge of H is drawn as a straight line segment in K_{2m+1}, denoting the resulting copy of H in K_{2m+1} as H_1. Hence $V(H_1) \subseteq \{v_0, v_1, \ldots, v_m\}$.

Each edge $v_s v_t$ of K_{2m+1} $(0 \leq s, t \leq 2m)$ is labeled $d_C(v_s, v_t)$, where then $1 \leq d_C(v_s, v_t) \leq m$. Consequently, K_{2m+1} contains exactly $2m+1$ edges labeled i for each i $(1 \leq i \leq m)$ and H_1 contains exactly one edge labeled i $(1 \leq i \leq m)$. Whenever an edge of H_1 is rotated through an angle (clockwise, say) of $2\pi k / (2m+1)$ radians, where $1 \leq k \leq m$, an edge of the same label is obtained. Denote the subgraph obtained by rotating H_1 through a clockwise angle of $2\pi k / (2m+1)$ radians by H_{k+1}. Then $H_{k+1} = H$ and a cyclic decomposition of K_{2m+1} into $2m+1$ copies of H results. ☐

As an illustration of Theorem 9.30, we consider the graceful graph $H = P_3$. A graceful labeling of H is shown in Figure 9.11 as well as the resulting cyclic H-decomposition of K_5.

Although K_{2m+1} has a cyclic decomposition into every graceful graph H of size m, it is not necessary for H to be graceful in order for K_{2m+1} to have a cyclic

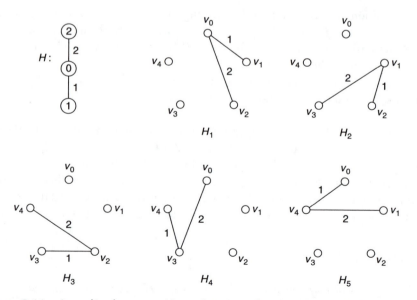

Figure 9.11   A cyclic decomposition of K_5 into the graceful graph P_3.

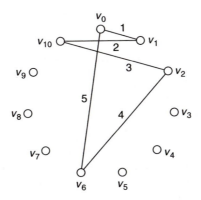

Figure 9.12 A cyclic decomposition of K_{11} into the nongraceful graph C_5.

H-decomposition. For example, we have seen that C_5 is not graceful; yet K_{11} has a cyclic C_5-decomposition. Such a decomposition is depicted in Figure 9.12.

It has been conjectured by Kotzig (see Rosa [R10]) that every nontrivial tree is graceful.

Kotzig's Conjecture

Every nontrivial tree is graceful.

Furthermore, the following conjecture concerning decompositions of complete graphs into trees has been made by Ringel [R6].

Ringel's Conjecture

For every tree T of size m, the complete graph K_{2m+1} is T-decomposable.

Of course, if Kotzing's conjecture is true, so is Ringel's. Indeed, the truth of Kotzig's conjecture implies the truth of the following conjecture, due jointly to G. Ringel and A. Kotzig.

The Ringel–Kotzig Conjecture

For every tree T of size m, K_{2m+1} can be cyclically decomposed into T.

We close this section by presenting another result involving cyclic decompositions. A *linear forest* is a forest each of whose components is a path. Since every 1-factor is a linear forest, the following result, due to Ruiz [R11], is a generalization of Theorem 9.19.

Theorem 9.31

If F is a linear forest of size k having no isolated vertices, then K_{2k} is F-decomposable.

Proof

Since the result is obvious for $k = 1$, we assume that $k \geq 2$. Let the vertex set of K_{2k} be denoted by $\{v_0, v_1, v_2, \ldots, v_{2k-1}\}$. Arrange the vertices $v_1, v_2, \ldots, v_{2k-1}$ cyclically in clockwise order about a regular $(2k-1)$-gon, calling the resulting cycle C, and place v_0 in the center of the $(2k-1)$-gon. Join every two vertices by a straight line segment to obtain the edges of K_{2k}. We label each edge that joins v_0 to a vertex of C by 0. There are $2k-1$ such edges. Every other edge of K_{2k} joins two vertices of C. If uv an a edge joining two vertices of C, then label uv by i if $d_C(u, v) = i$. Note that $1 \leq i \leq k-1$ and that for each $i = 1, 2, \ldots, k-1$, the graph K_{2k} contains $2k-1$ edges labeled i.

We now describe two paths P and Q of length k in K_{2k}. If k is even, then

$$P: v_0, v_1, v_{2k-1}, v_2, v_{2k-2}, v_3, \ldots, v_{k/2}, v_{3k/2}$$

and

$$Q: v_0, v_k, v_{k+1}, v_{k-1}, v_{k+2}, v_{k-2}, \ldots, v_{(k+2)/2}, v_{3k/2};$$

while if k is odd, then

$$P: v_0, v_1, v_{2k-1}, v_2, v_{2k-2}, v_3, \ldots, v_{(3k+1)/2}, v_{(k+1)/2}$$

and

$$Q: v_0, v_k, v_{k+1}, v_{k-1}, v_{k+2}, v_{k-2}, \ldots, v_{(3k-1)/2}, v_{(k+1)/2}.$$

Observe that, in either case, for $i = 1, 2, \ldots, k$, the ith edge of P and the ith edge of Q are labeled $i-1$.

Assume that the linear forest

$$F = P_{k_1+1} \cup P_{k_2+1} \cup \cdots \cup P_{k_t+1},$$

where then $\sum_{i=1}^{t} k_i = k$. We define a subgraph H of K_{2k} as follows. The edge set E of H consists of the first k_1 edges of P, edges $k_1 + 1$ through $k_1 + k_2$ of Q, edges $k_1 + k_2 + 1$ through $k_1 + k_2 + k_3$ of P, and so on until finally the last k_t edges of Q if t is even or the last k_t edges of P if t is odd. Define $H = \langle E \rangle$. Note that $H = F$ and that H contains exactly one edge labeled i for each $i = 0, 1, \ldots, k-1$.

Now for $j = 1, 2, \ldots, 2k-1$, define H_j to be the subgraph of K_{2k} obtained by revolving H about the $(2k-1)$-gon in a clockwise angle of $2\pi(j-1)/(2k-1)$

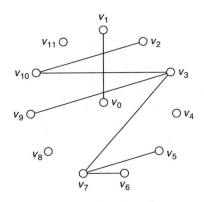

Figure 9.13 A step in the construction of an F-decomposition of K_{12} for $F = P_2 \cup P_3 \cup P_4$.

radians. Observe that for each $i = 0, 1, \ldots, k-1$ and each $j = 1, 2, \ldots, 2k-1$, the subgraph H_j contains exactly one edge labeled i. Since $H_j = F$ for each $j = 1, 2, \ldots, 2k-1$ and K_{2k} is decomposed into the subgraphs $H_1, H_2, \ldots, H_{2k-1}$, it follows that K_{2k} is F-decomposable. $\qquad\square$

The preceding theorem and its proof are illustrated in Figure 9.13 for $2k = 12$ and $F = P_2 \cup P_3 \cup P_4$. The labeling of the vertices of K_{12} is shown along with the subgraph H (or H_1).

EXERCISES 9.2

9.17 (a) Show that every bipartite graph G is a subgraph of a $\Delta(G)$-regular bipartite graph.

(b) Show that every bipartite graph G is of class one, that is, $\chi_1(G) = \Delta(G)$.

9.18 Give an example of a connected graph G of composite size having the property that whenever F is a factor of G and the size of F divides the size of G, then G is F-factorable.

9.19 (a) Prove that Q_n is 1-factorable for all $n \geq 1$.

(b) Prove that Q_n is k-factorable if and only if $k \mid n$.

9.20 Use the proof of Theorem 9.20 to give a 2-factorization of the graph of the octahedron (namely $K_{2,2,2}$).

9.21 Use the proof of Theorem 9.21 to produce a hamiltonian factorization of K_9.

9.22 Let k be a nonnegative even integer and $n \geq 5$ an odd integer with $k \leq n-3$. Prove that there exists a graph G of order n, all of whose vertices have degree k or $k+2$.

9.23 Prove Corollary 9.22.

9.24 Prove that K_{2k+1} cannot be factored into hamiltonian paths.

9.25 Give a constructive proof of Theorem 9.24.

9.26 Show that the graph of the octahedron is not $K_{1,4}$-decomposable.

9.27 (a) Use the fact that K_3 is graceful to find a K_3-decomposition of K_7.

     (b) Find a noncomplete regular K_3-decomposable graph.

9.28 Find an F-decomposition of K_{12} where $F = 2P_2 \cup 2P_3$.

9.29 Find a P_6-decomposition of K_{10}.

9.30 For each integer $k \geq 1$, show that

     (a) K_{2k+1} is $K_{1,k}$-decomposable.

     (b) K_{2k} is $K_{1,k}$-decomposable.

9.31 Use the drawing of the Petersen graph shown below to find cyclic decompositions into F_1, F_2 and F_3.

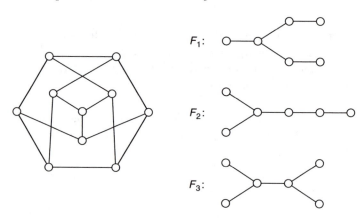

9.32 Find all graphs F of size 3 that are subgraphs of the Petersen graph P for which P is F-decomposable. (*Hint:* Use the drawing of the Petersen graph shown in Exercise 9.31.)

9.3 LABELINGS OF GRAPHS

In the previous section we discussed graceful labelings of graphs for the purpose of describing cyclic H-decompositions of certain complete graphs. In this section we discuss graceful labelings in more detail as well as describe another well-known labeling of graphs.

Recall that a graceful labeling of a graph G of size m is an assignment of distinct elements of the set $\{0, 1, \ldots, m\}$ to the vertices of G so that the edge labeling, which prescribes $|i - j|$ to the edge joining vertices labeled i and j, assigns the labels

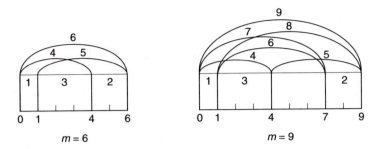

Figure 9.14 The Ruler Problem for 6 and 9 unit rulers.

$1, 2, \ldots, m$ to the edges of G. A graph possessing a graceful labeling is a graceful graph.

The topic of graceful labelings of graphs has a distinctive number theoretic flavor to it. Indeed, in number theory, a *restricted difference basis* (with respect to a positive integer m) is a set $\{a_1, a_2, \ldots, a_n\} \subseteq \{0, 1, \ldots, m\}$ such that every integer k with $1 \le k \le m$ can be represented in the form $k = a_j - a_i$. Hence a graph G of size m has a graceful labeling if the vertices of G can be labeled with the elements of a restricted difference basis in such a way that for each integer k with $1 \le k \le m$, there is a unique pair of adjacent vertices, labeled a_i and a_j say, so that $k = a_j - a_i$. A related problem is the *Ruler Problem*: For a given positive integer m, construct a ruler m units in length so that all integral distances from 1 to m can be measured and the ruler is marked at a minimum number of places. Rulers for $m = 6$ and $m = 9$ are shown in Figure 9.14.

Although there are no general sufficient conditions for a graph to be graceful, there are necessary conditions.

Theorem 9.32

If G is a graceful graph of size m, then there exists a partition of $V(G)$ into two subsets V_e and V_o such that the number of edges joining V_e and V_o is $\lceil m/2 \rceil$.

Proof

Let a graceful labeling of G be given. Denote the set of vertices labeled with an even integer by V_e and the set of vertices labeled with an odd integer by V_o. All edges labeled with an odd integer must then join a vertex of V_e and a vertex of V_o. Since there are $\lceil m/2 \rceil$ such edges, the result follows. □

A necessary condition for an eulerian graph to be graceful was discovered by Rosa [R10].

Theorem 9.33

If G is a graceful eulerian graph of size m, then $m \equiv 0 \pmod 4$ or $m \equiv 3 \pmod 4$.

Proof

Let $C: v_0, v_1, \ldots, v_{m-1}, v_m = v_0$ be an eulerian circuit of G, and let a graceful labeling of G be given that assigns the integer a_i ($0 \le a_i \le m$) to v_i for $0 \le i \le m$, where, of course, $a_i = a_j$ if $v_i = v_j$. Thus the label of the edge $v_{i-1} v_i$ is $|a_i - a_{i-1}|$. Observe that

$$|a_i - a_{i-1}| \equiv (a_i - a_{i-1}) \pmod 2$$

for $1 \le i \le m$. Thus the sum of the labels of the edges of G is

$$\sum_{i=1}^{m} |a_i - a_{i-1}| \equiv \sum_{i=1}^{m} (a_i - a_{i-1}) \equiv 0 \pmod 2,$$

that is, the sum of the edge labels of G is even. However, the sum of the edge labels is $\sum_{i=1}^{m} i = m(m+1)/2$; so $m(m+1)/2$ is even. Consequently, $4 \mid m(m+1)$, which implies that $4 \mid m$ or $4 \mid (m+1)$ so that $m \equiv 0 \pmod 4$ or $m \equiv 3 \pmod 4$. $\square$

We now determine which graphs in some well-known classes of graphs are graceful. Rosa [R10] determined the graceful cycles.

Theorem 9.34

The cycle C_n is graceful if and only if $n \equiv 0 \pmod 4$ or $n \equiv 3 \pmod 4$.

Proof

Since C_n is an eulerian graph, it follows by Theorem 9.33 that if $n \equiv 1 \pmod 4$ or $n \equiv 2 \pmod 4$, then C_n is not graceful; so it remains only to show that if $n \equiv 0 \pmod 4$ or $n \equiv 3 \pmod 4$, then C_n is graceful. Let $C_n: v_1, v_2, \ldots, v_n, v_1$. Assume first that $n \equiv 0 \pmod 4$. We assign v_i the label a_i, where

$$a_i = \begin{cases} (i-1)/2 & \text{if } i \text{ is odd} \\ n+1-i/2 & \text{if } i \text{ is even and } i \le n/2 \\ n-i/2 & \text{if } i \text{ is even and } i > n/2. \end{cases}$$

It remains to observe that this labeling is graceful. This labeling is illustrated in Figure 9.15 for $n = 12$.

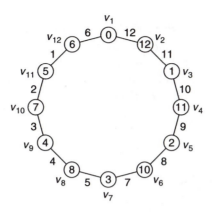

Figure 9.15 A graceful labeling of C_{12}.

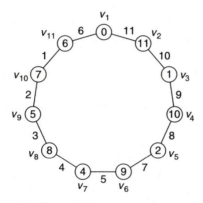

Figure 9.16 A graceful labeling of C_{11}.

Next, assume that $n \equiv 3 \pmod 4$. In this case we assign v_i the label b_i, where

$$b_i = \begin{cases} n+1-i/2 & \text{if } i \text{ is even} \\ (i-1)/2 & \text{if } i \text{ is odd and } i \le (n-1)/2 \\ (i+1)/2 & \text{if } i \text{ is odd and } i > (n-1)2 \end{cases}$$

This is a graceful labeling of C_n. An illustration is given in Figure 9.16 for $n = 11$. $\qquad\square$

If G is a graceful graph of order n and size m, then, of course, the vertices of G can be labeled with the elements of a set $\{a_1, a_2, \ldots, a_n\} \subseteq \{0, 1, \ldots, m\}$ so that the induced edge labels are precisely $1, 2, \ldots, m$. This means one vertex in some pairs of adjacent vertices is labeled 0 and the other vertex in the pair is labeled m. Also, if we were to replace each vertex label a_i by $m - a_i$, then we have a new graceful labeling, called the *complementary labeling*.

We saw in Figure 9.8 that the complete graphs K_3 and K_4 are graceful. It is very easy to show that K_2 is graceful. The following result of Golomb [G5] shows that there are no other graceful complete graphs.

Theorem 9.35

The complete graph K_n ($n \geq 2$) is graceful if and only if $n \leq 4$.

Proof

We have already observed that K_n is graceful if $2 \leq n \leq 4$. Assume then that $n \geq 5$ and suppose, to the contrary, that K_n is graceful. Hence there exists a graceful labeling of the vertices of K_n from an n-element subset of $\{0, 1, \ldots, m\}$, where $m = \binom{n}{2}$.

We have already seen that every graceful labeling of a graph of size m requires 0 and m to be vertex labels. Since some edge of K_n must be labeled $m - 1$, some vertex of K_n must be labeled 1 or $m - 1$. We may assume, without loss of generality, that a vertex of K_n is labeled 1; otherwise, we may use the complementary labeling.

To produce an edge labeled $m - 2$, we must have adjacent vertices labeled 0, $m - 2$ or 1, $m - 1$ or 2, m. If a vertex is labeled 2 or $m - 1$, then we have two edges labeled 1, which is impossible. Thus, some vertex of K_n must be labeled $m - 2$.

Since we now have vertices labeled 0, 1, $m - 2$ and m, we have edges labeled 1, 2, $m - 3$, $m - 2$, $m - 1$ and m. To have an edge labeled $m - 4$, we must have a vertex labeled 4 for all other choices result in two edges with the same label.

Now we have vertices labeled 0, 1, 4, $m - 2$ and m, which results in edges labeled 1, 2, 3, 4, $m - 6$, $m - 4$, $m - 2$, $m - 1$ and m. However, it is quickly seen that there is no vertex label that will produce the edge label $m - 5$ without also producing a duplicate edge label. Hence no graceful labeling of K_n exists. $\qquad \square$

Theorem 9.35 adds credence to the following conjecture [CHO1].

Conjecture

Graceful graphs with arbitrarily large chromatic numbers do not exist.

Unlike the classes of graphs we have considered, *every* complete bipartite graph is graceful.

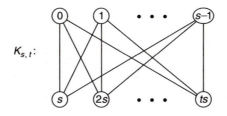

$K_{s,t}$:

Figure 9.17 A graceful labeling of $K_{s,t}$.

Theorem 9.36

Every complete bipartite graph is graceful.

Proof

Let $K_{s,t}$ have partite sets V_1 and V_2, where $|V_1| = s$ and $|V_2| = t$. Label the vertices of V_1 with $0, 1, \ldots, s-1$ and label the vertices of V_2 by $s, 2s, \ldots, (t-1)s, ts$ (see Figure 9.17). This is a graceful labeling. ☐

There is no result on graceful graphs as well-known as Kotzig's conjecture, which we recall.

Kotzig's conjecture

Every nontrivial tree is graceful.

Many classes of trees have been shown to be graceful. One of the most familiar of these is the class of paths.

Theorem 9.37

Every nontrivial path is graceful.

Proof

Let P: $v_0, v_1, \ldots, v_m$ be a path of size m. For i even, assign v_i the label $i/2$. If i is odd, then v_i is labeled $m - (i-1)/2$. It remains only to observe that this labeling is graceful (see Figure 9.18 for $m = 5$ and $m = 8$). ☐

Other familiar classes of graceful trees include stars, double stars and caterpillars (Exercise 9.35). Recall that a caterpillar is a tree the removal of whose

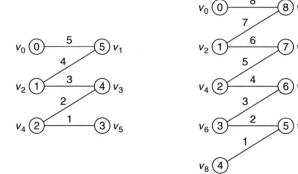

Figure 9.18 Graceful labelings of paths.

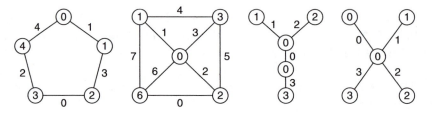

Figure 9.19 Harmonious graphs.

end-vertices produces a path. A *lobster* is a tree the removal of whose end-vertices produces a caterpillar. It is not known whether every lobster is graceful but, of course, it is conjectured that this is the case.

We now consider a graph labeling that is similar in nature to graceful labeling. A connected graph G of order n and size m with $m \geq n$ is *harmonious* if there exists a labeling $\phi: V(G) \to \mathbb{Z}_m$ of the vertices of G with distinct elements 0, $1, \ldots, m-1$ of $\mathbb{Z}_m$ such that each edge uv of G is labeled $\phi(u) + \phi(v)$ (addition in $\mathbb{Z}_m$) and the resulting edge labels are distinct. Such a labeling is called a *harmonious labeling*. If G is a tree (so that $m = n - 1$) exactly two vertices are labeled the same; otherwise, the definition is the same. Four harmonious graphs of order 5 (with harmonious labelings) are shown in Figure 9.19. Some examples of graphs that are not harmonious are shown in Figure 9.20.

We have now seen that C_5 is harmonious but C_4 is not. This serves as an illustration of a theorem of Graham and Sloane [GS1].

Theorem 9.38

The cycle C_n is harmonious if and only if n is odd.

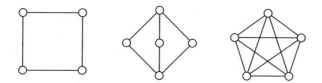

Figure 9.20 Three graphs that are not harmonious.

Proof

Assume first that n is odd and let C: $v_0, v_1, \ldots, v_{n-1}, v_0$ be a cycle of length n. The labeling that assigns v_i ($0 \leq i \leq n-1$) the label i is harmonious and hence C_n is harmonious if n is odd. (This labeling is illustrated for C_5 in Figure 9.19.)

Assume now that $n = 2k \geq 4$ is even and suppose, to the contrary, that C_n is harmonious. Let C_n: $v_0, v_1, \ldots, v_{n-1}, v_0$. Suppose that the labeling that assigns v_i the label a_i is harmonious. Consequently, the integers $a_0, a_1, \ldots, a_{n-1}$ are distinct and, in fact, $\{a_0, a_1, \ldots, a_{n-1}\} = \{0, 1, \ldots, n-1\}$. Therefore, the edge labels are $a_0 + a_1, a_1 + a_2, \ldots, a_{n-1} + a_0$ and, furthermore $\{a_0 + a_1, a_1 + a_2 \ldots, a_{n-1} + a_0\} = \{0, 1, \ldots, n-1\}$. Let $S = \sum_{i=0}^{n-1} a_i$. The sum of the edge labels of C_n is

$$(a_0 + a_1) + (a_1 + a_2) + \cdots + (a_{n-1} + a_0)$$

$$\equiv \sum_{i=0}^{n-1} i \equiv a_0 + a_1 + \cdots + a_{n-1} \pmod{n}.$$

Thus, $2S \equiv n(n-1)/2 \equiv S \pmod{n}$; so $S \equiv 0 \pmod{n}$ and $S \equiv k(n-1) \equiv k \pmod{n}$. Hence $k \equiv 0 \pmod{2k}$, which is impossible. $\square$

Although there is some similarity between the results for graceful cycles and harmonious cycles, there is no such similarity for complete bipartite graphs. The following result is due to Graham and Sloane [GS1].

Theorem 9.39

The complete bipartite graph $K_{s,t}$ is harmonious if and only if $s = 1$ or $t = 1$.

Proof

The labeling that assigns the central vertex of $K_{1,t}$ the label 0 and assigns the end-vertices the labels $0, 1, \ldots, t-1$ is harmonious. Consequently, every star is harmonious.

It remains to show that no other complete bipartite graph is harmonious. Suppose, to the contrary, that some complete bipartite graph $K_{s,t}$, where $s, t \geq 2$, is harmonious. Let the partite sets of $K_{s,t}$ be V_1 and V_2, where $|V_1| = s$ and $|V_2| = t$.

By assumption, there is a harmonious labeling of $K_{s,r}$. Suppose that this labeling assigns the integers $a_1, a_2, \ldots, a_s$ to the vertices of V_1 and $b_1, b_2, \ldots, b_t$ to the vertices of V_2. Thus, $A = \{a_1, a_2, \ldots, a_s\}$ and $B = (b_1, b_2, \ldots, b_t)$ are disjoint subsets of $\{0, 1, \ldots, st - 1\}$ and

$$\{a_i + b_j \,|\, 1 \leq i \leq s \quad \text{and} \quad 1 \leq j \leq t\} = \{0, 1, \ldots, st - 1\}.$$

Since for $(i, j) \neq (k, \ell)$, we have $a_i + b_j \neq a_k + b_\ell$, it follows that $a_i - b_\ell \neq a_k - b_j$ or, equivalently,

$$\left|\{a_i - b_j \,|\, 1 \leq i \leq s \quad \text{and} \quad 1 \leq j \leq t\}\right| = st.$$

Hence, for some i ($1 \leq i \leq s$) and j ($1 \leq j \leq t$), it follows that $a_i - b_j = 0$; so $a_i = b_j$, which contradicts the fact that A and B are disjoint. $\qquad\square$

We saw in Theorem 9.35 that the complete graph K_n ($n \geq 2$) is graceful if and only if $n \leq 4$. Harmonious complete graphs are characterized in exactly the same way.

Theorem 9.40

The complete graph K_n ($n \geq 2$) is harmonious if and only if $n \leq 4$.

We now turn our attention to trees. First, we show that every nontrivial path is harmonious. In the proof it is convenient to label some vertices by $-n + a$ rather than a, where $0 \leq a \leq n - 1$, which, of course, are equivalent in $\mathbb{Z}_n$.

Theorem 9.41

Every nontrivial path is harmonious.

Proof

Let P_n: $v_1, v_2, \ldots, v_n$, where $n \geq 2$. If n is even, write $n = 2k + 2$; while if n is odd, write $n = 2k + 3$. For an integer t, label the vertex v_i with a_i, where

$$a_i = \begin{cases} -t + (i - 1)/2 & \text{if } i \text{ is odd} \\ t + (i - 2)/2 & \text{if } i \text{ is even.} \end{cases}$$

If k and n are even, then let $t = k/2$; and if k is even and n is odd, the let $t = k/2 + 1$; while if k is odd, then let $t = (k + 1)/2$. In either case, this is a harmonious labeling of P_n, where if k and n are of the same parity, then t is the repeated label of P_n; while if k and n are of opposite parity, then $-t$ is the

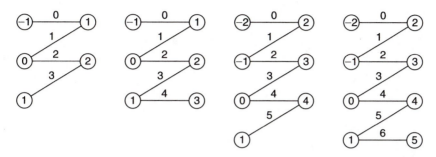

Figure 9.21 Harmonious labelings of P_n, $5 \leq n \leq 8$.

repeated label. Thus P_n is harmonious. (Harmonious labelings of P_n, $5 \leq n \leq 8$, are shown in Figure 9.21.) □

As with graceful labelings, many classes of trees have been shown to be harmonious, but whether all trees are harmonious is not known.

Graham–Sloane Conjecture

Every nontrivial tree is harmonious.

EXERCISES 9.3

9.33 Determine graceful labelings of C_{15} and C_{16}.

9.34 Determine graceful labelings of P_6, P_7, P_9, and P_{10}.

9.35 Show that the following classes of trees are graceful:
(a) stars, (b) double stars, (c) caterpillars.

9.36 Determine harmonious labelings of C_7 and C_9.

9.37 Use the proof of Theorem 9.37 to give harmonious labelings of P_9 and P_{10}.

9.38 Show that every double star is harmonious.

9.39 (a) Show that no disconnected forest is graceful.
(b) Give an example, with justification, of a disconnected graph without isolated vertices of order n and size m with $m \geq n-1$ is not graceful.

10

Domination in graphs

Next we turn to an area of graph theory that has received increased attention during recent decades.

10.1 THE DOMINATION NUMBER OF A GRAPH

A vertex v in a graph G is said to *dominate* itself and each of its neighbors, that is, v dominates the vertices in its *closed neighborhood* $N[v]$. A set S of vertices of G is a *dominating set* of G if every vertex of G is dominated by at least one vertex of S. Equivalently, a set S of vertices of G is a dominating set if every vertex in $V(G) - S$ is adjacent to at least one vertex in S. The minimum cardinality among the dominating sets of G is called the *domination number* of G and is denoted by $\gamma(G)$. A dominating set of cardinality $\gamma(G)$ is then referred to as a *minimum dominating set*. Although $\gamma(G)$ is the same notation that is used for the genus of a graph G, the notation in both instances is common and we will never discuss the domination number and the genus of a graph at the same time.

The sets $S_1 = \{v_1, v_2, y_1, y_2\}$ and $S_2 = \{w_1, w_2, x\}$ are both dominating sets in the graph G of Figure 10.1, indicated by solid circles. Since S_2 is a dominating set of minimum cardinality, $\gamma(G) = 3$.

Dominating sets appear to have their origins in the game of chess, where the goal is to cover or dominate various squares of a chessboard by certain chess pieces. In 1862 de Jaenisch [D1] considered the problem of determining the minimum number of queens (which can move either horizontally, vertically or diagonally over any number of unoccupied squares) that can be placed on a chessboard such that every square is either occupied by a queen or can be occupied by one of the queens in a single move. The minimum number of such queens is 5 and one possible placement of five such queens is shown in Figure 10.2.

Two queens on a chessboard are *attacking* if the square occupied by one of the queens can be reached by the other queen in a single move; otherwise, they are *nonattacking queens*. Clearly, every pair of queens on the chessboard of Figure 10.2 are attacking. The minimum number of nonattacking queens such that every square of the chessboard can be reached by one of the queens is also 5. A possible placement of five nonattacking queens is shown in Figure 10.3.

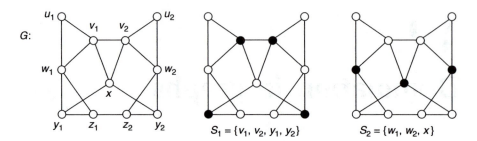

Figure 10.1 Dominating sets.

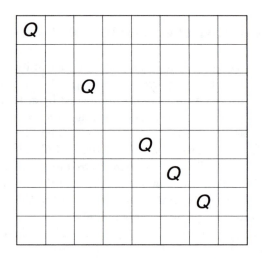

Figure 10.2 The minimum number of queens that dominate the squares of a chessboard.

The connection between the chessboard problem described above and dominating sets in graphs is immediate. The 64 squares of a chessboard are the vertices of a graph G and two vertices (squares) are adjacent in G if each square can be reached by a queen on the other square by a single move. The graph G is referred to as the *queen's graph*. Then the minimum number of queens that dominate all the squares of a chessboard is $\gamma(G)$. The minimum number of nonattacking queens that dominate all the squares of a chessboard is the minimum cardinality of an independent dominating set in G.

Domination as a theoretical area in graph theory was formalized by Berge in 1958 [see B6, p. 40] and Ore [O2, Chap. 13] in 1962. Since 1977, when Cockayne and Hedetniemi [CH3] presented a survey of domination results, domination theory has received considerable attention.

A *minimal dominating set* in a graph G is a dominating set that contains no dominating set as a proper subset. A minimal dominating set of minimum

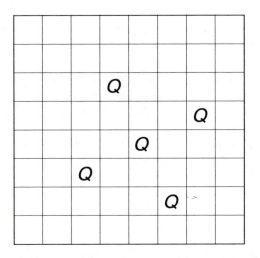

Figure 10.3 The minimum number of nonattacking queens that dominate the squares of a chessboard.

cardinality is, of course, a minimum dominating set and consists of $\gamma(G)$ vertices. For the graph G of Figure 10.1, the set $S_1 = \{v_1, v_2, y_1, y_2\}$ is a minimal dominating set that is not a minimum dominating set. Minimal dominating sets were characterized by Ore [O2, p. 206].

Theorem 10.1

A dominating set S of a graph G is a minimal dominating set of G if and only if every vertex v in S satisfies at least one of the following two properties:

(i) *there exists a vertex w in $V(G) - S$ such that $N(w) \cap S = \{v\}$*   (10.1)

(ii) *v is adjacent to no vertex of S.*   (10.2)

Proof

First, observe that if each vertex v in S has at least one of the properties (10.1) and (10.2), then $S - \{v\}$ is not a dominating set of G. Consequently, S is a minimal dominating set of G.

Conversely, assume that S is a minimal dominating set of G. Then certainly for each $v \in S$, the set $S - \{v\}$ is not a dominating set of G. Hence there is a vertex w in $V(G) - (S - \{v\})$ that is adjacent to no vertex of $S - \{v\}$. If $w = v$, then v is adjacent to no vertex of S. Suppose then that $w \neq v$. Since S is a dominating set of G and

$w \notin S$, the vertex w is adjacent to at least one vertex of S. However, w is adjacent to no vertex of $S - \{v\}$. Consequently, $N(w) \cap S = \{v\}$. ☐

Theorem 10.1 can be reworded as follows: A dominating set S of a graph G is a minimal dominating set of G if and only if for every vertex v in S either (1) v dominates some vertex of $V(G) - S$ that no other vertex of S dominates or (2) no other vertex of S dominates v.

The following result of Ore [O2, p. 207] gives a property of the complementary set of a minimal dominating set in a graph without isolated vertices.

Theorem 10.2

If S is a minimal dominating set of a graph G without isolated vertices, then $V(G) - S$ is a dominating set of G.

Proof

Let $v \in S$. Then v has at least one of the two properties (10.1) and (10.2) described in the statement of Theorem 10.1. Suppose first that there exists a vertex w in $V(G) - S$ such that $N(w) \cap S = \{v\}$. Hence v is adjacent to some vertex in $V(G) - S$. Suppose next that v is adjacent to no vertex in S. Then v is an isolated vertex of the subgraph $\langle S \rangle$. Since v is not isolated in G, the vertex v is adjacent to some vertex of $V(G) - S$. Thus $V(G) - S$ is a dominating set of G. ☐

For graphs G without isolated vertices, we now have an upper bound for $\gamma(G)$ in terms of the order of G.

Corollary 10.3

If G is a graph of order n without isolated vertices, then $\gamma(G) \leq n/2$.

Proof

Let S be a minimal dominating set of G. By Theorem 10.2, $V(G) - S$ is a dominating set of G. Thus

$$\gamma(G) \leq \min\{|S|, |V(G) - S|\} \leq n/2.$$ ☐

Nearly all connected graphs attaining the bound in Corollary 10.3 can be produced with the aid the following operation. The *corona* cor(H) of a graph H is that graph obtained from H by adding a pendant edge to each vertex of H. Let $G = \text{cor}(H)$, where G has order n. Then G has no isolated vertices and $\gamma(G) = n/2$. Indeed, Payan and Xuong [PX1] showed that every component of a graph G of order n without isolated vertices having $\gamma(G) = n/2$ is either C_4 or the corona of some (connected) graph.

Hence, if G is a graph of order n, then $\gamma(G) \leq n$; while, by Corollary 10.3, if $\delta(G) \geq 1$, then $\gamma(G) \leq n/2$. McCuaig and Shepherd [MS2] showed that if $\delta(G) \geq 2$ and G is not one of seven exceptional graphs, then $\gamma(G) \leq 2n/5$. Reed [R3] showed that if $\delta(G) \geq 3$, then $\gamma(G) \leq 3n/8$. A more general result is due to Payan [P1]. We delay its proof until Chapter 13 (Theorem 13.4) when the probabilistic method of proof is described.

Theorem 10.4

Let G be a graph of order n with $\delta = \delta(G) \geq 2$. Then

$$\gamma(G) \leq \frac{n(1 + \ln(\delta + 1))}{\delta + 1}.$$

Bollobás and Cockayne [BC2] showed that every graph without isolated vertices contains a minimum dominating set in which every vertex satisfies (10.1).

Theorem 10.5

Every graph G without isolated vertices contains a minimum dominating set S such that for every vertex v of S, there exists a vertex w of $G - S$ such that $N(w) \cap S = \{v\}$.

Proof

Among all minimum dominating sets of G, let S be one such that $\langle S \rangle$ has maximum size. Suppose, to the contrary, that S contains a vertex v that does not have the desired property. Then by Theorem 10.1, v is an isolated vertex in $\langle S \rangle$. Moreover, every vertex of $V(G) - S$ that is adjacent to v is adjacent to some other vertex of S as well. Since G contains no isolated vertices, v is adjacent to a vertex w in $V(G) - S$. Consequently, $(S - \{v\}) \cup \{w\}$ is a minimum dominating set of G whose induced subgraph contains at least one edge incident with w and hence has a greater size than $\langle S \rangle$. This produces a contradiction. □

Bounds for the domination number of a graph can be given in terms of the order and the maximum degree of the graph. The lower bound in the following theorem is due to Walikar, Acharya and Sampathkumar [WAS1], while the upper bound is due to Berge [B7].

Theorem 10.6

If G is a graph of order n, then

$$\left\lceil \frac{n}{1 + \Delta(G)} \right\rceil \leq \gamma(G) \leq n - \Delta(G).$$

Proof

We begin with the lower bound. Let S be a minimum dominating set of G. Then

$$V(G) - S \subseteq \bigcup_{v \in S} N(v),$$

implying that $|V(G) - S| \leq |S| \cdot \Delta(G)$. Therefore, $n - \gamma(G) \leq \gamma(G) \cdot \Delta(G)$ and so $\gamma(G) \geq \lceil n/(1 + \Delta(G)) \rceil$.

Next we establish the upper bound. Let v be a vertex of G with deg $v = \Delta(G)$. Then $V(G) - N(v)$ is a dominating set of cardinality $n - \Delta(G)$; so $\gamma(G) \leq n - \Delta(G)$. □

Since $\kappa(G) \leq \Delta(G)$ for every graph G, we have the following consequence of Theorem 10.6, due to Walikar, Acharya and Sampathkumar [WAS1].

Corollary 10.7

If G is a graph of order n, then

$$\gamma(G) \leq n - \kappa(G).$$

The domination number of a graph without isolated vertices is also bounded above by all of the covering and independence numbers.

Theorem 10.8

If G is a graph without isolated vertices, then

$$\gamma(G) \leq \min\{\alpha(G), \alpha_1(G), \beta(G), \beta_1(G)\}.$$

Proof

Since every vertex cover of a graph without isolated vertices is a dominating set, as is every maximal independent set of vertices, $\gamma(G) \le \alpha(G)$ and $\gamma(G) \le \beta(G)$. Let X be an edge cover of cardinality $\alpha_1(G)$. Then every vertex of G is incident with at least one edge in X.

Let S be a set of vertices, obtained by selecting an incident vertex with each edge in X. Then S is a dominating set of vertices and $\gamma(G) \le |S| \le |X| = \alpha_1(G)$.

Next, let M be a maximum matching in G. We construct a set S of vertices consisting of one vertex incident with an edge of M for each edge of M. Let $uv \in M$. The vertices u and v cannot be adjacent to distinct $\overline{M}$-vertices x and y, respectively; for otherwise, x, u, v, y is an M-augmenting path in G, contradicting Theorem 9.2. If u is adjacent to an $\overline{M}$-vertex, place u in S; otherwise, place v in S. This is done for each edge of M. Thus, S is a dominating set of G, and $\gamma(G) \le |S| = |M| = \beta_1(G)$. □

Vizing [V5] obtained an upper bound for the size of a graph in terms of its order and domination number.

Theorem 10.9

If G is a graph of order n and size m for which $\gamma = \gamma(G) \ge 2$, then

$$m \le \frac{(n - \gamma)(n - \gamma + 2)}{2}. \tag{10.3}$$

With the aid of Theorem 10.9, we can now supply bounds for the domination number of a graph in terms of its order and size. The lower bound is due to Berge [B7].

Theorem 10.10

If G is a graph of order n and size m, then

$$n - m \le \gamma(G) \le n + 1 - \sqrt{1 + 2m}.$$

Furthermore, $\gamma(G) = n - m$ if and only if each component of G is a star or an isolated vertex.

Proof

Rewriting the inequality (10.3) given in Theorem 10.9, we have

$$(n - \gamma(G))^2 + 2(n - \gamma(G)) - 2m \ge 0. \tag{10.4}$$

Solving the inequality (10.4) for $n - \gamma(G)$ and using the fact that $n - \gamma(G) \geq 0$, we have that

$$n - \gamma(G) \geq -1 + \sqrt{1 + 2m},$$

which establishes the desired upper bound.

Since $\gamma(G) \geq 1$, the lower bound is established when $m \geq n - 1$, which includes all connected graphs. Assume then that $m \leq n - 1$. Then G is a graph with at least $n - m$ components. The domination number of each component of G is at least 1; so $\gamma(G) \geq n - m$, with equality if and only if G has exactly $n - m$ components, each with domination number 1. This can occur only, however, if G is a forest with $n - m$ components, each of which is a star or an isolated vertex. □

The Nordhaus–Gaddum theorem (Theorem 8.15) provided sharp bounds on the sum and product of the chromatic numbers of a graph and its complement. We now present the corresponding result for the domination number. The following result is due to Jaeger and Payan [JP1], the proof of which is based on a proof by E. J. Cockayne.

Theorem 10.11

If G is a graph of order $n \geq 2$, then

    (i) $3 \leq \gamma(G) + \gamma(\overline{G}) \leq n + 1$,

    (ii) $2 \leq \gamma(G) \cdot \gamma(\overline{G}) \leq n$.

Proof

The lower bounds in (i) and (ii) follow immediately from the observation that if $\gamma(G) = 1$ or $\gamma(\overline{G}) = 1$, then $\gamma(\overline{G}) \geq 2$ or $\gamma(G) \geq 2$, respectively.

Next we verify the upper bound in (i). If G has an isolated vertex, then $\gamma(G) \leq n$ and $\gamma(\overline{G}) = 1$; while if $\overline{G}$ has an isolated vertex, then $\gamma(\overline{G}) \leq n$ and $\gamma(G) = 1$. So, in these cases, $\gamma(G) + \gamma(\overline{G}) \leq n + 1$. If neither G nor $\overline{G}$ has isolated vertices, then $\gamma(G) \leq n/2$ and $\gamma(\overline{G}) \leq n/2$ by Corollary 10.3 and so $\gamma(G) + \gamma(\overline{G}) \leq n$.

It remains then only to verify the upper bound in (ii). The upper bound is immediate if $\gamma(G) = 1$, so we assume that $\gamma(G) = k \geq 2$. Let $S = \{v_1, v_2, \ldots, v_k\}$ be a minimum dominating set of G and partition $V(G)$ into $\gamma(G) = k$ subsets V_1, $V_2, \ldots, V_k$ subject to the conditions that (a) $v_i \in V_i$ for $1 \leq i \leq k$ and all vertices in V_i are dominated by v_i and (b) the sum over all integers i ($1 \leq i \leq k$) of the number of vertices in V_i adjacent to all other vertices in V_i is a maximum.

We now show that each set V_i $(1 \leq i \leq k)$ is a dominating set of $\overline{G}$. Suppose that this is not the case. Then there exists a vertex $x \in V_t$ that is adjacent in $\overline{G}$ to no vertex of V_s for distinct integers s and t with $1 \leq s, t \leq k$. Then x is adjacent in G to *every* vertex of V_s. If $x = v_t$, then $S - \{v_t\}$ is a dominating set of G having cardinality less than $\gamma(G)$, which is impossible. Consequently, $x \in V_t - \{v_t\}$. If x is adjacent in G to every other vertex of V_t, then $(S - \{v_s, v_t\}) \cup \{x\}$ is a dominating set of G having cardinality less than $\gamma(G)$, which is again impossible. Therefore, x is adjacent in G to every vertex of V_s but *not* to every vertex of V_t.

Define $V_t' = V_t - \{x\}$ and $V_s' = V_s \cup \{x\}$. For $r \neq s, t$, define $V_r' = V_r$. Thus, we now have a partition of $V(G)$ into subsets $V_1', V_2', \ldots, V_k'$ such that $v_i \in V_i'$ for $1 \leq i \leq k$ and all vertices in V_i' are dominated by v_i. However, the sum over all subsets V_i' $(1 \leq i \leq k)$ of the number of vertices in V_i' adjacent to all other vertices of V_i' exceeds the corresponding sum for the partition $V_1, V_2, \ldots, V_k$, which is a contradiction.

Thus, as claimed, each subset V_i $(1 \leq i \leq k)$ is a dominating set in $\overline{G}$; so $\gamma(\overline{G}) \leq |V_i|$ for each i. Hence

$$n = \sum_{i=1}^{k} |V_i| \geq \gamma(G) \cdot \gamma(\overline{G}).$$

The upper bound in (i) in Theorem 10.11 can be restated as: If K_n $(n \geq 2)$ is factored into G_1 and G_2, then $\gamma(G_1) + \gamma(G_2) \leq n + 1$. Goddard, Henning and Swart [GHS1] obtained the corresponding upper bound for three factors.

Corollary 10.12

If K_n is factored into G_1, G_2 and G_3, then

$$\gamma(G_1) + \gamma(G_2) + \gamma(G_3) \leq 2n + 1.$$

Proof

Since $G_2 \oplus G_3 = \overline{G_1}$, it follows from Theorem 10.11 that $\gamma(G_1) + \gamma(G_2 \oplus G_3) \leq n + 1$. Now let S be a dominating set for $G_2 \oplus G_3$. Thus every vertex of $G_2 \oplus G_3$ is dominated by a vertex of S. Consequently, for each vertex v of $G_2 \oplus G_3$, the vertex v is not dominated by a vertex of S in at most one of G_2 and G_3. Thus, in extending S to dominating sets S_2 and S_3 for G_2 and G_3, respectively, each vertex of $G_2 \oplus G_3$ needs to be added at most once. So $\gamma(G_2) + \gamma(G_3) \leq \gamma(G_2 \oplus G_3) + n$. Therefore,

$$\gamma(G_1) + \gamma(G_2) + \gamma(G_3) \leq \gamma(G_1) + \gamma(G_2 \oplus G_3) + n \leq 2n + 1. \qquad \square$$

All of the bounds presented in Theorem 10.10 are sharp (Exercise 10.9); however, if neither G nor $\overline{G}$ has isolated vertices, then an improved upper bound for $\gamma(G) + \gamma(\overline{G})$, due to Joseph and Arumugam [JA1], can be given.

Theorem 10.13

If G is a graph of order $n \geq 2$ such that neither G nor $\overline{G}$ has isolated vertices, then

$$\gamma(G) + \gamma(\overline{G}) \leq \frac{n+4}{2}.$$

Proof

Since neither G nor $\overline{G}$ has isolated vertices, it follows from Corollary 10.3 that $\gamma(G) \leq n/2$ and $\gamma(\overline{G}) \leq n/2$. Hence if either $\gamma(G) = 2$ or $\gamma(\overline{G}) = 2$, then the proof is complete. If $\gamma(G) \geq 4$ and $\gamma(\overline{G}) \geq 4$, then by the upper bound in (ii) in Theorem 10.11, we have that $\gamma(G) \leq n/\gamma(\overline{G}) \leq n/4$ and $\gamma(\overline{G}) \leq n/\gamma(G) \leq n/4$; so $\gamma(G) + \gamma(\overline{G}) \leq n/2$. Hence we may assume that $\gamma(G) = 3$ or $\gamma(\overline{G}) = 3$, say the former. Thus $3 = \gamma(G) \leq n/2$, so $n \geq 6$. By Theorem 10.11, $\gamma(\overline{G}) \leq n/3$. Therefore

$$\gamma(G) + \gamma(\overline{G}) \leq 3 + \frac{n}{3} \leq 2 + \frac{n}{2}. \qquad \square$$

For the bound stated in Theorem 10.13 to be attained, either G or $\overline{G}$ must have domination number $n/2$. In the discussion following Corollary 10.3, graphs G of order n without isolated vertices and having $\gamma(G) = n/2$ were described.

We close this section with an unsolved problem, namely a conjecture due to V. G. Vizing.

Vizing's conjecture

For every two graphs G and H,

$$\gamma(G \times H) \geq \gamma(G) \cdot \gamma(H).$$

EXERCISES 10.1

10.1 Determine the domination numbers of the 3-cube Q_3 and the 4-cube Q_4.

10.2 (a) Determine (with proof) a formula for $\gamma(C_n)$.
  (b) Determine (with proof) a formula for $\gamma(P_n)$.

10.3 Obtain a sharp lower bound for the domination number of a connected graph G of order n and diameter k.

10.4 State and prove a characterization of those graphs G with $\gamma(G)=1$.

10.5 Investigate the sharpness of the bounds given in Theorem 10.6.

10.6 (a) Does there exist a graph G such that $\gamma(G)=\alpha(G)$ but $\gamma(G)$ is strictly less than each of the numbers $\alpha_1(G)$, $\beta(G)$ and $\beta_1(G)$?

(b) The question in (a) suggests three other questions. State and answer these questions.

10.7 Use Theorem 10.8 to give an alternative proof of Corollary 10.3.

10.8 Show that equality is possible for the upper bound given in Theorem 10.10.

10.9 Show that all bounds given in Theorem 10.11 are sharp.

10.2 THE INDEPENDENT DOMINATION NUMBER OF A GRAPH

It is not difficult to see that every maximal independent set of vertices in a graph G is a dominating set of G. Thus, $\gamma(G) \le i(G)$, where, recall, $i(G)$ is the lower independence number of G. Not every dominating set is independent, however. Indeed, not every minimum dominating set is independent. For example in the graph G of Figure 10.4, the set $S_1 = \{u_1, u_2, v_1, v_2, w_1, w_2\}$ is a maximal independent set (and consequently a dominating set) of G; while $S_2 = \{x, y, z\}$ is a minimum dominating set of G and certainly S_2 is not independent. (These sets are

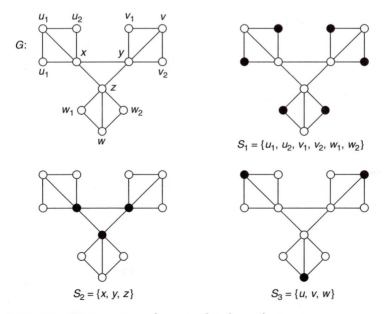

Figure 10.4 Dominating sets and maximal independent sets.

indicated in Figure 10.4 by solid circles.) However, G does contain a minimum dominating set of G that is independent, namely, $S_3 = \{u, v, w\}$.

Our attention now shifts in this section to dominating sets that are also maximal independent sets. A set S of vertices in a graph G is called an *independent dominating set* of G if S is both an independent and a dominating set of G. Thus the sets S_1 and S_3 in Figure 10.4 are independent dominating sets while S_2 and $S_4 = \{u_1, v_1, w_1\}$ are not. The *independent domination number* $i(G)$ of G is the minimum cardinality among all independent dominating sets of G. That this is precisely the notation used for the lower independence number of a graph is justified by the following observation of Berge [B7].

Theorem 10.14

A set S of vertices in a graph is an independent dominating set if and only if S is maximal independent.

Proof

We have already noted that every maximal independent set of vertices is a dominating set. Conversely, suppose that S is an independent dominating set. Then S is independent and every vertex not in S is adjacent to a vertex of S, that is, S is maximal independent. □

Another observation now follows.

Corollary 10.15

Every maximal independent set of vertices in a graph is a minimal dominating set.

Proof

Let S be a maximal independent set of vertices in a graph G. By Theorem 10.14, S is a dominating set. Since S is independent, certainly every vertex of S is adjacent to no vertex of S. Thus, every vertex of S satisfies property (ii) of Theorem 10.1. So, by Theorem 10.1, S is a minimal dominating set. □

Equality can hold since $\gamma(K_{1,t}) = i(K_{1,t}) = 1$ for every positive integer t. For $1 \leq s < t$, let H be the graph obtained from $K_{s,t}$ by adding a pendant edge to each vertex of the partite set of cardinality s. Then $\gamma(H) = i(H) = s$. For the queen's graph G, we also have $\gamma(G) = i(G) = 5$, as Figures 10.2 and 10.3 show. That the difference between the independent domination number and domination number of a graph can be arbitrarily large can be seen in the double star T containing two vertices of degree $k \geq 2$, where $i(T) = k$ and $\gamma(T) = 2$.

For some special classes of graphs, Bollobás and Cockayne [BC1] determined an upper bound for $i(G)$ in terms of $\gamma(G)$.

Theorem 10.16

If G is a $K_{1,k+1}$-free graph, where $k \geq 2$, then

$$i(G) \leq (k-1)\gamma(G) - (k-2).$$

Proof

Let S be a minimum dominating set of vertices of G and let S' be a maximal independent set of vertices of S in G. Thus, $|S| = \gamma(G)$ and $|S'| \geq 1$. Now, let T denote the set of all vertices in $V(G) - S$ that are adjacent in G to no vertex of S', and let T' be a maximal independent set of vertices in T. Certainly, then, $S' \cup T'$ is an independent set of vertices of G. Since every vertex of $V(G) - (S' \cup T)$ is adjacent to some vertex of S' and every vertex of $T - T'$ is adjacent to some vertex of T', it follows that $S' \cup T'$ is a maximal independent set of vertices. Thus, by Theorem 10.14, $S' \cup T'$ is an independent dominating set.

Observe that every vertex of $S - S'$ is adjacent to at most $k - 1$ vertices of T'; for if this were not the case, then some vertex v of $S - S'$ is adjacent to at least k vertices of T' and also at least one vertex of S', which contradicts the hypothesis that G contains no induced subgraph isomorphic to $K_{1,k+1}$. Also, observe that every vertex of T' is adjacent to some vertex of $S - S'$. Therefore,

$$|T'| \leq (k-1)|S - S'| = (k-1)(|S| - |S'|) = (k-1)(\gamma(G) - |S'|).$$

Consequently,

$$\begin{aligned} i(G) &\leq |S' \cup T'| = |S'| + |T'| \\ &\leq |S'| + (k-1)(\gamma(G) - |S'|) \\ &= (k-1)\gamma(G) - (k-2)|S'| \\ &\leq (k-1)\gamma(G) - (k-2). \end{aligned}$$ $\square$

The special case of Theorem 10.16 where $k = 2$ is of particular interest.

Corollary 10.17

If G is a claw-free graph, then $\gamma(G) = i(G)$.

The converse of Corollary 10.17 is certainly not true, though, since $\gamma(K_{1,3}) = i(K_{1,3}) = 1$. Since every line graph is claw-free (by Theorem 4.32), we have a consequence of Corollary 10.17.

Corollary 10.18

For every graph G,

$$\gamma(L(G)) = i(L(G)).$$

No forbidden subgraph characterization of graphs G for which $\gamma(G) = i(G)$ is possible; for suppose that H is a given graph and we define $G = K_1 + H$. Then $\gamma(G) = i(G) = 1$.

In Chapter 8, a graph G is defined to be perfect if $\chi(H) = \omega(H)$ for every induced subgraph H of G. A graph G is *domination perfect* if $\gamma(H) = i(H)$ for every induced subgraph H of G. A class of domination perfect graphs is provided by Corollary 10.17.

Corollary 10.19

Every claw-free graph is domination perfect.

EXERCISES 10.2

10.10 Show that a graph need not have any minimum dominating set that is independent.

10.11 Prove or disprove. If a graph G contains an independent minimum dominating set of vertices, then $\gamma(G) = i(G)$.

10.12 For each integer $k \geq 3$, show that there exists a graph G such that $i(G) = k$ and $\gamma(G) = 3$.

10.3 OTHER DOMINATION PARAMETERS

In the previous section we introduced a variant of the classical domination number, namely, the independent domination number. In this section, we describe several other domination parameters that have been the object of study.

For a set A of vertices in a graph G, the *closed neighborhood* $N[A]$ of A is defined by $N[A] = \cup_{v \in A} N[v]$. Equivalently, $N[A] = N(A) \cup A$. A set S of vertices in G is called an *irredundant set* of G if for every vertex $v \in S$, there exists a vertex $w \in N[v]$ such that $w \notin N[S - \{v\}]$. Equivalently, S is an irredundant set of vertices of G if $N[S - \{v\}] \neq N[S]$ for every vertex $v \in S$. Every vertex v with this property is an

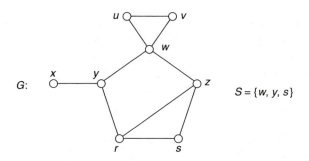

Figure 10.5 Irredundant sets of vertices.

irredundant vertex. Therefore, every vertex in an irredundant set is an irredundant vertex. A set S of vertices that is not irredundant is called *redundant*. Consequently, a set S of vertices in a graph G is redundant if and only if there exists a vertex v in S for which $N[S - \{v\}] = N[S]$. Such a vertex v is called a *redundant vertex* (with respect to S).

For the graph G of Figure 10.5, let $S = \{w, y, s\}$. Then S is an irredundant set of G. For example, $u \in N[S]$ but $u \notin N[S - \{w\}]$. Similarly $x \notin N[S - \{y\}]$ and $s \notin N[S - \{s\}]$.

A characterization of irredundant sets is presented next.

Theorem 10.20

A set S of vertices in a graph G is irredundant if and only if every vertex v in S satisfies at least one of the following two properties:

(i) *there exists a vertex w in $V(G) - S$ such that $N(w) \cap S = \{v\}$.* (10.5)

(ii) *v is adjacent to no vertex of S.* (10.6)

Proof

First, let S be a set of vertices of G such that for every vertex $v \in S$, at least one of the properties (10.5) and (10.6) is satisfied. If (10.5) is satisfied, then there exists a vertex $w \in N[v]$ such that $w \notin N[S - \{v\}]$. If (10.6) is satisfied, then $v \notin N[S - \{v\}]$. In either case, S is irredundant.

Conversely, let S be an irredundant set of vertices in G, and let $v \in S$. Since S is irredundant, there exists $w \in N[v]$ such that $w \notin N[S - \{v\}]$. If $w \neq v$, then (10.5) is satisfied; while if $w = v$, then (10.6) is satisfied. □

By Theorem 10.1, then, a minimal dominating set of vertices in a graph is an irredundant set. Hence, every graph has an irredundant dominating set of vertices.

If S is an irredundant set of vertices in a graph G, then for each $v \in S$, the set $N[v] - N[S - \{v\}]$ is nonempty. Each vertex in $N[v] - N[S - \{v\}]$ is referred to as a *private neighbor* of v. The vertex v may, in fact, be a private neighbor of itself. Consequently, a nonempty set S of vertices in a graph G is irredundant if and only if every vertex of S has a private neighbor. Certainly every nonempty subset of an irredundant set of vertices in a graph G is irredundant. Also, every independent set of vertices is an irredundant set.

The *irredundance number* $ir(G)$ of a graph G is the minimum cardinality among the maximal irredundant sets of vertices of G. Since the set $S = \{r, z\}$ is a maximal irredundant set of vertices of minimum cardinality for the graph G of Figure 10.5, it follows that for this graph, $ir(G) = 2$. To see that S is irredundant, observe that y is a private neighbor of r, and w is a private neighbor of z. To see that S is a *maximal* irredundant set, note that (1) $\{s, r, t\}$ is not irredundant since s would have no private neighbor, (2) $\{x, r, t\}$ and $\{y, r, t\}$ are not irredundant since r would have no private neighbor, and (3) $\{u, r, z\}$, $\{v, r, t\}$ and $\{w, r, z\}$ are not irredundant since z would have no private neighbor. Hence a maximal irredundant set need not be a dominating set and, strictly speaking, the irredundance number is not a domination parameter.

The next result summarizes how the parameters discussed thus far in this chapter are related.

Theorem 10.21

For every graph G,

$$ir(G) \leq \gamma(G) \leq i(G).$$

Proof

We have already observed that $\gamma(G) \leq i(G)$. The inequality $ir(G) \leq \gamma(G)$ is a consequence of the fact that every minimal dominating set of vertices of G is an irredundant set. □

That the inequality $ir(G) \leq \gamma(G)$ may be strict is illustrated in the graph G of Figure 10.5, where $\gamma(G) = 3$ and $ir(G) = 2$. Also, for the graph H of Figure 10.6, we have $\gamma(H) = 3$ and $ir(H) = 2$. The set $\{u_1, v_1\}$ is a maximal irredundant set of minimum cardinality in H. In order to see that $\{u_1, v_1\}$ is a maximal irredundant set in H, observe that (1) t has no private neighbor in $\{t, u_1, v_1\}$, (2) u_1 has no private neighbor in $\{u_1, v_1, u_2\}$ and $\{u_1, v_1, u_3\}$, and (3) v_1 has no private neighbor in $\{u_1, v_1, v_2\}$ and $\{u_1, v_1, v_3\}$. Moreover, for the tree T of Figure 10.6, $\gamma(T) = 5$ and $ir(T) = 4$. The set $\{w, x, y, z\}$ is a maximal irredundant set of minimum cardinality in T.

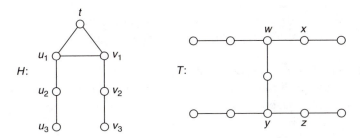

Figure 10.6 Graphs whose domination numbers exceed their irredundance numbers.

	minimum cardinality	maximum cardinality
maximal independent set of vertices in G	lower independence number $i(G)$	independence number $\beta(G)$
minimal dominating set of vertices in G	domination number $\gamma(G)$	upper domination number $\Gamma(G)$
maximal irredundant set of vertices in G	irredundance number $ir(G)$	upper irredundance number $IR(G)$

Figure 10.7 Summary of definitions of parameters.

We now introduce two other domination parameters — the so-called upper domination parameters. The *upper domination number* $\Gamma(G)$ of a graph G is the maximum cardinality of a minimal dominating set of G; while the *upper irredundance number* $IR(G)$ of G is the maximum cardinality of an irredundant set of G. We now summarize in Figure 10.7 the parameters we have introduced thus far in this chapter together with the vertex independence parameters we introduced earlier.

The six parameters described in Figure 10.7 make up a string of inequalities, which was first observed by Cockayne and Hedetniemi [CH3].

Theorem 10.22

For every graph G,

$$ir(G) \le \gamma(G) \le i(G) \le \beta(G) \le \Gamma(G) \le IR(G).$$

Proof

Since every minimal dominating set is an irredundant set, it follows that $\Gamma(G) \le IR(G)$. Moreover, every maximum independent set is a dominating set;

so $\beta(G) \leq \Gamma(G)$. Also, since an independent dominating set is independent, $i(G) \leq \beta(G)$. The result now follows from Theorem 10.21. $\qquad\square$

Cockayne, Favaron, Payan and Thomason [CFPT1] have shown that graphs exist having distinct values for all six parameters mentioned in Theorem 10.22. For bipartite graphs, however, three of these parameters must have the same value.

Theorem 10.23

For every bipartite graph G,

$$\beta(G) = \Gamma(G) = IR(G).$$

Proof

Let G be a bipartite graph with partite set U and W. Let S be a maximum irredundant set of vertices of G, and let T be the set of isolated vertices of $\langle S \rangle$. Furthermore, let

$$U_1 = T \cap U, \quad U_2 = (S \cap U) - T,$$
$$W_1 = (T \cap W), \quad W_2 = (S \cap W) - T,$$

where one or more of these sets may be empty. Each vertex $w \in W_2$ is irredundant in S. Since w is not isolated in $\langle S \rangle$, the vertex w is not its own private neighbor. However, since S is an irredundant set, w is a private neighbor of some vertex of $V(G) - S$. Hence for $w \in W_2$, there exists a vertex $w' \in V(G) - S$ such that $N(w') \cap S = \{w\}$. Moreover, since $w \in W$, it follows that $w' \in U$.

Let $A = \{w' | w \in W_2\}$. Then $|A| \geq |W_2|$ and $A \subset U$. Furthermore, no vertex of A is adjacent to a vertex of W_1. Consequently, $U_1 \cup U_2 \cup W_1 \cup A$ is independent in G. Hence

$$\beta(G) \geq |U_1| + |U_2| + |W_1| + |A| \geq |S| = IR(G).$$

The result now follows from Theorem 10.22. $\qquad\square$

We note, in closing, that the domination number may be defined for digraphs as well. In a digraph D, a vertex v *dominates* itself and all vertices adjacent from v. The *domination number* $\gamma(D)$ of D is the minimum cardinality of a set S of vertices of D such that every vertex of D is dominated by some vertex of S.

EXERCISES 10.3

10.13 (a) Prove that a set S of two or more vertices of G is irredundant if and only if it contains no redundant vertex.

(b) Characterize those graphs G of order $n \geq 2$ such that every set S of vertices of G with $|S| \geq 2$ contains a redundant vertex.

10.14 Verify that $\gamma(T) = 5$ and $ir(T) = 4$ for the tree T of Figure 10.6.

10.15 Let G be a graph for which $ir(G) = \gamma(G)$. Prove that if S is a minimal dominating set of G, then S is a maximal irredundant set.

10.16 Let G be a graph for which $ir(G) = \gamma(G)$. If S is a maximal irredundant set of vertices of G of minimum cardinality, is S a dominating set of G?

10.17 Give an example of an infinite class of graphs G for which $ir(G) < \gamma(G)$.

10.18 For a graph G, define

$$\text{dom } G = \min\{\gamma(D) \,|\, D \text{ is an orientation of } G\}$$

and

$$\text{DOM} G = \max\{\gamma(D) \,|\, D \text{ is an orientation of } G\}.$$

(a) Determine dom K_3 and DOM K_3.

(b) Show that dom $G = \gamma(G)$ for every graph G.

(c) Show that if k is an integer such that dom $G \leq k \leq$ DOM G, then there exists an orientation D' of G such that $\gamma(D') = k$.

11

Extremal graph theory

We have seen results which state that if a graph G of a fixed order n has a certain number $f(n)$ or more edges, then G contains a particular subgraph or G has some specified property. If the bound $f(n)$ on the number of edges is sharp, then there exists at least one graph of order n and size $f(n) - 1$ that doesn't contain the subgraph or doesn't possess the property involved. Such a graph is called an extremal graph. The problems of determining such sharp bounds $f(n)$ and resulting extremal graphs constitute a major part of an area of graph theory called extremal graph theory. Several problems of this type are considered in this chapter. There are also extremal problems that deal with determining the minimum order of a regular graph having a prescribed degree of regularity and girth. We discuss these as well.

11.1 TURÁN'S THEOREM

We have seen that if a graph G of order $n \geq 3$ has at least n edges, then G has a cycle. Indeed, if G has order $n \geq 3$ and at least $\binom{n-1}{2} + 2$ edges, then G has a hamiltonian cycle (Exercise 4.17). Both bounds are sharp since every tree of order n has size $n - 1$ and certainly contains no cycles; while the graph of order n obtained by adding a pendant edge to K_{n-1} has $\binom{n-1}{2} + 1$ edges but is not hamiltonian. Furthermore, if a graph G of order $n \geq 2$ has at least $\binom{n-1}{2} + 1$ edges, then G is connected; indeed, G has a hamiltonian path. Moreover, if n is even, then G contains a 1-factor. The graph $K_{n-1} \cup K_1$ shows that all of these bounds are sharp. These observations lead us to the main topic of this section and the next.

For a graph F of order k and an integer n with $n \geq k$, the *extremal number* $ex(n; F)$ of F is the maximum number of edges in a graph of order n that does not contain F as a subgraph. Consequently, every graph of order n and size $ex(n; F) + 1$ contains F as a subgraph. The graphs of order n and size $ex(n; F)$ not containing F as a subgraph are the *extremal graphs*. From the discussion above, $ex(n; C_n) = \binom{n-1}{2} + 1$ for $n \geq 3$. We now determine $ex(n; F)$ for some 'small' graphs F. If $F = K_2$, then $ex(n; F) = 0$ for $n \geq 2$; while if $F = P_3$, then $ex(n; F) = \lfloor n/2 \rfloor$ for $n \geq 3$. Furthermore, if $F = 2K_2$, then $ex(n; F) = n - 1$ for $n \geq 4$ (Exercises 11.1 and 11.2).

Of all the extremal numbers $ex(n; F)$ that have been investigated, the best known ones have been when F is complete. We begin with the case $F = K_3$. The following result is due to Turán [T6].

Theorem 11.1

Every graph of order $n \geq 3$ and size at least $\lfloor n^2/4 \rfloor + 1$ contains a triangle.

Proof

We proceed by induction on n. For $n = 3$, the only graph of order n and size at least $\lfloor n^2/4 \rfloor + 1$ is K_3, which, of course, is a triangle. For $n = 4$, the only graphs with the given conditions are $K_4 - e$ and K_4, both of which contain triangles. Thus the result is true for $n = 3$ and $n = 4$.

Assume that every graph of order k and size at least $\lfloor k^2/4 \rfloor + 1$ contains a triangle for each integer k with $3 \leq k < n$, where $n \geq 5$. Now let G be a graph of order n containing at least $\lfloor n^2/4 \rfloor + 1$ edges. Let u and v be adjacent vertices of G and define $H = G - u - v$. If u and v are mutually adjacent to a vertex of H, then G contains a triangle. Otherwise, each vertex of H is adjacent to at most one of u and v, and the size of H is at least

$$\left\lfloor \frac{n^2}{4} \right\rfloor + 1 - (n - 1) = \left\lfloor \frac{n^2 - 4n + 4}{4} \right\rfloor + 1 = \left\lfloor \frac{(n-2)^2}{4} \right\rfloor + 1.$$

By the inductive hypothesis, H contains a triangle and, consequently, so does G. $\square$

That the bound presented in Theorem 11.1 is best possible follows from the fact that for $n \geq 3$, the graph $K_{\lfloor n/2 \rfloor, \lceil n/2 \rceil}$ has order n, has size $\lfloor n^2/4 \rfloor$ and, of course, is triangle-free. This verifies that $ex(n; K_3) = \lfloor n^2/4 \rfloor$ for $n \geq 3$. We shall soon see that $K_{\lfloor n/2 \rfloor, \lceil n/2 \rceil}$ is, in fact, the unique extremal graph.

By Theorem 11.1, it follows, of course, that if G is a graph of order $n \geq 3$ and size at least $(n^2/4) + 1$, then G contains K_3 as a subgraph. This result is now extended to complete graphs of any order $r \geq 2$.

Theorem 11.2

Let r and n be positive integers, where $n \geq r \geq 2$. Then every graph of order n and size at least

$$\left(\frac{r-2}{2r-2} \right) n^2 + 1$$

contains K_r as a subgraph.

Proof

First observe that the result is true if $r=2$. Hence it suffices to assume that $n \geq r \geq 3$. We now proceed by induction on r. The result follows for $r=3$ and all integers $n \geq 3$ by Theorem 11.1. Assume, for an integer $r-1 \geq 3$ and all integers $n \geq r-1$, that every graph of order n and size at least

$$\left(\frac{r-3}{2r-4}\right)n^2 + 1$$

contains K_{r-1} as a subgraph.

It remains to show that every graph G of order n and size m, where $n \geq r$ and

$$m \geq \left(\frac{r-2}{2r-2}\right)n^2 + 1$$

contains K_r as a subgraph. We verify this by induction on n. For $n=r$, we have

$$m \geq \left(\frac{n-2}{2n-2}\right)n^2 + 1 \geq \binom{n}{2}.$$

Thus, $G=K_n=K_r$ and the result follows.

Assume now that every graph H of order k, where $r \leq k < n$, and size at least

$$\left(\frac{r-2}{2r-2}\right)k^2 + 1$$

contains K_r as a subgraph. Let G be a graph of order n and size m, where

$$m \geq \left(\frac{r-2}{2r-2}\right)n^2 + 1.$$

We show that G contains K_r as a subgraph. Since

$$m \geq \left(\frac{r-2}{2r-2}\right)n^2 + 1 \geq \left(\frac{r-3}{2r-4}\right)n^2 + 1,$$

it follows from the inductive hypothesis that G contains K_{r-1} as a subgraph. Let U be the vertex set of a subgraph of G that is isomorphic to K_{r-1}, and define $H=G-U$. If some vertex of H is adjacent to all vertices of U, then G contains K_r as a subgraph. Otherwise, every vertex of H is adjacent to at most $r-2$ vertices of U. Thus, the size of G is at most

$$\binom{r-1}{2} + (n-r+1)(r-2) + \binom{n-r+1}{2}.$$

If $n-r+1<r$, then $n\le 2(r-1)$. However, the inequalities $r\le n\le 2(r-1)$ are equivalent to the inequality

$$\binom{r-1}{2}+(n-r+1)(r-2)+\binom{n-r+1}{2}\le\left(\frac{r-2}{2r-2}\right)n^2,$$

which contradicts the fact that the size of G is at least

$$\left(\frac{r-2}{2r-2}\right)n^2+1.$$

Thus $n-r+1\ge r$. Since H has order $n-r+1\ge r$ and size at least

$$\left(\frac{r-2}{2r-2}\right)n^2+1-\binom{r-1}{2}-(n-r+1)(r-2)$$

$$=\left(\frac{r-2}{2r-2}\right)(n-r+1)^2+1,$$

it follows by the inductive hypothesis that H contains K_r as a subgraph. Therefore, G contains K_r as a subgraph. ☐

By Theorem 11.2, it follows that for $n\ge r$,

$$ex(n;K_r)\le\left(\frac{r-2}{2r-2}\right)n^2.$$

Also, by Theorem 11.2, for $r=4$ and $n=10$, every graph of order 10 and size at least 35 contains K_4 as a subgraph. However, every graph of order 10 and size 34 also contains K_4 as a subgraph. Hence the bound presented in Theorem 11.2 is not sharp for $r=4$. The exact value of $ex(n;K_r)$ for all integers n and r with $n\ge r\ge 2$ is due to Turán [T6].

Prior to presenting this more general result, we introduce some terminology and notation that will be useful in its proof. A *near regular complete multipartite graph* is a complete multipartite graph, the cardinalities of whose partite sets differ by at most 1. Thus the degrees of the vertices of a near regular complete multipartite graph have at most two values. A near regular complete k-partite graph of order n is unique and we denote this graph by $R(n,k)$. If $q=\lfloor n/k\rfloor$ is the quotient obtained when n is divided by k, then, by the division algorithm, $n=qk+r$, where $0\le r<k$. Necessarily, then, r of the partite sets in $R(n,k)$ contain $q+1$ vertices, while the remaining $k-r$ partite sets contain q vertices. Thus the size of $\overline{R(n,k)}$ is $r\binom{q+1}{2}+(k-r)\binom{q}{2}$ and the size of $R(n,k)$ is

$$\binom{n}{2}-r\binom{q+1}{2}-(k-r)\binom{q}{2}.$$

We denote the size of $R(n, k)$ by $m(n, k)$. Consequently,

$$m(n, k) = \binom{n}{2} - r\binom{q+1}{2} - (k - r)\binom{q}{2}.$$

The following proof of Turán's theorem is based on one due to A. J. Schwenk.

Theorem 11.3

Let n and p be integers with $2 \leq n \leq p$. Every graph of order p and size at least $m(p, n-1) + 1$ contains K_n as a subgraph. Furthermore, the only K_n-free graph of order p and size $m(p, n-1)$ is $R(p, n-1)$.

Proof

We proceed by induction on $n (\geq 2)$. For $n = 2$ and $p \geq 2$, the graph $R(p, n-1) = R(p, 1) = \overline{K_p}$; so $m(p, 1) = 0$. Consequently, every graph of order p and size at least $m(p, 1) + 1$ contains K_2 as a subgraph. Certainly, $\overline{K_p}$ is the unique graph of order p containing no edges. Therefore, the result is true for $n = 2$.

Assume, for $n \geq 3$, that every graph of order s ($\geq n-1$) and size at least $m(s, n-2) + 1$ contains K_{n-1} as a subgraph and that $R(s, n-2)$ is the only graph of order s and size $m(s, n-2)$ that does not contain K_{n-1} as a subgraph. For $p \geq n$, let G be a K_n-free graph of maximum size having order p.

Let v be a vertex of G such that $\deg_G v = \Delta(G) = \Delta$. Since G does not contain K_n as a subgraph, the subgraph $\langle N(v) \rangle$ induced by the neighbors of v does not contain K_{n-1}.

Next we show that $\Delta \geq n - 1$; for suppose, to the contrary, that $\Delta \leq n - 2$. Since G has order p and $p \geq n$, it follows that there is a vertex u ($\neq v$) in G such that u is not adjacent to v. Since G is a K_n-free graph of maximum size having order p, the graph $G + uv$ contains a subgraph F isomorphic to K_n. With the possible exception of u and v, all of the vertices of $G + uv$ have degree at most $n - 2$. However, F is $(n-1)$-regular and has order at least 3. This produces a contradiction; so $\Delta \geq n - 1$, as claimed.

Since $\langle N(v) \rangle$ does not contain K_{n-1} as a subgraph, the size of $\langle N(v) \rangle$ does not exceed the size $m(\Delta, n-2)$ of the graph $R(\Delta, n-2)$.

Let $U = \{u_1, u_2, \ldots, u_t\}$ denote the vertex set of the graph $G - N[v]$. Since each vertex u_i ($1 \leq i \leq t$) has degree at most Δ in G, it follows that

$$|E(G)| \leq (t + 1)\Delta + m(\Delta, n - 2).$$

If, in fact, $|E(G)| = (t+1)\Delta + m(\Delta, n - 2)$, then $\langle N(v) \rangle = R(\Delta, n - 2)$. Define

$$G' = R(\Delta, n - 2) + \overline{K}_{t+1}.$$

Thus, G' is a complete $(n-1)$-partite graph of order p and size $(t+1)\Delta + m(\Delta, n-2)$. Since G' is $(n-1)$-partite, it does not contain K_n as a subgraph. Therefore.

$$(t+1)\Delta + m(\Delta, n-2) = \left|E(G')\right| \le \left|E(G)\right| \le (t+1)\Delta + m(\Delta, n-2).$$

Consequently, G has size $(t+1)\Delta + m(\Delta, n-2)$.

Next we show that $G = G'$, that is, G' is the unique K_n-free graph of order p and size $(t+1)\Delta + m(\Delta, n-2)$. The degree in G of every vertex of U is Δ; for otherwise $|E(G)| < |E(G')|$. Moreover, U is independent; otherwise, $|E(G)| < |E(G')|$. Therefore, $U \cup \{v\}$ is independent and $G = G'$, as claimed.

Since $G = R(\Delta, n-2) + \overline{K}_{t+1}$, it follows that $G = K(t+1, p_1, p_2, \ldots, p_{n-2})$, where we may assume that $p_1 \le p_2 \le \cdots \le p_{n-2}$. It remains only to show that G is near regular. By the induction hypothesis, $R(\Delta, n-2) = K(t+1, p_1, p_2, \ldots, p_{n-2})$ is near regular, so $p_{n-2} \le p_1 + 1$. Since v is a vertex of maximum degree in G, it follows that $t+1 \le p_1$.

Hence, it remains to show that $p_{n-2} \le t+2$. Suppose, to the contrary, that $p_{n-2} \ge t+3$. Let $H = K(t+2, p_1, p_2, \ldots, p_{n-3}, p_{n-2}-1)$. Thus

$$\left|E(H)\right| - \left|E(G)\right| = (p_{n-2}-1) - (t+1) \ge 1,$$

which contradicts the defining property of G. Therefore, $p_{n-2} \le t+2$ and $G = R(p, n-1)$. $\qquad\square$

EXERCISES 11.1

11.1 Show that every graph of order $n \ge 3$ and size $\lfloor n/2 \rfloor + 1$ contains P_3 as a subgraph. Describe the extremal graphs.

11.2 Show that every graph of order $n \ge 4$ and size n contains $2K_2$ as a subgraph. Describe the extremal graphs.

11.3 For $n \ge 4$, determine $ex(n; K_{1,3})$ and all extremal graphs.

11.2   EXTREMAL RESULTS ON GRAPHS

In this section, we consider a variety of other extremal results in graph theory. By Turán's theorem, $ex(n; K_4) = m(n, 3)$. Dirac [D7] obtained a related result.

Theorem 11.4

Every graph of order $n \ge 4$ and size at least $2n - 2$ contains either K_4 or a subdivision of K_4 as a subgraph.

Mader [M1] established the following result which had been a longstanding conjecture of G. A. Dirac.

Theorem 11.5

Every graph of order $n \geq 5$ and size at least $3n-5$ contains either K_5 or a subdivision of K_5 as a subgraph.

We have already seen that the minimum size which guarantees that every graph of order n contains a cycle is n. Although barely a teenager at the time, Pósa (see Erdös [E4]) determined the minimum size of a graph G of order $n \geq 6$ which guarantees that G contains two disjoint cycles.

Theorem 11.6

Every graph of order $n \geq 6$ and size at least $3n-5$ contains two disjoint cycles.

Proof

It suffices to show that every graph of order n and size $3n-5$ contains two disjoint cycles for $n \geq 6$. We employ induction on n. There are only two graphs of order 6 and size 13, one obtained by removing two nonadjacent edges from K_6 and the other obtained by removing two adjacent edges from K_6. In both cases, the graph has two disjoint triangles. Thus, the result is true for $n=6$.

Assume for all k with $6 \leq k < n$ that every graph of order k and size $3k-5$ contains two disjoint cycles. Let G be a graph of order n and size $3n-5$. Since

$$\sum_{v \in V(G)} \deg v = 6n - 10,$$

there exists a vertex v_0 of G such that $\deg v_0 \leq 5$. Assume first that $\deg v_0 = 5$, and $N(v_0) = \{v_1, v_2, \ldots, v_5\}$. If $\langle N[v_0] \rangle$ contains 13 or more edges, then we have already noted that $\langle N[v_0] \rangle$ has two disjoint cycles, implying that G has two disjoint cycles. If, on the other hand, $\langle N[v_0] \rangle$ contains 12 or fewer edges, then, since $\deg v_0 = 5$, some neighbor of v_0, say v_1, is not adjacent with two other neighbors of v_0, say v_2 and v_3. Add to G the edges $v_1 v_2$ and $v_1 v_3$ and delete the vertex v_0, obtaining the graph G'; that is, $G' = G + v_1 v_2 + v_1 v_3 - v_0$. The graph G' is a graph of order $n-1$ and size $3n-8$ and, by the inductive hypothesis, contains two disjoint cycles C_1 and C_2. At least one of these cycles, say C_1, does not contain the vertex v_1 and thus contains neither the edge $v_1 v_2$ nor the edge $v_1 v_3$. Hence C_1 is a cycle of G. If C_2 contains neither $v_1 v_2$ nor $v_1 v_3$, then C_1 and C_2 are disjoint cycles of G. If C_2 contains $v_1 v_2$ but not $v_1 v_3$, then by removing $v_1 v_2$ and adding v_0, $v_0 v_1$ and $v_0 v_2$, we produce a cycle of G that is disjoint from C_1. The procedure is similar if C_2 contains v_1v_3 but not v_1v_2. If C_2 contains both v_1v_2 and v_1v_3, then by removing v_1 from C_2 and adding v_0, v_0v_2 and v_0v_3, a cycle of G disjoint from C_1 is produced.

Suppose next that $\deg v_0 = 4$, where $N(v_0) = \{v_1, v_2, v_3, v_4\}$. If $\langle N[v_0] \rangle$ is not complete, then some two vertices of $N(v_0)$ are not adjacent, say v_1 and v_2. By adding v_1v_2 to G and deleting v_0, we obtain a graph G' of order $n-1$ and size $3n-8$, which by the inductive hypothesis contains two disjoint cycles. We may proceed as before to show now that G has two disjoint cycles. Assume then that $\langle N[v_0] \rangle$ is a complete graph of order 5. If some vertex of $V(G) - N[v_0]$ is adjacent to two or more neighbors of v_0, then G contains two disjoint cycles. Hence we may assume that no vertex of $V(G) - N[v_0]$ is adjacent to more than one vertex of $N(v_0)$. Remove the vertices v_0, v_1, v_2 from G, and note that the resulting graph G'' has order $n-3$ and contains at least $(3n-5) - (n-5) - 9 = 2n-9$ edges. However, $n \geq 6$ implies that $2n-9 \geq n-3$; so G'' contains at least one cycle C. The cycle C and the cycle v_0, v_1, v_2, v_0 are disjoint and belong to G.

Finally, we assume that $\deg v_0 \leq 3$. The graph $G - v_0$ is a graph of order $n-1$ and size m, where $m \geq 3n-8$. Hence by the inductive hypothesis, $G - v_0$ (and therefore G) contains two disjoint cycles. ◻

To see that the bound $3n-5$ presented in Theorem 11.6 is sharp, observe that the complete 4-partite graph $K_{1,1,1,n-3}$ has order n and size $3n-6$, and that every cycle contains at least two of the three vertices having degree $n-1$. Thus no two cycles of $K_{1,1,1,n-3}$ are disjoint.

For a graph of order n to contain two edge-disjoint cycles, only $n+4$ edges are required. This result is also due to Pósa (see Erdös [E4]). The proof we present establishes the result for multigraphs. It is a simple observation that every multigraph of order n and size n contains a cycle.

Theorem 11.7

Every multigraph of order $n \geq 2$ and size at least $n+4$ contains two edge-disjoint cycles.

Proof

We employ induction on n. Let G be a multigraph of order 2 and size at least 6. Then clearly G has two edge-disjoint cycles.

Assume that all multigraphs of order $n-1$ and size at least $n+3$, where $n \geq 3$, contain two edge-disjoint cycles, and let G be a multigraph of order n and size at least $n+4$. If G has a vertex v of degree at most 1, then $G-v$ has order $n-1$ and size at least $n+3 = (n-1)+4$. By the inductive hypothesis, $G-v$ contains two edge-disjoint cycles and so G does as well. Suppose next that G has a vertex v of degree 2, whose neighbors are u and w. Consider the multigraph G' obtained from G by removing v and adding the edge uw. Then G' has order $n-1$ and size at least $n+3$ and so, by the inductive hypothesis, contains two edge-disjoint cycles. But then G, too, contains two edge-disjoint cycles. Thus we may assume that $\delta(G) \geq 3$.

Let C be a shortest cycle G. If $G-E(C)$ contains a cycle, then G contains two edge-disjoint cycles. Thus we may assume that $G-E(C)$ is a forest. Let v be an end-vertex of $G-E(C)$. Since $\delta(G) \geq 3$, v is adjacent to at least two vertices of C. This implies that C has length at most 4 or else C is not a shortest cycle of G. But then $G-E(C)$ has order n and size at least n and so $G-E(C)$ contains a cycle, which prduces a contradiction. Thus G contains two edge-disjoint cycles. □

Theorem 11.1 could very well be interpreted as a result concerning cycles rather than a result concerning complete graphs. From this point of view, we know that if G is a graph of order $n \geq 3$ and size m, where $m \geq (n^2/4)+1$, then G contains a 3-cycle. We now turn to 4-cycles.

In order to present a proof of the next result, it is convenient to be acquainted with an inequality involving nonincreasing sequences of integers. Let $a_1, a_2, \ldots, a_n$ and $b_1, b_2, \ldots, b_n$ be nonincreasing sequences of integers. Then

$$\sum_{1 \leq i < j \leq n} (a_i - a_j)(b_i - b_j) \geq 0. \tag{11.1}$$

By rearranging the terms in (11.1), we arrive at

$$(n-1)\sum_{i=1}^{n} a_i b_i \geq \sum_{1 \leq i \neq j \leq n} a_i b_j. \tag{11.2}$$

Adding $\sum_{i=1}^{n} a_i b_i$ to both sides of (11.2), we obtain

$$n\sum_{i=1}^{n} a_i b_i \geq \sum_{1 \leq i,j \leq n} a_i b_j = \left(\sum_{i=1}^{n} a_i\right)\left(\sum_{i=1}^{n} b_i\right)$$

or, equivalently,

$$\sum_{i=1}^{n} a_i b_i \geq n\left(\frac{1}{n}\sum_{i=1}^{n} a_i\right)\left(\frac{1}{n}\sum_{i=1}^{n} b_i\right). \tag{11.3}$$

That is, the inequality (11.3) states that the sum of the integers $a_i b_i$ ($1 \leq i \leq n$) is at least n times the product of the averages of $a_1, a_2, \ldots, a_n$ and of $b_1, b_2, \ldots, b_n$.

Suppose then that G is a graph of order n and size m with vertex set $V(G) = \{v_1, v_2, \ldots, v_n\}$ such that $\deg v_i = d_i$ ($1 \leq i \leq n$) and $d_1, d_2, \ldots, d_n$ is a nonincreasing sequence. Then, of course, $\sum_{i=1}^{n} d_i = 2m$. From inequality (11.3), it follows that

$$\sum_{i=1}^{n} \binom{d_i}{2} = \frac{1}{2} \sum_{i=1}^{n} d_i(d_i - 1) \geq \frac{n}{2}\left(\frac{2m}{n}\right)\left(\frac{2m-n}{n}\right). \tag{11.4}$$

We are now prepared to present the aforementioned result dealing with 4-cycles (see Lovász [L5, p. 69]).

Theorem 11.8

If G is a graph of order n and size m with $n \geq 4$ and

$$m \geq \frac{n + n\sqrt{4n-3}}{4} + 1,$$

then G contains a 4-cycle.

Proof

Suppose that G is a graph of order n and size m with $n \geq 4$ that contains no 4-cycles. For a vertex v of G, the number of distinct pairs of vertices that are mutually adjacent to v is $\binom{\deg v}{2}$. However, since G contains no 4-cycles, each pair of vertices that are mutually adjacent to another vertex is counted exactly once in the sum $\sum_{v \in V(G)} \binom{\deg v}{2}$. Hence

$$\sum_{v \in V(G)} \binom{\deg v}{2} \leq \binom{n}{2}. \tag{11.5}$$

Denote the degrees of the vertices of G by $d_1, d_2, \ldots, d_n$. Then

$$\sum_{v \in V(G)} \binom{\deg v}{2} = \sum_{i=1}^{n} \binom{d_i}{2} \geq \frac{n}{2}\left(\frac{2m}{n}\right)\left(\frac{2m-n}{n}\right),$$

where the inequality follows from (11.4). Combining (11.4) and (11.5), we have

$$\frac{m(2m - n)}{n} \leq \frac{n(n - 1)}{2}. \tag{11.6}$$

Solving inequality (11.6) for m gives us

$$m \leq \frac{n + n\sqrt{4n-3}}{4},$$

which completes of proof. $\qquad\qquad\square$

Somewhat fewer edges guarantee that a graph contains *either* a 3-cycle or a 4-cycle.

Theorem 11.9

If G is a graph of order $n \geq 4$ and size

$$m \geq \frac{n\sqrt{n-1}}{2} + 1,$$

then G contains a 3-cycle or a 4-cycle.

Proof

Suppose that G is a graph with $n \geq 4$ vertices and m edges that contains no 3-cycle or 4-cycle. We proceed as in the proof of Theorem 11.8. Since G contains no 4-cycles, each pair of vertices that are mutually adjacent to another vertex is counted exactly once in the sum $\sum_{v \in V(G)} \binom{\deg v}{2}$. Since G contains no 3-cycles, each pair of vertices that are mutually adjacent to another vertex are themselves not adjacent. Hence

$$\sum_{v \in V(G)} \binom{\deg v}{2} \leq \binom{n}{2} - m. \tag{11.7}$$

Denote the degrees of the vertices of G by $d_1, d_2, \ldots, d_n$. Applying the inequality (11.4), we obtain

$$\sum_{v \in V(G)} \binom{\deg v}{2} = \sum_{i=1}^{n} \binom{d_i}{2} \geq \frac{n}{2} \left(\frac{2m}{n} \right) \left(\frac{2m-n}{n} \right). \tag{11.8}$$

Combining (11.7) and (11.8), we have

$$m^2 \leq \frac{n^2(n-1)}{4},$$

which yields the desired result. □

Letting $n = 5$ in Theorem 11.9, we find that if the size of a graph G of order 5 is at least 6, then G contains a 3-cycle or a 4-cycle. This cannot be improved because of the 5-cycle. For $n = 10$ it follows that if a graph G of order 10 has at least 16 edges, then G has a 3-cycle or a 4-cycle. This too cannot be improved since the Petersen graph has order 10, size 15, and contains no 3-cycles or 4-cycles.

We now turn to the problem of determining the number of edges that a graph G of order n must have to guarantee that G contains a subgraph with a specified minimum degree.

Theorem 11.10

Let k and n be integers with $1 \le k < n$. Every graph of order n and size at least

$$(k-1)n - \binom{k}{2} + 1$$

contains a subgraph with minimum degree k.

Proof

We proceed by induction on $n \ge k+1$. First, assume that $n = k+1$. Let G be a graph of order n and size at least

$$(k-1)n - \binom{k}{2} + 1 = (n-2)n - \binom{n-1}{2} + 1 = \binom{n}{2}.$$

Then $G = K_n = K_{k+1}$ and so G itself is a graph with minimum degree k.

Assume that every graph of order $n-1 \ge k+1$ and size at least

$$(k-1)(n-1) - \binom{k}{2} + 1$$

contains a subgraph with minimum degree k. Let G be a graph of order n and size m, where

$$m \ge (k-1)n - \binom{k}{2} + 1.$$

We show that G contains a subgraph with minimum degree k. If G itself is not such a graph, then G contains a vertex v with $\deg v \le k-1$. Then the order of $G - v$ is $n-1$ and its size is at least

$$m - \deg v \ge (k-1)n - \binom{k}{2} + 1 - (k-1) = (k-1)(n-1) - \binom{k}{2} + 1.$$

By the induction hypothesis, $G - v$, and therefore G as well, contains a subgraph with minimum degree k. $\qquad\square$

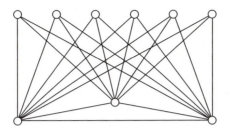

Figure 11.1 A graph of order 9 and size 21 containing no subgraph with minimum degree 4.

The bound given in Theorem 11.10 cannot be improved. If $k=1$, then the graph $\overline{K}_n$ contains no subgraph with minimum degree 1. More generally, for $2 \leq k < n$, the graph $\overline{K}_{n-k+1} + K_{k-1}$ has order n and size $(k-1)n - \binom{k}{2}$ but contains no subgraph with minimum degree k. For $n=9$ and $k=4$, the graph $\overline{K}_{n-k+1} + K_{k-1}$ is shown in Figure 11.1.

By Theorem 3.8, if G is a graph such that $\delta(G) \geq k$ for some positive integer k, then G contains every tree of size k as a subgraph. Combining this result with Theorem 11.10 gives us the following corollary.

Corollary 11.11

Let k and n be integers with $1 \leq k < n$. If G is a graph of order n and size at least

$$(k-1)n - \binom{k}{2} + 1,$$

then G contains every tree of size k as a subgraph.

Corollary 11.11 is not best possible however. A longstanding conjecture due to P. Erdös and V. Sós is that if k and n are integers with $1 \leq k < n$ and G is a graph of order n and size greater than $n(k-1)/2$, then G contains every tree of size k as a subgraph.

In 1959 Erdös and Gallai [EG2] determined the maximum number of edges in a graph of order $n \geq 2k$ that contains no matching of size k; that is, for $n \geq 2k$, they determined $ex(n; kK_2)$. In 1968 Moon [M10] obtained a more general result. It is Moon's proof of the formula for $ex(n; kK_2)$ that we present here.

Theorem 11.12

For positive integers n and k with $n \geq 2k$,

$$ex(n; kK_2) = \max\left\{ (k-1)n - \binom{k}{2}, \binom{2k-1}{2} \right\}.$$

Proof

Let G be a graph of order n and size m containing a maximum matching M of size $k-1$, and let U denote the set of vertices of G that are incident with no edge of M. Since M is a maximum matching, U is an independent set of vertices. Moreover, U does not contain two vertices that are adjacent to distinct vertices of $V(G) - U$ that are joined by an edge of M; otherwise, G contains an M-augmenting path, which is impossible since M is a maximum matching (Theorem 9.2).

We now partition the edges of M into two subsets A and B. The set A consists of those edges xy of M such that one of x and y, say y, is adjacent to at least two vertices of U. The vertex x of each such edge xy of A is therefore not adjacent to any vertex of U. The set B then consists of the remaining edges of M. Hence if uv is an edge of B, then there is at most one vertex of U that is adjacent to u or v (or both). Let $a = |A|$ and $b = |B|$; so $a + b = k - 1$.

Now observe that if $x_1 y_1$ and $x_2 y_2$ are edges of A, where each of y_1 and y_2 is adjacent to at least two vertices of U, then $x_1 x_2 \notin E(G)$; for otherwise G contains an M-augmenting path. Note that for every two edges e_1 and e_2 of A, at least one vertex incident with e_1 is not adjacent to a vertex incident with e_2; so the size of $\langle A \rangle$ is at most $\binom{2a}{2} - \binom{a}{2}$. For each edge e of A, there is a vertex incident with e that is adjacent to no vertex of U. Thus the number of edges joining a vertex of $\langle A \rangle$ and a vertex of U is at most $a(n - 2k + 2)$. Therefore, the number of edges that are incident with two vertices of A or with a vertex of A and a vertex of U is at most

$$\binom{2a}{2} - \binom{a}{2} + a(n - 2k + 2).$$

Next observe that if $x_1 y_1$ and $x_2 y_2$ are edges of A, where y_1 and y_2 are adjacent to two or more vertices of U, then x_1 and x_2 are not adjacent to distinct incident vertices of an edge uv of B; for otherwise G contains an M-augmenting path. Moreover, if u is adjacent to x_1, say, then v is not adjacent to any vertex of U, for once again an M-augmenting path results. This implies that the number of edges of G incident with at least one vertex of B is at most

$$\binom{2b}{2} + (2b)a + ba + 2b.$$

Since every edge of G is one of the types described above, it follows that

$$m \le \binom{2a}{2} - \binom{a}{2} + \binom{2b}{2} + a(n - 2k + 2) + 3ab + 2b$$

$$= \binom{2k-1}{2} + \frac{a(2n - 5k + 2)}{2} - \frac{ab}{2} \tag{11.9}$$

$$\leq \binom{2k-1}{2} + \frac{a(2n-5k+2)}{2}$$
$$= (k-1)n - \binom{k}{2} - \frac{b(2n-5k+2)}{2}. \tag{11.10}$$

If $2n - 5k + 2 = 0$, then, of course, $\binom{2k-1}{2} = (k-1)n - \binom{k}{2}$. If $2n - 5k + 2 > 0$, then (11.9) and (11.10) attain their maximum value when $b = 0$; while if $2n - 5k + 2 < 0$, then (11.9) and (11.10) attain their maximum values when $a = 0$. Thus

$$m \leq \max\left\{(k-1)n - \binom{k}{2}, \binom{2k-1}{2}\right\}$$

and so

$$ex(n; kK_2) \leq \max\left\{(k-1)n - \binom{k}{2}, \binom{2k-1}{2}\right\}.$$

The graph $K_{2k-1} \cup \overline{K}_{n-2k+1}$ has order n, size $\binom{2k-1}{2}$, and a maximum matching of size $k-1$. Moreover, the graph $K_{k-1} + \overline{K}_{n-k+1}$ has order n, size $(k-1)n - \binom{k}{2}$, and a maximum matching of size $k-1$. Therefore,

$$ex(n; kK_2) \geq \max\left\{(k-1)n - \binom{k}{2}, \binom{2k-1}{2}\right\},$$

which yields the desired result. □

An equivalent statement of Theorem 11.12 is that for a positive integer k, every graph of order $n \geq 2k$ and size at least

$$\max\left\{1 + (k-1)n - \binom{k}{2}, 1 + \binom{2k-1}{2}\right\}$$

contains a matching of size k. From Corollary 11.11, we know that if G is a graph of order $n > k \geq 1$ and size at least $(k-1)n - \binom{k}{2} + 1$, then G contains every tree of size k. Brandt [B10] obtained a generalization of both Corollary 11.11 and Theorem 11.12.

Theorem 11.13

Let n and k be positive integers with $n \geq 2k$. Every graph of order n and size at least

$$\max\left\{1 + (k-1)n - \binom{k}{2}, 1 + \binom{2k-1}{2}\right\}$$

contains every forest of size k without isolated vertices as a subgraph.

EXERCISES 11.2

11.4 Prove Theorem 11.4.

11.5 For $n \geq 9$ determine the smallest positive integer m such that every graph of order n and size m contains three pairwise disjoint cycles.

11.6 Let n and k be positive integers such that $n \geq (5k-2)/2$. Prove that if G is a graph of order n and size at least $(k-1)n - \binom{k}{2} + 1$, then G contains every forest of size k and without isolated vertices as a subgraph.

11.7 Let G be a graph containing a subgraph H of order at least $2k$ such that $\delta(H) \geq k$. Prove that G contains every forest of size k and without isolated vertices as a subgraph.

11.3 CAGES

We close this chapter with a different type of extremal topic. Recall that the length of a smallest cycle in a graph G that contains cycles is called the girth of G which we denote by $g(G)$. Therefore, $g(K_n) = 3$ for $n \geq 3$, $g(K_{s,t}) = 4$ for $s, t \geq 2$, and $g(C_n) = n$ for $n \geq 3$. We are interested in the smallest order of an r-regular graph of girth g for given integers r and g. Thus, for positive integers $r \geq 2$ and $g \geq 3$, we define $f(r, g)$ as the smallest positive integer n for which there exists an r-regular graph of girth g having order n. The r-regular graphs of order $f(r, g)$ with girth g have been the object of many investigations; such graphs are called (r, g)-*cages*. The $(3, g)$-cages are commonly referred to simply as g-*cages*. We introduce the notation $[r, g]$-*graph* to indicate an r-regular graph having girth g. Thus, an (r, g)-cage is an $[r, g]$-graph; indeed, it is one of minimum order.

It is clear that $f(r, g) \geq \max\{r+1, g\}$. Thus, $f(2, g) = g$ since C_g is a 2-regular graph with girth g. Likewise, $f(r, 3) = r+1$ since K_{r+1} is an r-regular graph having girth 3. In fact, the complete graph K_4 is the unique 3-cage. A lower bound for the order of any $[r, g]$-graph is presented next (see Holton and Sheehan [HS3, p. 184]). For $r, g \geq 3$, we define

$$f_0(r, g) = \begin{cases} 1 + \frac{r[(r-1)^{(g-1)/2} - 1]}{r-2} & \text{if } g \text{ is odd} \\ \frac{2[(r-1)^{g/2} - 1]}{r-2} & \text{if } g \text{ is even.} \end{cases}$$

Theorem 11.14

If G is an $[r, g]$-graph of order n, then $n \geq f_0(r, g)$.

Proof

First, suppose that g is odd. Then $g = 2k+1$ for some positive integer k. Let $v \in V(G)$. For $1 \leq i \leq k$, the number of vertices at distance i from v is $r(r-1)^{i-1}$.

Hence

$$n \geq 1 + r + r(r-1) + r(r-1)^2 + \cdots + r(r-1)^{k-1} = 1 + \frac{r[(r-1)^k - 1]}{r-2}$$

Next, suppose that g is even. Then $g = 2\ell$, where $\ell \geq 2$. Let $e = uv \in E(G)$. For $1 \leq i \leq \ell - 1$, the number of vertices at distance i from u or v is $2(r-1)^i$. Thus

$$n \geq 2 + 2(r-1) + 2(r-1)^2 + \cdots + 2(r-1)^{\ell-1} = 2\left[\frac{(r-1)^\ell - 1}{r-2}\right]. \qquad \square$$

Consequently, if for integers $r, g \geq 3$, there exists an (r, g)-cage, then $f(r, g) \geq f_0(r, g)$. We now show that for every pair r, $g \geq 3$ of integers, there is at least one (r, g)-cage. The proof of the following result is due to Erdös and Sachs [ES1].

Theorem 11.15

For every pair r, g of integers at least 3, the number $f(r, g)$ exists and, in fact,

$$f(r, g) \leq \left(\frac{r-1}{r-2}\right)[(r-1)^{g-1} + (r-1)^{g-2} + (r-4)].$$

Proof

Since

$$\sum_{i=1}^{g-1} (r-1)^i = \left(\frac{r-1}{r-2}\right)[(r-1)^{g-1} - 1]$$

and

$$\sum_{i=1}^{g-2} (r-1)^i = \left(\frac{r-1}{r-2}\right)[(r-1)^{g-2} - 1],$$

it follows that

$$\left(\frac{r-1}{r-2}\right)[(r-1)^{g-1} + (r-1)^{g-2} + (r-4)]$$

is an integer. Denote this integer by n, and let $\mathcal{S}$ be the set of all graphs H of order n such that $g(H) = g$ and $\Delta(H) \leq r$. Note that $n \geq g$. The set $\mathcal{S}$ is nonempty since the graph consisting of a g-cycle and $n-g$ isolated vertices belongs to $\mathcal{S}$. For each $H \in \mathcal{S}$, define

$$M(H) = \{v \in V(H) \mid \deg v < r\}.$$

If for some $H \in \mathcal{S}$, $M(H) = \emptyset$, then we have the desired result; thus we assume for all $H \in \mathcal{S}$, $M(H) \neq \emptyset$. For $H \in \mathcal{S}$, we define $d(H)$ to be the maximum distance between two vertices of $M(H)$. (We define $d(u_1, u_2) = +\infty$ if u_1 and u_2 are not connected.)

Let $\mathcal{S}_1$ be those graphs in $\mathcal{S}$ containing the maximum number of edges, and denote by $\mathcal{S}_2$ the set of all those graphs H of $\mathcal{S}_1$ for which $|M(H)|$ is maximum. Now among the graphs of $\mathcal{S}_2$, let G be chosen so that $d(G)$ is maximum.

Let $u, v \in M(G)$ such that $d(u, v) = d(G)$. Suppose that $d(G) \geq g-1 \geq 2$. By adding the edge uv to G, we obtain a graph G' of order n having $g(G') = g$ and $\Delta(G') \leq r$. Hence $G' \in \mathcal{S}$; However, G' has more edges than G, and this produces a contradiction. Therefore, $d(G) \leq g-2$ and $d(u, v) \leq g-2$. (The vertices u and v may not be distinct.)

Denote by W the set of all those vertices w of G such that $d(u, w) \leq g-2$ or $d(v, w) \leq g-1$. From our earlier remark, it follows that $u, v \in W$. The number of vertices different from u at a distance at most $g-2$ from u cannot exceed

$$\sum_{i=1}^{g-2} (r-1)^i = \left(\frac{r-1}{r-2}\right)[(r-1)^{g-2} - 1];$$

while the number of vertices different from v and at a distance at most $g-1$ from v cannot exceed

$$\sum_{i=1}^{g-1} (r-1)^i = \left(\frac{r-1}{r-2}\right)[(r-1)^{g-1} - 1].$$

Hence the number of elements in W is at most

$$\left(\frac{r-1}{r-2}\right)[(r-1)^{g-2} - 1] + \left(\frac{r-1}{r-2}\right)[(r-1)^{g-1} - 1];$$

however,

$$\left(\frac{r-1}{r-2}\right)[(r-1)^{g-1} + (r-1)^{g-2} - 2] = n - r + 1 < n.$$

Therefore, there is a vertex $w_1 \in V(G) - W$, so $d(u, w_1) \geq g-1$ and $d(v, w_1) \geq g$. Since $d(u, w_1) > d(G)$ and $u \in M(G)$, it follows that $w_1 \notin M(G)$ and $\deg w_1 = r \geq 3$. Therefore, there exists an edge e incident with w_1 whose removal from G results in a graph having girth g. Suppose that $e = w_1 w_2$. Clearly, $d(v, w_2) \geq g - 1$, so $w_2 \notin M(G)$ and $\deg w_2 = r$.

We now add the edge uw_1 to G and delete the edge $w_1 w_2$, producing the graph G_1. The graph G_1 also belongs to S and, in fact, belongs to S_1. The set $M(G_1)$ contains all the members of $M(G)$ except possibly u and, in addition, contains w_2. From the manner in which G was chosen, $|M(G_1)| \leq |M(G)|$; so $u \notin M(G_1)$ and $|M(G_1)| = |M(G)|$. Therefore, $\deg u = r$ in G_1, implying that, in G, $\deg u = r - 1$. Furthermore, G_1 belongs to S_2.

We now show that u is not the only vertex of $M(G)$, for suppose that it is. Since there is an even number of odd vertices, we must have r and n odd; however, this cannot occur since n is even when r is odd. We conclude that u and v are distinct vertices of $M(G)$.

The vertices v and w_2 are distinct vertices of $M(G_1)$. If there exists no v–w_2 path in G_1, then $d(G_1) = +\infty$, and this is contrary to the fact that $d(G_1) \leq d(G)$. Thus v and w_2 are connected in G_1. Let P be a shortest v–w_2 path in G_1. If P is also in G, then P has length at least $d_G(v, w_2)$ in G, but

$$d_G(v, w_2) \geq g - 1 > d(G),$$

which is impossible. If P is not in G, then P contains the edge uw_1 and a u–v path of length $d_G(u, v)$ as a subpath. Hence P has length exceeding $d_G(u, v) = d(G)$, again a contradiction.

It follows for some H in S that $M(H) = \emptyset$, that is, H is an r-regular graph of order n having girth g. ☐

We now determine the value of the number $f(r, 4)$.

Theorem 11.16

For $r \geq 2$, $f(r, 4) = 2r$. Furthermore, there is only one $(r, 4)$-cage, namely $K_{r,r}$.

Proof

By Theorem 11.14, $f(r, 4) \geq 2r$. Obviously, the graph $K_{r,r}$ is r-regular, has girth 4, and has order $2r$, thus implying that $f(r, 4) = 2r$.

To show that $K_{r,r}$ is the only $(r, 4)$-cage, let G be an $[r, 4]$-graph of order $2r$, and let $u_1 \in V(G)$. Denote by $v_1, v_2, \ldots, v_r$ the vertices of G adjacent to u_1. Since $g(G) = 4$, v_1 is adjacent to none of the vertices v_i, $2 \leq i \leq r$; hence G contains $r - 1$ additional vertices $u_2, u_3, \ldots, u_r$. Since every vertex has degree r and G contains no triangle, each vertex u_i ($1 \leq i \leq r$) is adjacent to every vertex v_j ($1 \leq j \leq r$); therefore, $G = K_{r,r}$. ☐

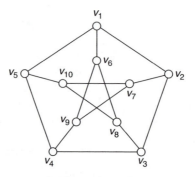

Figure 11.2 The Petersen graph: the unique 5-cage.

An $[r, g]$-graph of order n is called a *Moore graph* if $n = f(r, g) = f_0(r, g)$. Hence, a Moore graph is an (r, g)-cage of order $f_0(r, g)$. Consequently, the graphs $K_{r,r}$ for $r \geq 3$ are Moore graphs. The best known Moore graph is the Petersen graph (Figure 11.2). It is not difficult to verify that the Petersen graph is a 5-cage. That it is the *only* 5-cage is verified next.

Theorem 11.17

The Petersen graph is the unique 5-cage.

Proof

As mentioned earlier, it is not difficult to show that the Petersen graph is a 5-cage. In order to see that it is unique, assume that G is a [3, 5]-graph of order 10. We show that G is isomorphic to the Petersen graph.

Let $v_1 \in V(G)$, and suppose v_2, v_3 and v_4 are the vertices adjacent to v_1. Since $g(G) = 5$, each vertex v_1, $i = 2, 3, 4$, is adjacent to two new vertices of G. Let v_5 and v_6 be adjacent with v_2; v_7 and v_8 with v_3; and v_9 and v_{10} with v_4. Hence $V(G) = \{v_1 \mid i = 1, 2, \ldots, 10\}$. The fact that the girth of G is 5 and that every vertex of G has degree 3 implies that v_5 is adjacent with one of v_7 and v_8 and one of v_9 and v_{10}. Without loss of generality, we assume v_5 to be adjacent to v_7 and v_9. We must now have v_6 adjacent to v_8 and v_{10}. Therefore, the edges $v_7 v_{10}$ and $v_8 v_9$ are also present and no others. Thus, G is isomorphic to the Petersen graph. □

Since $f_0, (r, 5) = r^2 + 1$, a Moore graph of girth 5 has order $r^2 + 1$. We have seen that the Petersen graph is the only cubic Moore graph of girth 5. However, the Petersen graph is one of only two (or possibly three) Moore graphs of girth 5. This fact was established by Hoffman and Singleton [HS2].

Theorem 11.18

If G is an r-regular Moore graph (r ≥ 3) of girth 5, then r = 3, r = 7 or, possibly, r = 57.

A graph referred to as the *Hoffman–Singleton graph* is the 7-regular Moore graph of girth 5 (and of order 50). Its construction is described in Holton and Sheehan [HS3, p. 202].

Moore graphs can be considered from a different point of view. Consider integers $r, g \geq 3$, where $g = 2k + 1$ is odd. If G is an r-regular graph of order n and diameter k, then by the proof of Theorem 11.14, it follows that $f_0(r, g) \geq n$. Hence a Moore graph of odd girth has the maximum order consistent with its degree and diameter constraints and the minimum order consistent with its degree and girth constraints. Such a statement also applies to Moore graphs with even girth.

We now summarize the information concerning Moore graphs; namely, r-regular Moore graphs of odd girth g exist when

- $g = 3$, $r \geq 3$, and K_{r+1} is the unique Moore graph;
- $g = 5$, $r = 3$, and the Petersen graph is the unique Moore graph;
- $g = 5$, $r = 7$, and the Hoffman-Singleton graph is the unique Moore graph; and
- $g = 5$ and $r = 57$ is undecided.

Furthermore, r-regular Moore graphs of even girth g exist when

- $g = 4$, $r \geq 4$, and $K_{r,r}$ is the unique Moore graph;
- $g = 6$ and for all r for which there exists a projective plane of order $r - 1$;

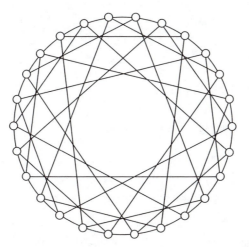

Figure 11.3 The 4-regular Moore graph of girth 6.

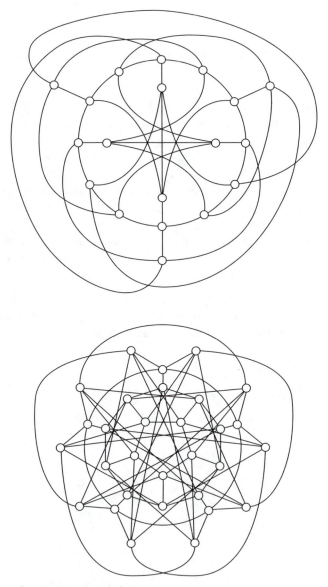

Figure 11.4 The (4, 5)-cage and (5, 5)-cage.

- $g = 8$ and for all r for which there exists a certain projective geometry; and
- $g = 12$ and for all r for which there exists a certain projective geometry.

As an additional example of a Moore graph referred to above, there is a unique 4-regular Moore graph of girth 6 and order $f_0(4, 6) = 26$. This graph is shown in Figure 11.3.

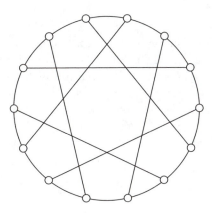

Figure 11.5 The Heawood graph: the unique 6-cage.

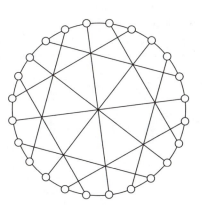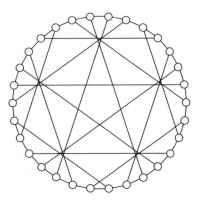

Figure 11.6 The 7-cage and 8-cage.

	$g=3$	$g=4$	$g=5$	$g=6$	$g=7$	$g=8$	$g=9$	$g=10$	$g=11$	$g=12$
$r=3$	4	6	10	14	24	30	58	70	112	126
$r=4$	5	8	19	26	?	80	?	?	?	728
$r=5$	6	10	30	42	?	170	?	?	?	2730
$r=6$	7	12	40	62	?	312	?	?	?	7812
$r=7$	8	14	50	90	?	?	?	?	?	?
$r=8$	9	16	?	114	?	800	?	?	?	39216
$r=9$	10	18	?	146	?	1170	?	?	?	74898
$r=10$	11	20	?	182	?	1640	?	?	?	132860
$r=11$	12	22	?	?	?	?	?	?	?	?
$r=12$	13	24	?	266	?	2928	?	?	?	354312
$r=13$	14	26	?	?	?	?	?	?	?	?
$r=14$	15	28	?	366	?	4760	?	?	?	804468

Figure 11.7 Some known values of $f(r,g)$.

We now return to cages that are not necessarily Moore graphs. The $(4, 5)$-cage and $(5, 5)$-cage are shown in Figure 11.4. The $(6, 5)$-cage has order 40 while the $(7, 5)$-cage, as mentioned earlier, is known to have order 50. There is only one 6-cage, referred to as the *Heawood graph*, and this is shown in Figure 11.5. There are only a few known g-cages, $g \geq 7$. The 7-cage (known as the *McGee graph*) and the 8-cage (the so-called *Tutte–Coxeter graph*) are shown in Figure 11.6. The 11-cage has order 112 and the 12-cage has order 126. Some other known values of $f(r, g)$ are shown in Figure 11.7.

EXERCISES 11.3

11.8 Let G be a connected graph with cycles. Show that $g(G) \leq 2 \operatorname{diam}(G) + 1$.

11.9 (a) Prove that $f(3, 6) = 14$.

(b) Prove that the Heawood graph is the only 6-cage.

11.10 Let G be an $[r, g]$-graph ($r \geq 2, g \geq 3$) of order $f(r, g)$; that is, let G be an (r, g)-cage. Prove that if $H = G \times K_2$ is an $[s, g]$-graph, then H cannot be an (s, g)-cage.

12

Ramsey theory

Probably the best known and most studied area within extremal graph theory is Ramsey theory. We begin this study with the classical Ramsey numbers.

12.1   CLASSICAL RAMSEY NUMBERS

For positive integers s and t, the *Ramsey number* $r(s, t)$ is the least positive integer n such that for every graph G of order n, either G contains K_s as a subgraph or $\overline{G}$ contains K_t as a subgraph; that is, G contains either s mutually adjacent vertices or an independent set of t vertices. The Ramsey number is named for Frank Ramsey [R1], who studied this concept in a set theoretic framework and essentially verified the existence of Ramsey numbers. Since $\overline{\overline{G}} = G$ for every graph G, it follows that the Ramsey number $r(s, t)$ is symmetric in s and t and $r(s, t) = r(t, s)$.

It is straightforward to show that $r(s, t)$ exists if at least one of s and t does not exceed 2 and that

$$r(1, t) = 1 \quad \text{and} \quad r(2, t) = t.$$

The degree of difficulty in determining the values of other Ramsey numbers increases sharply as s and t increase, and no general values like the above are known.

It is sometimes convenient to investigate Ramsey numbers from an 'edge coloring' point of view. For every graph G of order n, the edge sets of G and $\overline{G}$ partition the edges of K_n. Thus, $r(s, t)$ can be thought of as the least positive integer n such that if every edge of K_n is arbitrarily colored red or blue (where, of course, adjacent edges may receive the same color), then there exists either a complete subgraph of order s, all of whose edges are colored red, or a complete subgraph of order t, all of whose edges are colored blue. In the first case, we say that there is a red K_s; in the second case, a blue K_t. We call the coloring a *red–blue coloring* of K_n. For example, for $t \geq 2$, $r(2, t) > t - 1$ since if all $\binom{t-1}{2}$ edges of K_{t-1} are colored blue, then K_{t-1} contains neither a red K_2 nor a blue K_t. However, $r(2, t) \leq t$ since in an arbitrary red–blue coloring of K_t, either all the edges are blue and we have a blue K_t, or at least one edge is red and we have a red K_2. Thus, $r(2, t) = t$.

Theorem 12.1 gives the value of the first nontrivial Ramsey number $r(3, 3)$.

317

Theorem 12.1

The Ramsey number $r(3,3) = 6$.

Proof

Since neither C_5 nor $\overline{C}_5$ contains K_3 as a subgraph, $r(3,3) \geq 6$. Consider any red–blue coloring of K_6 and let v be a vertex of K_6. Then, v is incident with at least three edges of the same color. Without loss of generality, we may assume that vv_1, vv_2 and vv_3 are red edges. If any of v_1v_2, v_1v_3 and v_2v_3 is a red edge, then there is a red K_3; otherwise, these three edges are blue and we have a blue K_3. Thus, $r(3,3) \leq 6$. Combining the two inequalities, we have $r(3,3) = 6$. □

Before proceeding further, we show that all Ramsey numbers exist and, at the same time, establish an upper bound for $r(s,t)$, which was discovered originally by Erdös and Szekeres [ES2]. In the proof of Theorem 12.2 we use the definition of the Ramsey number directly, rather than the equivalent edge coloring point of view.

Theorem 12.2

For every two positive integers s and t, the Ramsey number $r(s,t)$ exists; moreover,

$$r(s,t) \leq \binom{s+t-2}{s-1}.$$

Proof

We proceed by induction on $k = s + t$. Note that we have equality for $s = 1$ or $s = 2$, and arbitrary t; and for $t = 1$ or $t = 2$, and arbitrary s. Hence the result is true for $k \leq 5$. Furthermore, we may assume that $s \geq 3$ and $t \geq 3$.

Assume that $r(s',t')$ exists for all positive integers s' and t' with $s' + t' < k$, where $k \geq 6$, and that

$$r(s',t') \leq \binom{s'+t'-2}{s'-1}.$$

Let s and t be positive integers such that $s + t = k$, $s \geq 3$ and $t \geq 3$. By the inductive hypothesis, it follows that $r(s-1,t)$ and $r(s,t-1)$ exist, and that

$$r(s-1,t) \leq \binom{s+t-3}{s-2} \quad \text{and} \quad r(s,t-1) \leq \binom{s+t-3}{s-1}.$$

Since

$$\binom{s+t-3}{s-2} + \binom{s+t-3}{s-1} = \binom{s+t-2}{s-1},$$

we have that

$$r(s-1, t) + r(s, t-1) \le \binom{s+t-2}{s-1}. \tag{12.1}$$

Let G be a graph of order $r(s-1, t) + r(s, t-1)$. We show that either G contains K_s as a subgraph or $\overline{G}$ contains K_t as a subgraph. Let $v \in V(G)$. We consider two cases.

Case 1. Assume that $\deg_G v \ge r(s-1, t)$. Thus if S is the set of vertices adjacent to v in G, then either $\langle S \rangle_G$ contains K_{s-1} as a subgraph or $\overline{\langle S \rangle}_G = \langle S \rangle_{\overline{G}}$ contains K_t as a subgraph. If $\langle S \rangle_{\overline{G}}$ contains K_t as a subgraph, then so does $\overline{G}$. If $\langle S \rangle_G$ contains K_{s-1} as a subgraph, then G contains K_s as a subgraph since in G, the vertex v is adjacent to each vertex in S. Hence in this case, $K_s \subseteq G$ or $K_t \subseteq \overline{G}$.

Case 2. Assume that $\deg_G v < r(s-1, t)$. Then $\deg_{\overline{G}} v \ge r(s, t-1)$. Thus if T denotes the set of vertices adjacent to v in $\overline{G}$, then $|T| \ge r(s, t-1)$ and either $\langle T \rangle_G$ contains K_s as a subgraph or $\langle T \rangle_{\overline{G}}$ contains K_{s-1} as a subgraph. It follows, as in Case 1, that either $K_s \subseteq G$ or $K_t \subseteq \overline{G}$.

Since G was an arbitrary graph of order $r(s-1, t) + r(s, t-1)$, we conclude that $r(s, t)$ exists and that

$$r(s, t) \le r(s-1, t) + r(s, t-1). \tag{12.2}$$

Combining (12.1) and (12.2), we obtain the desired result. □

The proof of Theorem 12.2 gives a potentially improved upper bound for $r(s, t)$. This is stated next, together with another interesting fact.

Corollary 12.3

For integers $s \ge 2$ *and* $t \ge 2$,

$$r(s, t) \le r(s-1, t) + r(s, t-1). \tag{12.3}$$

Moreover, if $r(s-1, t)$ *and* $r(s, t-1)$ *are both even, then strict inequality holds in* (12.3).

Proof

The inequality in (12.3) follows from the proof of Theorem 12.2.

In order to complete the proof of the corollary, assume that $r(s-1,t)$ and $r(s,t-1)$ are both even, and let G by any graph of order $r(s-1,t)+r(s,t-1)-1$. We show that either G contains K_s as a subgraph or $\overline{G}$ contains K_t as a subgraph.

Since G has odd order, some vertex v of G has even degree. If $\deg_G v \geq r(s-1,t)$, then, as in Case 1 of Theorem 12.2, either G contains K_s as a subgraph or $\overline{G}$ contains K_t as a subgraph. If, on the other hand, $\deg_G < r(s-1,\ t)$, then $\deg_G v \leq r(s-1,t)-2$ since $\deg_G v$ and $r(s-1,t)$ are both even. But then $\deg_{\overline{G}} v \geq r(s,t-1)$, and we may proceed as in Case 2 of Theorem 12.2. $\qquad\square$

As we have already noted, the bound given in Theorem 12.2 for $r(s,t)$ is exact if one of s and t is 1 or 2. The bound is also exact for $s=t=3$. By Theorem 12.2,

$$r(3,t) \leq \frac{t^2+t}{2}.$$

An improved bound for $r(3,t)$ is now presented.

Theorem 12.4

For every integer $t \geq 3$,

$$r(3,t) \leq \frac{t^2+3}{2}. \tag{12.4}$$

Proof

We proceed by induction on t. For $t=3$, $r(3,t)=6$ while $(t^2+3)/2=6$, so that (12.4) holds if $t=3$. Assume that $r(3,t-1) \leq ((t-1)^2+3)/2$, for some $t \geq 4$, and consider $r(3,t)$. By Corollary 12.3,

$$r(3,t) \leq t + r(3,t-1). \tag{12.5}$$

Moreover, strict inequality holds if t and $r(3,t-1)$ are both even.

Combining (12.5) and the inductive hypothesis, we have

$$r(3,t) \leq t + \frac{(t-1)^2+3}{2} = \frac{t^2+4}{2}. \tag{12.6}$$

To complete the proof, it suffices to show that the inequality given in (12.6) is strict.

If t is odd, then $r(3, t) < (t^2 + 4)/2$ since $t^2 + 4$ is odd. Thus we may assume that t is even. If $r(3, t-1) < ((t-1)^2 + 3)/2$, then clearly the inequality in (12.6) is strict. If, on the other hand, $r(3, t-1) = ((t-1)^2 + 3)/2 = t^2/2 - t + 2$, then $r(3, t-1)$ is even since t is even. Therefore the inequality in (12.5) is strict, which implies the desired result. □

According to Theorem 12.4, $r(3, 4) \leq 9$ and $r(3, 5) \leq 14$. Actually, equality holds in both of these cases. The equality $r(3, 5) = 14$ follows since there exists a graph G of order 13 containing neither a triangle nor an independent set of five vertices; that is, $K_3 \not\subseteq G$ and $K_5 \not\subseteq \overline{G}$. The graph G is shown in Figure 12.1. On the other hand, $r(3, 6) \leq 19$ by Theorem 12.4, but the value of $r(3, 6)$ is 18.

Theorem 12.2 gives an upper bound for the 'diagonal' Ramsey numbers $r(s, s)$, namely $r(s, s) \leq \binom{2s-2}{s-1}$. There are three ways in which lower bounds for $r(s, s)$ have been obtained: the constructive method, a counting method and the probabilistic method. In the constructive method, a lower bound for $r(s, s)$ is established by explicitly constructing a graph G of an appropriate order such that neither G nor $\overline{G}$ contains K_s as a subgraph. Better lower bounds, however, have been obtained using a counting method, which we describe here briefly. (The probabilistic method will be discussed in Chapter 13.) Suppose that we wish to establish the existence of a graph G of order n having some given property P. If we can estimate the number of graphs of order n that do not have property P and we can show that this number is strictly less than the total number of graphs of order n, then there must exist a graph G of order n having property P. Of course, this procedure offers no method for constructing G. In 1947, in one of the first applications of a counting method, Erdös [E2] established the following bound.

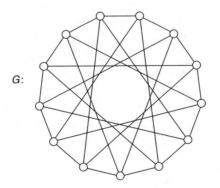

Figure 12.1   An extremal graph showing $r(3, 5) \geq 14$.

Theorem 12.5

For every integer $t \geq 3$,

$$r(t, t) > \lfloor 2^{t/2} \rfloor.$$

Proof

Let $n = \lfloor 2^{t/2} \rfloor$. We demonstrate the existence of a graph G of order n such that neither G nor $\overline{G}$ contains K_t as a subgraph.

There are $2^{\binom{n}{2}}$ distinct labeled graphs of order n with the same vertex set V. For each subset S of V with $|S| = t$, the number of these graphs in which S induces a complete graph is $2^{\binom{n}{2} - \binom{t}{2}}$. Thus, if M denotes the number of graphs with vertex set V that contain a subgraph isomorphic to K_t, then

$$M \leq \binom{n}{t} 2^{\binom{n}{2} - \binom{t}{2}} < \frac{n^t}{t!} 2^{\binom{n}{2} - \binom{t}{2}}. \tag{12.7}$$

By hypothesis, $n \leq 2^{t/2}$. Thus, $n^t \leq 2^{t^2/2}$. Since $t \geq 3$, we have $2^{t^2/2} < (\frac{1}{2}) t! 2^{\binom{t}{2}}$, and so

$$n^t < \left(\frac{1}{2}\right) t! 2^{\binom{t}{2}}. \tag{12.8}$$

Combining (12.7) and (12.8), we conclude that

$$M < \left(\frac{1}{2}\right) 2^{\binom{n}{2}}.$$

If we list the M graphs with vertex set V that contain a subgraph isomorphic to K_t, together with their complements, then there are at most $2M < 2^{\binom{n}{2}}$ graphs in the list. Since there are $2^{\binom{n}{2}}$ graphs with vertex set V, we conclude that there is a graph G with vertex set V such that neither G nor $\overline{G}$ appears in the afore-mentioned list, that is, neither G nor $\overline{G}$ contains a subgraph isomorphic to K_t. $\square$

By Theorems 12.2 and 12.5, we have $4 < r(4, 4) \leq 20$. Actually $r(4, 4) = 18$ (Exercise 12.4); in fact, the only known Ramsey numbers $r(s, t)$ for $3 \leq s \leq t$ are

$$r(3, 3) = 6 \qquad r(3, 6) = 18 \quad r(3, 9) = 36$$
$$r(3, 4) = 9 \qquad r(3, 7) = 23 \quad r(4, 4) = 18$$
$$r(3, 5) = 14 \qquad r(3, 8) = 28 \quad r(4, 5) = 25.$$

EXERCISES 12.1

12.1 Show that if G is a graph of order $r(s,t) - 1$, then

(a) $K_{s-1} \subseteq G$ or $K_t \subseteq \overline{G}$,

(b) $K_s \subseteq G$ or $K_{t-1} \subseteq \overline{G}$.

12.2 If $2 \le s' \le s$ and $2 \le t' \le t$, then prove that $r(s',t') \le r(s,t)$. Furthermore, prove that equality holds if and only if $s' = s$ and $t' = t$.

12.3 Show that $r(3,4) = 9$.

12.4 The accompanying graph has order 17 and contains neither four mutually adjacent vertices nor an independent set of four vertices. Thus, $r(4,4) > 17$. Show that $r(4,4) \le 18$.

12.5 The value of the Ramsey number $r(5,5)$ is unknown. Establish upper and lower bounds (with explanations) for this number.

12.2 GENERALIZED RAMSEY THEORY

For positive integers s_1 and s_2, the classical Ramsey number $r(s_1, s_2)$, discussed in Section 12.1, may be defined as the least positive integer n such that for any factorization $K_n = F_1 \oplus F_2$ (therefore, $F_2 = \overline{F}_1$), either $K_{s_1} \subseteq F_1$ or $K_{s_2} \subseteq F_2$. Defining the Ramsey number in this manner suggests a variety of interesting generalizations. In this section we consider one of these generalizations.

Let $G_1, G_2, \ldots, G_k$ $(k \ge 2)$ be graphs. The (generalized) *Ramsey number* $r(G_1, G_2, \ldots, G_k)$ is the least positive integer n such that for any factorization

$$K_n = F_1 \oplus F_2 \oplus \cdots \oplus F_k,$$

the graph G_i is a subgraph of F_i for at least one $i = 1, 2, \ldots, k$. Hence, $r(K_{s_1}, K_{s_2}) = r(s_1, s_2)$. Furthermore, we denote $r(K_{s_1}, K_{s_2}, \ldots, K_{s_k})$ by $r(s_1, s_2, \ldots, s_k)$. The existence of such Ramsey numbers is guaranteed by the existence of the classical Ramsey numbers, as we now see.

Theorem 12.6

Let the graphs $G_1, G_2, \ldots, G_k$ $(k \ge 2)$ be given. Then the Ramsey number $r(G_1, G_2, \ldots, G_k)$ exists.

Proof

It suffices to show that if $s_1, s_2, \ldots, s_k$ are positive integers, then $r(s_1, s_2, \ldots, s_k)$ exists; for suppose that $G_1, G_2, \ldots, G_k$ have orders $s_1, s_2, \ldots, s_k$, respectively, and that $r(s_1, s_2, \ldots, s_k)$ exists. If $F_1 \oplus F_2 \oplus \cdots \oplus F_k$ is any factorization of the complete graph of order $r(s_1, s_2, \ldots, s_k)$, then $K_{s_i} \subseteq F_i$ for some i, $1 \le i \le k$. Since $G_i \subseteq K_{s_i}$, it follows that $G_i \subseteq F_i$. Thus $r(G_1, G_2, \ldots, G_k)$ exists and $r(G_1, G_2, \ldots, G_k) \le r(s_1, s_2, \ldots, s_k)$.

We proceed by induction on k, where $r(s_1, s_2)$ exists for all positive integers s_1 and s_2 by Theorem 12.2. Assume that $r(s_1, s_2, \ldots, s_{k-1})$ exists ($k \ge 3$) for any $k-1$ positive integers $s_1, s_2, \ldots, s_{k-1}$, and let $s_1, s_2, \ldots, s_k$ be k positive integers. We show that $r(s_1, s_2, \ldots, s_k)$ exists. By the inductive hypothesis, $r(s_1, s_2, \ldots, s_{k-1})$ exists; say $r(s_1, s_2, \ldots, s_{k-1}) = n_0$. Let $r(n_0, s_k) = n$. We now verify that $r(s_1, s_2, \ldots, s_k) \le n$, thereby establishing the required existence.

Let $K_n = F_1 \oplus F_2 \oplus \cdots \oplus F_k$ be an arbitrary factorization of K_n into k factors. We show that $K_{s_i} \subseteq F_i$ for at least one i, $1 \le i \le k$. Let $H = F_1 \oplus F_2 \oplus \cdots \oplus F_{k-1}$; hence $K_n = H \oplus F_k$. Since $r(n_0, s_k) = n$, it follows that $K_{n_0} \subseteq H$ or $K_{s_k} \subseteq F_k$.

Suppose that $K_{n_0} \subseteq H$. Let V_0 be a set of n_0 mutually adjacent vertices of H, and define $F_i' = \langle V_0 \rangle_{F_i}$ for $i = 1, 2, \ldots, k-1$. Since $H = F_1 \oplus F_2 \oplus \cdots \oplus F_{k-1}$, it follows that $K_{n_0} = F_1' \oplus F_2' \oplus \cdots \oplus F_{k-1}'$. However, $r(s_1, s_2, \ldots, s_{k-1}) = n_0$; so $K_{s_i} \subseteq F_i'$ for some i, $1 \le i \le k-1$. Because $F_i' \subseteq F_i$ for all i, $1 \le i \le k-1$, the graph K_{s_i} is a subgraph of F_i for at least one i, $1 \le i \le k-1$.

Hence, we may conclude that $K_{s_i} \subseteq F_i$ for some i, $1 \le i \le k$. □

While it is known that $r(3, 3, 3) = 17$, no other nontrivial numbers of the type $r(s_1, s_2, \ldots, s_k)$, $k \ge 2$, have been evaluated except those mentioned in the preceding section. It may be surprising that there has been considerably more success in evaluating the numbers $r(G_1, G_2, \ldots, G_k)$ when not all the graphs G_i are complete. One of the most interesting results in this direction is due to Chvátal [C6], who determined the Ramsey number $r(T_s, K_t)$, where T_s is an arbitrary tree of order s. This very general result has a remarkably simple proof.

Theorem 12.7

Let T_s be any tree of order $s \ge 1$ and let t be a positive integer. Then

$$r(T_s, K_t) = 1 + (s-1)(t-1).$$

Proof

For $s = 1$ or $t = 1$, $r(T_s, K_t) = 1 = 1 + (s-1)(t-1)$. Thus, we may assume that $s \ge 2$ and $t \ge 2$.

The graph $F = (t-1)K_{s-1}$ does not contain T_s as a subgraph since each component of F has order $s-1$. The complete $(t-1)$-partite graph $\overline{F} = K(s-1, s-1, \ldots, s-1)$ does not contain K_t as a subgraph. Therefore, $r(T_s, K_t) \ge 1 + (s-1)(t-1)$.

Let F be any graph of order $1+(s-1)(t-1)$. We show that $T_s \subseteq F$ or $K_t \subseteq \overline{F}$, implying that $r(T_s, K_t) \le 1+(s-1)(t-1)$ and completing the proof. If K_t is not a subgraph of $\overline{F}$, then $\beta(F) \le t-1$. Therefore, since F has order $1+(s-1)(t-1)$ and $\beta(F) \le t-1$, it follows that $\chi(F) \ge s$ (Exercise 8.1). Let H be a subgraph of F that is critically s-chromatic. By Corollary 8.3, $\delta(H) \ge s-1$. Now applying Theorem 3.8, we have that $T_s \subseteq H$, so that $T_s \subseteq F$. $\square$

For $k \ge 3$, the determination of Ramsey numbers $r(G_1, G_2, \ldots, G_k)$ has proved to be quite difficult, for the most part. For only a very few classes of graphs has any real progress been made. One such example, however, is where each G_i, $1 \le i \le k$, is a star. The following result is by Burr and Roberts [BR3].

Theorem 12.8

Let $s_1, s_2, \ldots, s_k$ ($k \ge 2$) be positive integers, t of which are even. Then

$$r(K_{1, s_1}, K_{1, s_2}, \ldots, K_{1, s_k}) = \sum_{i=1}^{k} (s_i - 1) + \theta_t,$$

where $\theta_t = 1$ if t is positive and even and $\theta_t = 2$ otherwise.

Proof

Let $r(K_{1, s_1}, K_{1, s_2}, \ldots, K_{1, s_k}) = n$, and let $\sum_{i=1}^{k} s_i = N$. First, we show that $n \le N - k + \theta_t$. Since each vertex of K_{N-k+2} has degree $N-k+1 = \sum_{i=1}^{k}(s_i - 1) + 1$, any factorization

$$K_{N-k+2} = F_1 \oplus F_2 \oplus \cdots \oplus F_k$$

necessarily has $K_{1, s_i} \subseteq F_i$ for at least one i, $1 \le i \le k$. Thus, $n \le N - k + 2$. To complete the proof of the inequality $n \le N - k + \theta_t$, it remains to show that $n \le N - k + 1$ if t is positive and even. Observe that, in this case, $N - k + 1$ is odd. Suppose, to the contrary, that there exists a factorization

$$K_{N-k+1} = F_1 \oplus F_2 \oplus \cdots \oplus F_k$$

such that K_{1, s_i} is not a subgraph of F_i for each $i = 1, 2, \ldots, k$. Since each vertex of K_{N-k+1} has degree $N - k = \sum_{i=1}^{k}(s_i - 1)$, this implies that F_i is an $(s_i - 1)$-factor of K_{N-k+1} for each $i = 1, 2, \ldots, k$. However, $N - k + 1$ is odd and $s_j - 1$ is odd for some j ($1 \le j \le k$); thus, F_j contains an odd number of odd vertices, which is impossible.

Next we show that $n \ge N - k + \theta_t$. If $t = 0$, then each integer s_i is odd as is $N - k + 1$. By Theorem 9.21, the complete graph K_{N-k+1} can be factored into $(N-k)/2$ hamiltonian cycles. For each $i = 1, 2, \ldots, k$, let F_i be the union of $(s_i - 1)/2$ of these cycles, so F_i is an $(s_i - 1)$-factor of K_{N-k+1}. Hence there exists a factorization $K_{N-k+1} = F_1 \oplus F_2 \oplus \cdots \oplus F_k$ such that K_{1,s_i} is not a subgraph of F_i, for each i $(1 \le i \le k)$. This implies that $n \ge N - k + 2$ if $t = 0$.

Assume that t is odd. Then $N - k + 1$ is even. By Theorem 9.19, K_{N-k+1} is 1-factorable and therefore factors into $N - k$ 1-factors. For $i = 1, 2, \ldots, k$, let F_i be the union of $s_i - 1$ of these 1-factors, so each F_i is an $(s_i - 1)$-factor of K_{N-k+1}. Thus, there exists a factorization $K_{N-k+1} = F_1 \oplus F_2 \oplus \cdots \oplus F_k$ such that K_{1,s_i} is not a subgraph of F_i for each i. Thus $n \ge N - k + 2$ if t is odd.

Finally, assume that t is even and positive, and suppose that s_1, say, is even. Then there is an odd number of even integers among $s_1 - 1, s_2, \ldots, s_k$, which implies by the previous remark that

$$n \ge r(K_{1,s_1-1}, K_{1,s_2}, \ldots, K_{1,s_k}) \ge N - k + 1.$$

Hence, in all cases $n \ge N - k + \theta_t$, so that $n = N - k + \theta_t$. $\qquad\qquad\square$

For $k = 2$ in Theorem 12.8, we have the following.

Corollary 12.9

Let s and t be positive integers. Then

$$r(K_{1,s}, K_{1,t}) = \begin{cases} s + t - 1 & \text{if } s \text{ and } t \text{ are both even} \\ s + t & \text{otherwise.} \end{cases}$$

EXERCISES 12.2

12.6 Show that $r(3, 3, 3) \le 17$.

12.7 Show for graphs $G_1, G_2, \ldots, G_k$ $(k \ge 2)$ that

$$r(G_1, G_2, \ldots, G_k, K_2) = r(G_1, G_2, \ldots, G_k).$$

12.8 Show for positive integers $t_1, t_2, \ldots, t_k$ $(k \ge 2)$ that

$$r(K_{t_1}, K_{t_2}, \ldots, K_{t_k}, T_s) = 1 + (r - 1)(s - 1),$$

where T_s is any tree of order $s \ge 1$ and $r = r(t_1, t_2, \ldots, t_k)$.

12.9 Let s and t be integers with $s \ge 3$ and $t \ge 1$. Show that

$$r(C_s, K_{1,t}) = \begin{cases} 2t + 1 & \text{if } s \text{ is odd and } s \le 2t + 1 \\ s & \text{if } s \ge 2t. \end{cases}$$

(Note that this does not cover the case where s is even and $s < 2t$.)

12.10 Let G_1 be a graph whose largest component has order s, and let G_2 be a graph with $\chi(G_2) = t$. Prove that $r(G_1, G_2) \ge 1 + (s - 1)(t - 1)$.

12.11   Show for positive integers ℓ and t, that $r(K_\ell + \overline{K}_t, T_s) \leq \ell(s-1) + t$, where T_s is any tree of order $s \geq 1$.

12.12   Let s and t be positive integers, and recall that $a(G)$ denotes the vertex-arboricity of a graph G. Determine a formula for $a(s, t)$, where $a(s, t)$ is the least positive integer n such that for any factorization $K_n = F_1 \oplus F_2$, either $a(F_1) \geq s$ or $a(F_2) \geq t$.

12.3   RAINBOW RAMSEY NUMBERS

In this section we consider an example of a Ramsey-like number. We will investigate these numbers from an edge coloring point of view.

The *rainbow Ramsey number* $RR(G_1, G_2)$ of graphs G_1 and G_2 is the least positive integer n such that if the edges of the complete graph K_n are colored with *any* number of colors the resulting graph contains a subgraph isomorphic to G_1, all of whose edges are the same color, or a subgraph isomorphic to G_2, all of whose edges are different colors. There is no reason to expect $RR(G_1, G_2)$ and $RR(G_2, G_1)$ to be the same number. In fact, as we shall see, one may exist while the other does not.

For simplicity, we say that a graph is *monochromatic* if all of its edges are colored the same color, and we say that a graph is *rainbow* if all of its edges are colored with different colors. Thus $RR(G_1, G_2)$ is the least positive integer n such that if the edges of K_n are colored with any number of colors, the resulting graph contains a monochromatic subgraph isomorphic to G_1 or a rainbow subgraph isomorphic to G_2.

Theorem 12.10 characterizes those pairs G_1 and G_2 for which $RR(G_1, G_2)$ exists. Its proof uses the observation that if $RR(G_1, G_2)$ exists, and H_1 is a subgraph of G_1 and H_2 is a subgraph of G_2, then $RR(H_1, H_2)$ exists and $RR(H_1, H_2) \leq RR(G_1, G_2)$. The proof of Theorem 12.10 follows that given in Eroh [E5]. The existence of rainbow Ramsey numbers was also established in Jamison, Jiang, and Ling [JJL1].

Theorem 12.10

The rainbow Ramsey number $RR(G_1, G_2)$ exists if and only if G_1 is a star or G_2 is a forest.

Proof

We first show that if G_1 is not a star and G_2 is not a forest, then $RR(G_1, G_2)$ does not exist. For any integer n, label the vertices of K_n with the integers $1, 2, \ldots, n$

and color edge ij with color min (i,j). For any color i, every edge of color i is incident with vertex i. Thus any monochromatic subgraph must be a star.

Suppose that K_n contains a rainbow subgraph isomorphic to G_2. Since G_2 is not a forest, it must contain a cycle C_k. Thus, K_n contains a rainbow subgraph isomorphic to C_k. Let C_k: $v_1, v_2, \ldots, v_k, v_1$. We may assume, without loss of generality, that $v_1 \leq v_i$ for each i, $2 \leq i \leq k$. But then by the definition of the coloring, edge $v_k v_1$ and edge $v_1 v_2$ are both colored with the same color v_1. This contradiction implies that there is no rainbow subgraph isomorphic to G_2.

Since this coloring may be used for any integer n, the rainbow ramsey number does not exist if G_1 is not a star and G_2 is not a forest.

The remainder of the proof consists of three cases. Case 1 serves as a lemma for Case 2. Theorem 12.10 follows from Case 2 and Case 3.

Case 1. Assume $G_1 = K_\ell$ and $G_2 = K_{1,m}$ for some positive integer ℓ and m. Let

$$n = \frac{(m-1)^{(\ell-2)(m-1)+2} - 1}{m-2} = \sum_{i=0}^{(\ell-2)(m-1)+1} (m-1)^i.$$

Color the edges of K_n with any number of colors. Choose an arbitrary vertex v_1. If m or more colors appear on the edges incident with v_1, then we have a rainbow copy of $K_{1,m}$. Otherwise, at most $m-1$ colors appear, so there must be at least

$$\frac{n-1}{m-1} = \sum_{i=0}^{(\ell-2)(m-1)} (m-1)^i$$

edges incident with v_1 which are colored with the same color, say color 1. Consider the edges colored 1 and the set W_1 of vertices incident with them.

Now, choose any vertex v_2 in W_1. Again, if m or more colors appear among the edges between v_2 and the other vertices of W_1, then there is a rainbow copy of $K_{1,m}$. Otherwise, at most $m-1$ colors appear, so there exists a set of at least

$$\frac{|W_1| - 1}{m-1} \geq \sum_{i=0}^{(\ell-2)(m-1)-1} (m-1)^i$$

edges having the same color, say color 2, between v_2 and the other vertices of W_1, where the colors 1 and 2 are not necessarily distinct.

Continuing in this fashion, we may assume that we have a sequence v_1, $v_2, \ldots, v_{(\ell-2)(m-1)+2}$ such that every edge $v_i v_j$, for $1 \leq i < j \leq (\ell-2)(m-1)+2$ is colored i, where the colors i and j are not necessarily distinct for $i \neq j$. If m or more of these colors are distinct, say $i_1, i_2, \ldots, i_m$, then the vertex $v_{(\ell-2)(m-1)+2}$ is the central vertex of a rainbow $K_{1,m}$ with end-vertices $v_{i_1}, v_{i_2}, \ldots, v_{i_m}$. Otherwise, there are at most $m-1$ distinct colors appearing in the subgraph. Thus, of the

$(\ell - 2)(m-1) + 1$ colors appearing, there must be a set of at least $\ell - 1$ colors which are identical, say i_1, $i_2, \ldots, i_{\ell-1}$. In this case, the subgraph induced by $v_{i_1}, v_{i_2}, \ldots, v_{i_{\ell-1}}$ and the vertex $v_{(\ell-2)(m-1)+2}$ is a monochromatic complete subgraph of order ℓ.

Case 2: Assume $G_1 = K_\ell$ and G_2 is a tree of order m for some positive integers ℓ and m.

We proceed by induction on the order m of the tree. Since a tree of order 2 or 3 is a star, the base case is included in Case 1.

Suppose for some integer m the rainbow Ramsey number $RR(K_\ell, T)$ exists for any tree T of order $m-1$. Let T' be a tree of order m with an end-vertex v adjacent to a vertex u. Let $RR(K_\ell, T' - v) = t$. From Case 1 we know that the rainbow Ramsey number $RR(K_\ell, K_{1,m-1})$ exists; suppose it is n.

Consider the complete graph K_{nt} of order nt. Suppose the edges are colored with any number of colors. We may divide the vertices of K_{nt} into n disjoint sets of t vertices each. By the inductive hypothesis, the subgraph induced by each set of t vertices contains either a monochromatic copy of K_ℓ or a rainbow copy of $T' - v$. If a monochromatic copy of K_ℓ appears, we are done, so we may assume that we have n rainbow copies of $T' - v$. Let $u_1, u_2, \ldots, u_n$ be the corresponding copies of the vertex u. Now, the graph induced by $u_1, u_2, \ldots, u_n$ contains either a monochromatic copy of K_ℓ or a rainbow copy of $K_{1,m-1}$. If it contains a monochromatic copy of K_ℓ, the proof is complete. Suppose, then, that u_i is the central vertex of a rainbow copy of $K_{1,m-1}$. One of the $m - 1$ colors appearing in this star is different than the $m - 2$ colors appearing in the ith rainbow copy of $T' - v$. Thus, we may add the edge in this color to the ith rainbow copy of $T' - v$ to produce a rainbow copy of T'.

Since the rainbow Ramsey number $RR(G_1, G_2)$ exists when G_1 is complete and G_2 is a tree, the rainbow ramsey number $RR(G_1, G_2)$ exists for any graph G_1 provided G_2 is a forest.

Case 3: Assume $G_1 = K_{1,\ell}$ and $G_2 = K_m$ for some positive integers ℓ and m.

In what follows, for an integer p let

$$p^{(k)} = p(p - 1) \cdots (p - k + 1) = \frac{p!}{(p - k)!}$$

Choose an integer n divisible by $\ell - 1$ so that

$$n \geq 3 + \frac{(\ell - 1)(m + 1)^{(4)}}{8} \tag{12.9}$$

and color the edges of K_n with any number of colors. Assume that there is no monochromatic copy of $K_{1,\ell}$ in K_n. We show there is a rainbow copy of K_m. Notice the total number of different copies of K_m in K_n is $\binom{n}{m}$.

We wish to bound the number of copies of K_m that are not rainbow. First, consider the number of copies of K_m that contain adjacent edges uv and uw which are the same color. There are n choices for the vertex u. Suppose there are a_i edges of color i incident with u, where $1 \le i \le k$. Then $\sum_{i=1}^{k} a_i = n - 1$ where $1 \le a_i \le \ell - 1$ for each i, and the number of different choices for v and w is $\sum_{i=1}^{k} \binom{a_i}{2}$. The maximum occurs when each a_i is as large as possible so there are at most

$$\sum_{i=1}^{(n-1)/(\ell-1)} \binom{\ell - 1}{2} = \frac{n - 1}{\ell - 1}\binom{\ell - 1}{2}$$

choices for v and w. Then there are at most $\binom{n-3}{m-3}$ choices for the remaining vertices of K_m. Thus there are at most

$$n\frac{n - 1}{\ell - 1}\binom{\ell - 1}{2}\binom{n - 3}{m - 3}$$

copies of K_m of this type.

Now consider copies of K_m in which two nonadjacent edges have the same color. There are $\binom{n}{2}$ choices for the first edge, and $n - 2$ ways to choose one end-vertex for the second edge. This vertex is incident with at most $\ell - 1$ edges which are the same color as the first edge and not adjacent to that edge. Since neither the order in which the edges are chosen nor the order in which the end-vertices of the second edge are chosen is important, we are counting each pair of edges at least four times this way. Thus, there are at most $n(n-1)(n-2)(\ell-1)/8$ ways to choose two nonadjacent edges of the same color, and $\binom{n-4}{m-4}$ ways to choose the remaining vertices of K_m. The edge-colored copy of K_n can contain at most

$$\frac{n(n - 1)(n - 2)(\ell - 1)}{8}\binom{n - 4}{m - 4}$$

copies of K_m of this type.

Thus, there are at most

$$n\frac{n - 1}{\ell - 1}\binom{\ell - 1}{2}\binom{n - 3}{m - 3} + \frac{n(n - 1)(n - 2)(\ell - 1)}{8}\binom{n - 4}{m - 3}$$

$$= \binom{n}{m}\left[\frac{(\ell - 2)m^{(3)}}{2(n - 2)} + \frac{(\ell - 1)m^{(4)}}{8(n - 3)}\right]$$

$$< \binom{n}{m}\left[\frac{(\ell - 1)m^{(3)}}{2(n - 3)} + \frac{(\ell - 1)m^{(4)}}{8(n - 3)}\right]$$

$$= \binom{n}{m}\left[\frac{(\ell - 1)m^{(3)}(4 + m - 3)}{8(n - 3)}\right] = \binom{n}{m}\left[\frac{(\ell - 1)(m + 1)^{(4)}}{8(n - 3)}\right] \le \binom{n}{m}$$

nonrainbow copies of K_m in K_n, which means there is at least one rainbow copy. The last inequality follows from (12.9).

Since $RR(K_{1,\ell}, K_m)$ exists, the rainbow Ramsey number $RR(K_{1,\ell}, G_2)$ exists for any graph G_2 of order m. □

Rainbow Ramsey numbers also have a strong relationship with generalized Ramsey numbers, as the next theorem illustrates. In the statement of the theorem, $r(G; m-1) = r(G, G, \ldots, G)$, the generalized Ramsey number for a monochromatic subgraph isomorphic to G when the edges of a complete graph are colored with $m-1$ colors.

Theorem 12.11

For any integer $m \geq 2$ and any graph G,

$$r(G; m - 1) \leq RR(G, mK_2) \leq r(G; m - 1) + 2(m - 1)$$

Proof

For the lower bound with $m-1$ colors, suppose $n = r(G; m-1)$. Consider a coloring of K_{n-1} with $m-1$ colors that does not contain a monochromatic copy of G. Since K_{n-1} is colored with fewer than m colors, it does not contain a rainbow copy of mK_2. Thus $RR(G, mK_2) \geq n$.

For the upper bound, let $M = r(G; m-1) + 2(m-1)$. Color the edges of K_M with any number of colors. If at most $m-1$ colors are used, then there is a monochromatic copy of G. Thus we may assume that at least m colors appear. Choose an edge in, say, color 1 and remove the two vertices incident with this edge. If at most $m-1$ colors are used on the remaining K_{M-2}, then there is a monochromatic copy of G. Thus we may assume that at least m colors appear on the remaining portion of the graph and there must be some edge in a color other than 1. Remove the edges adjacent with this edge. We may repeat this argument until we have removed the vertices incident with $m-1$ edges in $m-1$ different colors, leaving a complete graph of order $r(G; m-1)$. If this complete graph is edge-colored with $m-1$ or fewer colors, then it contains a monochromatic copy of G. Otherwise, it contains at least m colors, including a color distinct from the colors used on the $m-1$ edges already removed. In that case, we have a rainbow copy of mK_2. Thus, $RR(G, mK_2) \leq M$. □

EXERCISES 12.3

12.13 Show that $RR(K_{1,n}, K_{1,m}) \leq (n-1)(m-1) + 2$.

12.14 Show that $RR(K_{1,n}, K_{1,m}) > (n-1)(m-1) + 1$.

12.15 Show that $R(K_{1,n}, mK_2) \geq (n-1)(m-1) + 2$ if n is odd or m is even.

12.16 Show that $RR(K_{1,n}, mK_2) \geq (n-1)(m-1) + 1$ if n is even and m is odd.

12.17 Show that $RR(nK_2, mK_2) \geq m \ (n-1) + 2$.

13

The probabilistic method in graph theory

In this chapter we investigate a powerful, nonconstructive proof technique known as the probabilistic method and then study properties of random graphs.

13.1 THE PROBABILISTIC METHOD

In Chapter 12 we showed that for every integer $t \geq 3$, the Ramsey number $r(t, t) > \lfloor 2^{t/2} \rfloor$. We did so by proving the existence of a graph G of order $n = \lfloor 2^{t/2} \rfloor$ such that neither G nor $\overline{G}$ contains K_t as a subgraph. More specifically, we counted the number of different labeled graphs of order n that contain a subgraph isomorphic to K_t, together with their complements, and showed that there were fewer than $2^{\binom{n}{2}}$ of these graphs. Here we revisit this proof from a probabilistic point of view. Recall that the assignment of the colors red or blue to the edges of a graph G is called a red–blue coloring of G.

Theorem 13.1

For every integer $t \geq 3$,

$$r(t, t) > \left\lfloor 2^{t/2} \right\rfloor.$$

Proof

Let $n = \left\lfloor 2^{t/2} \right\rfloor$. We show that there exists a red–blue coloring of K_n that contains no monochromatic K_t, that is, there is neither a red K_t nor a blue K_t in K_n.

Consider the probability space whose elements are the red–blue colorings of K_n, where $V(K_n) = \{v_1, v_2, \ldots, v_n\}$. The probabilities are defined by setting

$$P[v_i v_j \text{ is red}] = P[v_i v_j \text{ is blue}] = \tfrac{1}{2}$$

for each pair v_i, v_j of distinct vertices of K_n (where $P[E]$ denotes the probability of event E) and letting these events be mutually independent. Thus each of the $2^{\binom{n}{2}}$ red–blue colorings is equally likely with probability $2^{-\binom{n}{2}}$.

For a fixed t-element set $S \subseteq \{v_1, v_2, \ldots, v_n\}$, let A_S denote the event that the subgraph induced by S in K_n is a red K_t or a blue K_t. Then

$$P[A_S] = \left(\frac{1}{2}\right)^{\binom{t}{2}} + \left(\frac{1}{2}\right)^{\binom{t}{2}} = 2^{1-\binom{t}{2}}$$

since the $\binom{t}{2}$ edges joining the vertices of S must be all red or all blue.

Consider the event $\vee A_S$, the disjunction over all t-element subsets S of $\{v_1, v_2, \ldots, v_n\}$. Since there are $\binom{n}{t}$ such subsets,

$$P[\vee A_S] \leq \sum P[A_S] = \binom{n}{t}(2^{1-\binom{t}{2}}) < \left(\frac{n^t}{t!}\right)(2^{1-\binom{t}{2}}).$$

Since $n \leq 2^{t/2}$, we have $n^t \leq 2^{t^2/2}$. Furthermore, since $t \geq 3$, it follows that $2^{t^2/2} < (\frac{1}{2})t!2^{\binom{t}{2}}$. Thus

$$P[\vee A_S] \leq \frac{2^{t^2/2}}{t!}(2^{1-\binom{t}{2}}) < \left(\frac{1}{2}\right)2^{\binom{t}{2}}(2^{1-\binom{t}{2}}) = 1.$$

Since $P[\vee A_S] < 1$, it follows that $P[\overline{\vee A_S}] > 0$, that is, $P[\wedge \overline{A_S}] > 0$. Thus $\wedge \overline{A_S}$ is not the null event and so there is a point in the probability space for which $\wedge \overline{A_S}$ holds. Such a point, however, is a red–blue coloring of K_n with no monochromatic K_t and the proof is complete. $\qquad\square$

The proof of Theorem 13.1 illustrates the basic technique of the probabilistic method. An appropriate probability space is defined on a set of objects (in our case, red–blue colorings of K_n). An event A is then defined representing the desired structure. In the proof of Theorem 13.1, this event $A = \wedge \overline{A_S}$. We then show that A has positive probability so that an object with the desired characteristics or structure must exist.

Before presenting another example of the probabilistic method we introduce some standard terminology. In the proof of Theorem 13.1 we defined a probability space whose objects consisted of all red–blue colorings of K_n in which each such coloring was equally likely. In such a case we refer to a *random red–blue coloring* of K_n.

Our second example of the probabilistic method involves tournaments. A tournament T of order $n \geq 2$ has property S_k ($1 \leq k \leq n-1$) if for every set S of k vertices of T there is a vertex $w \notin S$ such that $(w, v) \in E(T)$ for every v in S, that is, there is a vertex $w \notin S$ that is adjacent to every vertex of S. Using the probabilistic method, we show that for every such integer k there is a tournament T of order n having property S_k for all sufficiently large n.

Theorem 13.2

For every positive integer k and sufficiently large integer n, there is a tournament T of order n with property S_k.

Proof

For a fixed integer n, consider a random tournament T on n vertices. More specifically, consider the probability space whose elements are the $2^{\binom{n}{2}}$ different labeled tournaments T with vertex set $\{v_1, v_2, \ldots, v_n\}$. The probabilities are defined by setting

$$P[(v_i, v_j) \in E(T)] = P[(v_j, v_i) \in E(T)] = \tfrac{1}{2},$$

and then letting these events be mutually independent. Thus each of these $2^{\binom{n}{2}}$ different labeled tournaments is equally likely.

For a fixed k-element set $S \subseteq \{v_1, v_2, \ldots, v_n\}$, let A_S denote the event that there is no $w \in V(T) - S$ that is adjacent to every vertex of S. Each vertex $w \in V(T) - S$ has probability $\left(\tfrac{1}{2}\right)^k$ of being adjacent to every vertex of S, and there are $n - k$ such vertices w, all of whose chances are mutually independent. Thus,

$$P[A_S] = (1 - 2^{-k})^{n-k},$$

and so

$$P[\vee A_S] \le \sum P[A_S] = \binom{n}{k}(1 - 2^{-k})^{n-k}.$$

Thus, if we choose n so that $\binom{n}{k}(1 - 2^{-k})^{n-k} < 1$, then $P[\vee A_S] < 1$. For such an integer n, it follows that $P[\overline{\vee A_S}] = P[\wedge \overline{A_S}] > 0$. Thus there is a point in the probability space for which $\wedge \overline{A_S}$ is true, that is, there exists a tournament T with property S_k. □

Observe again that we have defined an appropriate probability space and event A. This is done so that A has positive probability and, consequently, the desired object (in this case, a tournament of order n with property S_k) exists.

For a probability space $\mathcal{S}$, a *random variable X on $\mathcal{S}$* is a real-valued function on $\mathcal{S}$. The *expected value* $E[X]$ of X is the weighted average

$$E[X] = \sum k P[X = k],$$

where the sum is taken over all possible values k of X. It is easy to see that expectation is *linear*, that is, if X_1, X_2 and X are random variables on a probability space $\mathcal{S}$ and $X = X_1 + X_2$, then $E[X] = E[X_1] + E[X_2]$. Furthermore, if $E[X] = t$, then $X(s_1) \geq t$ and $X(s_2) \leq t$ for some elements s_1 and s_2 of $\mathcal{S}$. This second observation will prove to be very powerful, as indicated in the proof of Theorem 13.3. This result of Szele [S9] is often considered the first use of the probabilistic method.

Theorem 13.3

For each positive integer n there is a tournament of order n with at least $n!2^{-(n-1)}$ hamiltonian paths.

Proof

Consider a random tournament T of order n, and let X be the number of hamiltonian paths in T. For each of the $n!$ permutations σ of $V(T)$, let X_σ be the *indicator random variable* for σ giving a hamiltonian path, that is, X_σ is 1 or 0 depending on whether σ does or does not describe a hamiltonian path in T. Then $P[X_\sigma = 1] = \left(\frac{1}{2}\right)^{n-1}$ (since each of the $n-1$ arcs in the potential hamiltonian path must be correct) and so $E[X_\sigma] = \left(\frac{1}{2}\right)^{n-1}$. Let $X = \sum X_\sigma$, where the summation is taken over all permutations σ of $V(T)$. Then X gives the number of hamiltonian paths in T and

$$E[X] = E\left[\sum X_\sigma\right] = n!2^{-(n-1)}.$$

Hence there is a point in the probability space, namely a specific tournament T, for which X exceeds or equals its expectation. This T has at least $n!2^{-(n-1)}$ hamiltonian paths. □

In our fourth example of the probabilistic method in graph theory, the objects in the sample space under consideration are the vertex subsets of a fixed graph G. Here we obtain an upper bound on the domination number $\gamma(G)$ of G, due to Payan [P1], in terms of the minimum degree of G (Theorem 10.4).

Theorem 13.4

Let G be a graph of order n with $\delta = \delta\ (G) \geq 2$. Then

$$\gamma(G) \leq \frac{n(1 + \ln(\delta + 1))}{\delta + 1}.$$

Proof

Set $p = (\ln(\delta + 1))/(\delta + 1)$ and consider a random set $S \subseteq V(G)$ whose vertices are chosen independently with probability p. That is, consider the probability space whose elements are the 2^n subsets of $V(G) = \{v_1, v_2, \ldots, v_n\}$. The probabilities are assigned by setting $P[v_i \in S] = p$, and letting these events be mutually independent. Thus a subset $S \subseteq V(G)$ occurs with probability $p^{|S|}(1 - p)^{n-|S|}$. For a random set S, let $Y = Y_S$ be the set of vertices not in S having no neighbors in S. Then $S \cup Y_S$ is a dominating set of G. We show that

$$E[|S| + |Y|] \leq \frac{n(1 + \ln(\delta + 1))}{\delta + 1}.$$

Certainly, the expected value of $|S|$ is np. Now, for each $v \in V(G)$,

$$P[v \in Y_S] = P[v \text{ and its neighbors are not in } S] \leq (1 - p)^{\delta + 1}$$

since v has degree at least $\delta = \delta(G)$. Furthermore, since $1 - p \leq e^{-p}$, we have that $P[v \in Y_S] \leq e^{-p(\delta + 1)}$. Since the expected value of a sum of random variables is the sum of their expectations, and since $|Y|$ can be written as a sum of n indicator variables X_v, where $v \in V(G)$, and $X_v = 1$ if $v \in Y$ and $X_v = 0$ otherwise, we conclude that the expected value of $|S| + |Y|$ is at most

$$np + ne^{-p(\delta + 1)} = \frac{n(1 + \ln(\delta + 1))}{\delta + 1}.$$

Thus, for some set S, we have

$$|S| + |Y| \leq \frac{n(1 + \ln(\delta + 1))}{\delta + 1},$$

that is, we have a dominating set $S \cup Y_S$ of G whose cardinality is at most $(n(1 + \ln(\delta + 1)))/(\delta + 1)$. □

One important new idea is involved in the previous proof. The random choice did not give the required dominating set immediately; it gave us a set S which then needed to be altered (in this case, by adding Y_S) to obtain the desired dominating set. The proof of Theorem 13.5 employs the same technique of alteration. This proof also uses Markov's inequality, which states that for a random variable X and positive number t,

$$P[X \geq t] \leq \frac{E[X]}{t}.$$

Theorem 13.5 was first stated as Theorem 8.14 without proof.

Theorem 13.5

For every two integers $k \geq 2$ and $\ell \geq 3$ there exists a k-chromatic graph whose girth exceeds ℓ.

Proof

For $k=2$, any even cycle of length greater than ℓ has the desired properties. Assume then that $k \geq 3$. Let $0 < \theta < 1/\ell$ and, for a fixed positive integer n, let $p = n^{\theta - 1}$. Consider a random graph G of order n whose edges are chosen independently with probability p; that is, consider the probability space whose elements are the $2^{\binom{n}{2}}$ different labeled graphs G with vertex set $\{v_1, v_2, \ldots, v_n\}$. The probabilities are defined by setting $P[v_i v_j$ is an edge of $G] = p$, and then letting these events be mutually independent. Thus each of the different labeled graphs of size m occurs with probability $p^m (1 - p)^{\binom{n}{2} - m}$.

Let X be the random variable that gives the number of cycles of length at most ℓ. For a fixed i, $3 \leq i \leq \ell$, there are $\binom{n}{i}$ i-element subsets of $\{v_1, v_2, \ldots, v_n\}$. For each such set S there are $i!/(2i) = (i-1)!/2$ different cyclic orderings of the vertices in S. Thus there are

$$\binom{n}{i} \frac{(i-1)!}{2} = \frac{n(n-1)\ldots(n-i+1)}{2i}$$

potential cycles of length i, and so X is the sum of $\sum_{i=3}^{\ell} (n(n-1)\ldots(n-i+1))/2i$ indicator variables. Furthermore, since a cycle of length i occurs with probability p^i, the linearity of expectation gives

$$E[X] = \sum_{i=3}^{\ell} \frac{n(n-1)\ldots(n-i+1)}{2i} p^i. \tag{13.1}$$

Since $p = n^{\theta - 1}$, it follows from (13.1) that

$$E[X] \leq \sum_{i=3}^{\ell} \frac{n^{\theta i}}{2i}. \tag{13.2}$$

By the choice of θ, it follows that $\theta \ell = 1 - \varepsilon$ for some real number ε with $0 < \varepsilon < 1$. Thus

$$\frac{E[X]}{n/2} \leq \sum_{i=3}^{\ell} \frac{n^{\theta i}}{ni} \leq \sum_{i=3}^{\ell} \frac{n^{\theta \ell}}{ni} = \frac{K}{n^{\varepsilon}},$$

where $K = \sum_{i=3}^{\ell} 1/i$, and so $\lim_{n \to \infty} E[X]/(n/2) = 0$.

By Markov's inequality,

$$P[X \geq n/2] \leq \frac{E[X]}{n/2}.$$

Thus, for n sufficiently large, $P[X \geq n/2] < 0.5$.

Let $t = \lceil 3(\ln n)/p \rceil$. The probability that a given t-element subset of $\{v_1, v_2, \ldots, v_n\}$ is independent is $(1-p)^{\binom{t}{2}}$. Since there are $\binom{n}{t}$ such sets, it follows that

$$P[\beta(G) \geq t] \leq \binom{n}{t}(1-p)^{\binom{t}{2}}.$$

However, $1 - p < e^{-p}$, and so

$$P[\beta(G) \geq t] < \binom{n}{t}e^{-p\binom{t}{2}} < (ne^{-p(t-1)/2})^t.$$

Since $ne^{-p(t-1)/2} < 1$ for n sufficiently large, it follows that we can choose n so that $P[\beta(G) \geq t] < 0.5$ and $P[X \geq n/2] < 0.5$. For such an n, $P[\beta(G) < t$ and $X < n/2] > 0$. Thus there is a specific graph G with $\beta(G) < t$ of order n having fewer than $n/2$ cycles of length at most ℓ. Since $t = \lceil 3(\ln n)/p \rceil = \lceil 3(\ln n)n^{1-\theta} \rceil$, we may assume that n is sufficiently large to ensure that $t < n/2k$.

Remove one vertex from each cycle of G of length at most ℓ, denoting the resulting graph by G^*. Then G^* has girth greater than ℓ and $\beta(G^*) \leq \beta(G) < n/2k$. Furthermore,

$$\chi(G^*) \geq \frac{|V(G^*)|}{\beta(G^*)} \geq \frac{n/2}{n/2k} = k.$$

Finally, we remove vertices from G^*, if necessary, to produce a graph G^{**} with girth greater than ℓ and $\chi(G^{**}) = k$. □

EXERCISES 13.1

13.1 (a) Show that if $\binom{n}{t}2^{1-\binom{t}{2}} < 1$, then $r(t, t) > n$.

(b) Stirling's formula states that $\lim_{t \to \infty} t!/(t/e)^t = \sqrt{2\pi t}$. Use this fact to prove that

$$r(t, t) > \frac{t2^{t/2}}{e\sqrt{2}}.$$

13.2 (a) Show, without probabilistic techniques, that every graph of order n and size m contains a bipartite subgraph with at least $m/2$ edges.

(b) Give a probabilistic proof that every graph G of order n and size m contains a bipartite subgraph with at least $m/2$ edges. (*Hint*: Consider the probability space whose elements are the 2^n subsets of $V(G) = \{v_1, v_2, \ldots, v_n\}$. For a random set $S \subseteq V(G)$, the probabilities are assigned by setting $P[v_i \in S] = 0.5$ and letting these events be mutually independent. Let X be the random variable defined so that $X(S)$ is the number of edges incident with exactly one vertex of S, and consider the expected value of X.)

13.3 Give a probabilistic proof that there is a red–blue coloring of K_n with at most $\binom{n}{a} \cdot 2^{1-\binom{a}{2}}$ monochromatic copies of K_a.

13.4 Show that if $\binom{n}{s} \cdot 2^{-\binom{s}{2}} + \binom{n}{t} \cdot 2^{-\binom{t}{2}} < 1$, then $r(s, t) > n$.

13.2 RANDOM GRAPHS

In Section 13.1 we found it useful to define appropriate probability spaces in order to prove the *existence* of graphs with desired properties. In this section we give a formal model for a random graph and answer questions about the *probability* that a random graph has certain properties such as nonplanarity or k-connectedness.

For a positive integer n and positive real number p less than 1, the *random graph* $G(n, p)$ denotes the probability space whose elements are the $2^{\binom{n}{2}}$ different labeled graphs with vertex set $\{v_1, v_2, \ldots, v_n\}$. The probabilities are determined by setting $P[v_i v_j \in E(G)] = p$, with these events mutually independent, so that the probability of any specific graph with m edges is $p^m (1 - p)^{\binom{n}{2} - m}$. Although we refer to the 'random graph $G(n, p)$', it is important to remember that we are, in fact, referring to an element selected from the probability space $G(n, p)$.

In this section we discuss properties shared by almost all graphs. Specifically, given a graph theoretic property Q, we say that *almost all graphs* (in $G(n, p)$) have property Q if $\lim_{n \to \infty} P[G \in G(n, p)$ has property $Q] = 1$. A useful technique to establish that almost all graphs have property Q is to define a nonnegative integer-valued random variable X on $G(n, p)$ so that G has property Q if $X = 0$. Then

$$P[X = 0] \leq P[G \in G(n, p) \text{ has property } Q];$$

so that if $\lim_{n \to \infty} P[X = 0] = 1$, then we also know that $\lim_{n \to \infty} P[G \in G(n, p)$ has property $Q] = 1$. Since X is an integer-valued function, $\lim_{n \to \infty} P[X = 0] = 1$ if and only if $\lim_{n \to \infty} P[X \geq 1] = 0$. Using Markov's inequality, we see that since $P[X \geq 1] \leq E[X]$ it follows that if $\lim_{n \to \infty} E[X] = 0$, then $\lim_{n \to \infty} P[X \geq 1] = 0$ and so almost all graphs have property Q.

Our first result shows that for a constant real number p $(0 < p < 1)$, almost all graphs are connected with diameter 2. This strengthens an earlier result of Gilbert [G4] that almost all graphs are connected.

Theorem 13.6

For any fixed positive real number $p < 1$, almost all graphs are connected with diameter 2.

Proof

For each graph G in $G(n, p)$, let the random variable $X(G)$ be the number of (unordered) pairs of distinct vertices of G with no common adjacency. Certainly, if $X(G) = 0$ then G is connected with diameter 2 (or G is the single exception K_n). Thus (by Markov's inequality), it suffices to show that $\lim_{n \to \infty} E[X] = 0$.

List the $\binom{n}{2}$ pairs of vertices of G. Then X can be written as the sum of $\binom{n}{2}$ indicator variables X_i, $1 \le i \le \binom{n}{2}$, where $X_i = 1$ if the ith pair has no common adjacency and 0, otherwise. Then $X = X_1 + X_2 + \cdots + X_{\binom{n}{2}}$ and, by the linearity of expectation, $E[X] = \sum_{i=1}^{\binom{n}{2}} E[X_i]$. If the ith pair is u, v, then $P[X_i = 1]$ is the probability that no other vertex is adjacent to u and v. For a fixed vertex z $(\ne u, v)$, the probability that z is not adjacent to both u and v is $1 - p^2$. This probability is independent of the probability that any other vertex is not adjacent to u and v. Thus the probability that none of the $n - 2$ vertices z $(\ne u, v)$ is adjacent to both u and v is $(1 - p^2)^{n-2}$ and so $E[X_i = 1] = (1 - p^2)^{n-2}$. It follows then that

$$E[X] = \binom{n}{2}(1 - p^2)^{n-2}$$

and, clearly, $\lim_{n \to \infty} E[X] = 0$. ☐

The basic idea used to define the random variable X in the proof of Theorem 13.6 was generalized by Blass and Harary [BH3] in order to study other properties of almost all graphs.

Theorem 13.7

For fixed nonnegative integers k and ℓ and a positive real number $p < 1$, almost all graphs have the property that if S and T are disjoint k-element and ℓ-element subsets of vertices, then there is a vertex $z \notin S \cup T$ that is adjacent to every vertex of S and to no vertex of T.

Proof

Define a pair S, T of disjoint k-element and ℓ-element subsets of $V(G)$ to be *bad* if no vertex $z \notin S \cup T$ is adjacent to every vertex of S and to no vertex of T. For each G in $G(n, p)$, let $X(G)$ be the number of such bad pairs S, T. We wish to show that almost all graphs have no bad pairs of sets and, as in the proof of Theorem 13.6, we need only show that $\lim_{n\to\infty} E[X] = 0$. The variable X can be written as the sum of indicator variables X_i, where $X_i = 1$ if the ith pair S, T is bad, and $X_i = 0$, otherwise. Then $P[X_i = 1] = (1 - p^k(1 - p)^\ell)^{n-k-\ell}$. Since the number of pairs S, T is $N = \binom{n}{k}\binom{n-k}{\ell} = n!/(k!\ell!(n - k - \ell)!)$, it follows that

$$E[X] = \sum_{i=1}^{N} E[X_i] = \frac{n!}{k!\ell!(n - k - \ell)!}(1 - p^k(1 - p)^\ell)^{n-k-\ell}.$$

As n tends to infinity, the first factor in the expression for $E[X]$ tends to infinity (as a polynomial in n) and the second factor tends to 0 exponentially. Thus, $\lim_{n\to\infty} E[X] = 0$, and the proof is complete. □

In the case $k = 2$ and $\ell = 0$, Theorem 13.7 reduces to Theorem 13.6.

For fixed nonnegative integers k and ℓ, let $Q_{k,\ell}$ denote the property that if S and T are disjoint sets of vertices of a graph with $|S| \leq k$ and $|T| \leq \ell$, then there is a vertex $z \notin S \cup T$ that is adjacent to every vertex of S and to no vertex of T.

Corollary 13.8

For fixed nonnegative integers k and ℓ and a positive real number $p < 1$, almost all graphs have property $Q_{k,\ell}$.

If Q_1 and Q_2 are graphical properties such that almost all graphs (in $G(n, p)$) have property Q_1 and almost all graphs have property Q_2, then almost all graphs have both properties Q_1 and Q_2 (Exercise 13.5).

Corollary 13.9

For fixed nonnegative integers k and ℓ and a positive real number $p < 1$, let Q be a graphical property deducible from finitely many applications of Corollary 13.8. Then almost all graphs have property Q.

As an example of the use of Corollary 13.9 to show that almost all graphs have property Q, we prove that for each graph H and fixed real number p $(0 < p < 1)$, almost all graphs (in $G(n, p)$) contain H as an induced subgraph.

Theorem 13.10

For each graph H and fixed positive real number $p < 1$, almost all graphs contain H as an induced subgraph.

Proof

Let $k = |V(H)|$. We proceed by induction on k. For $k = 1$, all graphs contain H as an induced subgraph since $H = K_1$. Assume that for every graph H' of order $k - 1 \geq 1$, almost all graphs contain H' as an induced subgraph, and consider a graph H of order k. Select a vertex v of H and let $H' = H - v$. Then, by the inductive hypothesis, almost all graphs contain H' as an induced subgraph. Furthermore, if v is adjacent to exactly s vertices of H' in H, then since almost all graphs have property $Q_{k,k}$, it follows from Corollary 13.9 that almost all graphs contain H as an induced subgraph. □

If Q is a property like planarity that implies certain graphs (such as K_5 and $K_{3,3}$) do not exist as induced subgraphs, then Theorem 13.10 immediately implies that for p fixed, almost no graph in $G(n, p)$ has property Q. Here, of course, we mean that $\lim_{n \to \infty} P[G \in G(n, p)$ has property $Q] = 0$.

Corollary 13.11

For any fixed positive real number $p < 1$, almost no graphs are planar.

Corollary 13.12

For any positive integer k and fixed positive real number $p < 1$, almost no graphs are k-colorable.

Corollary 13.13

For any positive integer k and fixed positive real number $p < 1$, almost no graphs have genus k.

Other results can be obtained in a manner similar to that used in the proof of Theorem 13.10.

Theorem 13.14

For any fixed positive integer k and positive real number p < 1, almost all graphs are k-connected.

It should be noted that for a fixed real number p ($0 < p < 1$), there are interesting properties of almost all graphs that *cannot* be proved by applying Corollary 13.9. For example, Blass and Harary [BH3] showed that almost all graphs are hamiltonian; however, Corollary 13.9 cannot be used to establish this result.

If p is fixed and X is the random variable defined on $G(n, p)$ by $X(G) = |E(G)|$, then the expected value of X is $p^{\binom{n}{2}}$, and consequently we are dealing with dense graphs. So, in some sense, the preceding results of this section are not surprising. We next briefly consider $G(n, p(n))$, that is, $G(n, p)$ where p is not fixed and $p = p(n)$ is a function of n. We begin with an example involving complete subgraphs.

Let Q be the property that a graph G has clique number $\omega(G) < 4$, and let $p(n)$ be a function of n. For each graph G in $G(n, p(n))$, let the random variable $X(G)$ denote the number of copies of K_4 in G. If $X(G) = 0$, then G has property Q. Thus, by Markov's inequality, if $\lim_{n \to \infty} E[X] = 0$, then almost every graph in $G(n, p)$ has clique number less than 4. For each 4-element subset S of $V(G)$, let X_S be the indicator variable with $X_S = 1$ if $\langle S \rangle$ is complete and $X_S = 0$, otherwise. Then $X = \sum X_S$, where the sum is taken over all 4-element subsets of $V(G)$. Furthermore, $E[X_S] = P[X_S] = (p(n))^6$. By the linearity of expectation, then,

$$E[X] = \sum E[X_S] = \binom{n}{4}(p(n))^6 < n^4(p(n))^6.$$

If $\lim_{n \to \infty} E[X] = 0$, then almost all graphs in $G(n, p(n))$ have clique number less than 4. Consequently, if $\lim_{n \to \infty} p(n)/n^{-2/3} = 0$, then almost all graphs have clique number less than 4. We can think of this result as saying that if $p(n)$ is 'significantly smaller' than $n^{-2/3}$, then almost all graphs in $G(n, p)$ have clique number less than 4. The surprising fact is that, using the second moment method of probability theory, it can be shown that if $\lim_{n \to \infty} p(n)/n^{-2/3} = \infty$, then almost no graph has clique number less than 4. Thus $n^{-2/3}$ can be thought of as a threshold for clique number less than 4. Equivalently, if $\lim_{n \to \infty} p(n)/n^{-2/3} = 0$, then almost no graph G has clique number $\omega(G) \geq 4$ while if $\lim_{n \to \infty} p(n)/n^{-2/3} = \infty$, then almost every graph G has $\omega(G) \geq 4$.

Generally, let Q be a graph theoretic property that is not destroyed by the addition of edges to a graph. A function $r(n)$ is called a *threshold function for Q* if $\lim_{n \to \infty} p(n)/r(n) = 0$ implies that almost no graph has property Q, that is, $\lim_{n \to \infty} P[G \in G(n, p(n))$ has property $Q] = 0$ and $\lim_{n \to \infty} p(n)/r(n) = \infty$ implies that almost every graph has property Q, that is, $\lim_{n \to \infty} P[G \in G(n, p(n))$ has property $Q] = 1$.

Property	Threshold
Contains a path of length k	$r(n) = n^{-(k+1)/k}$
Is not planar	$r(n) = 1/n$
Contains a hamiltonian path	$r(n) = (\ln n)/n$
Is connected	$r(n) = (\ln n)/n$
Contains a copy of K_k	$r(n) = n^{-2(k-1)}$

Figure 13.1   Threshold functions.

Figure 13.1 indicates some of the properties Q for which a threshold function $r(n)$ exists and is known (see [S6, p. 17]).

EXERCISES 13.2

13.5 (a) Show that $P[A \text{ and } B] \geq 1 - (P[\bar{A}] + P[\bar{B}])$.

      (b) Show that if almost all graphs (in $G(n, p)$) have property Q_1 and almost all graphs have property Q_2, then almost all graphs have both properties Q_1 and Q_2.

13.6 Prove Corollary 13.13.

13.7 Prove Theorem 13.14.

13.8 Without using the results given in Figure 13.1, show that if $\lim_{n \to \infty} p(n)/n^{-2/(k-1)} = 0$, then almost no graph in $G(n, p(n))$ contains a copy of K_k.

13.9 For p fixed, let $T\left(n, \frac{1}{2}\right)$ denote the probability space consisting of the $2^{\binom{n}{2}}$ different labeled tournaments T of order n with vertex set $\{v_1, v_2, \ldots, v_n\}$, where the probabilities are defined by setting $P[(v_i, v_j) \in E(T)] = P[(v_j, v_i) \in E(T)] = \frac{1}{2}$. Show that for a fixed positive integer k, almost all tournaments have property S_k (Theorem 13.2).

Glossary of symbols

Symbol	Meaning	Page
$A(G)$	adjacency matrix	1
$\mathrm{Aut}(G)$	automorphism group	38
$a(G)$	vertex-arboricity	66
$a_1(G)$	(edge)-arboricity	68
$B(G)$	incidence matrix	2
C_n	cycle of order n	18
$C(G)$	closure	95
$C_{n+1}(G)$	$(n+1)$-closure	98
$\mathrm{Cen}(G)$	center	20
D	digraph	25
$D(G)$	degree matrix	63
$D_\Delta(G)$	Cayley color graph	43
$d(u, v)$	distance	19
$\deg_G v$ or $\deg v$	degree	2
$\mathrm{diam}\, D$	diameter of a digraph	29
$\mathrm{diam}\, G$	diameter of a graph	20
$E(G)$	edge set of graph	1
$E(D)$	edge set of digraph	25
$e(v)$	eccentricity of a vertex in a digraph	29
$e(v)$	eccentricity of a vertex in a graph	20
$E[X]$	expected value	335
$\mathrm{ex}(n; F)$	extremal number	293
$f(r, g)$	smallest order of an r-regular graph with girth g	308
G	graph	1
$g(G)$	girth	204, 308
$G(n, p)$	random graph	340
$\mathrm{grac}(G)$	gracefulness	255
$\mathrm{id}\, v$	indegree	26
$i(G)$	lower independence number	244

$i(G)$	independent domination number	284
$i_1(G)$	lower edge independence number	245
$IR(G)$	upper irredundance number	289
$ir(G)$	irredundance number	288
K_n	complete graph of order n	7
$K_n{}^*$	complete symmetric digraph of order n	28
$K(r,s)$ or $K_{r,s}$	complete bipartite graph	9
$K(n_1, n_2, \ldots, n_k)$	complete k-partite graph	9
$K_{n_1, n_2, \ldots, n_k}$	complete k-partite graph	9
$k(G)$	number of components	18
$k_0(G)$	number of odd components	237
$M(G)$	matching graph	246
m or $m(G)$	size	1
$\mathrm{Med}(G)$	median	22
n or $n(G)$	order	1
$N(U)$	neighborhood	235
$N[v]$, $N[A]$	closed neighborhood	273, 286
od v	outdegree	26
P_n	path of order n	18
$\mathrm{Per}(G)$	periphery	21
$\mathrm{P}[E]$	probability	334
Q_n	n-cube	10
r	number of regions	128
rad D	radius of a digraph	29
rad G	radius of a graph	20
$r(n)$	threshold function	344
$R(n, k)$	near regular complete multipartite graph	296
$\mathrm{RR}(G_1, G_2)$	rainbow Ramsey number	327
$r(s, t)$	Ramsey number	317
$r(G_1, G_2, \ldots, G_k)$	(generalized) Ramsey number	323
(r, g)-cage	smallest $[r, g]$-graph	308
$[r, g]$-graph	r-regular graph with girth g	308
S_k	surface of genus k	170
$t(G)$	toughness	81
$td(u)$	total distance	22
$V(G)$	vertex set of graph	1
$V(D)$	vertex set of digraph	25
$\alpha(G)$	vertex covering number	241
$\alpha_1(G)$	edge covering number	241
$\beta(G)$	independence number	82
$\beta_1(G)$	edge independence number	211
$\Gamma(G)$	upper domination number	289
$\gamma(G)$	domination number	273

$\gamma(G)$	genus	161
$\gamma_M(G)$	maximum genus	186
$\Delta(G)$	maximum degree	2
$\delta(G)$	minimum degree	2
$\theta_1(G)$	edge-thickness or thickness	158
$\kappa(G)$	vertex-connectivity	70
$\kappa_1(G)$	edge-connectivity	70
$\nu(G)$	crossing number	150
$\bar{\nu}$	rectilinear crossing number	152
$\xi_0(G)$	number of components of odd size	189
$\xi(G)$	minimum $\xi_0(G)$	189
$\chi(G)$	chromatic number	193
$\chi_1(G)$	edge chromatic number or chromatic index	209
$\chi_2(G)$	total chromatic number	215
χ_ℓ	list chromatic number	206
$\chi^*(G)$	region chromatic number	218
$\chi(S_k)$	chromatic number of a surface	229
$\omega(G)$	clique number	199
$\overline{G}$	complement	8
G_k	kth power	105
G^*	symmetric digraph	28
G_d	dual	218
$G_1 = G_2$	isomorphic	3
$G_1 \cup G_2$	union	9
$G_1 + G_2$	join	9
$G_1 \times G_2$	Cartesian product	10
$G_1 \oplus G_2$	factorization, decomposition	246, 252
$H \subseteq G$	subgraph	5
$\langle U \rangle$	subgraph induced by a set U of vertices	6
$\langle X \rangle$	subgraph induced by a set X of edges	6
$G - v$	deletion of a vertex	5
$G - e$	deletion of an edge	5
$G + f$	addition of an edge	6
$\tilde{T}$	associated tournament	115

Graph theory books 1936–2004

1936

König, D. (1936) *Theorie der endlichen und unendlichen Graphen*, Teubner, Leipzig.

1958

Berge, C. (1958) *Théorie des Graphes et Ses Applications*, Dunod, Paris.

1959

Ringel, G. (1959) *Färbungsprobleme auf Flächen und Graphen*, Veb Deutscher Verlag der Wissenschaften, Berlin.

1961

Seshu, S. and Reed, M. B. (1961) *Linear Graphs and Electrical Networks*, Addison-Wesley, Reading, MA.

1962

Berge, C. (1962) *The Theory of Graphs and its Applications*, Methuen, London.
Ford, L. R. and Fulkerson, D. R. (1962) *Flows in Networks*, Princeton University Press, Princeton.
Ore, O. (1962) *Theory of Graphs*, **38** Amer. Math. Soc. Colloq. Pub., Providence, RI.

1963

Ore, O. (1963) *Graphs and Their Uses*, Random House, New York.

1964

Grossman, I. and Magnus, W. (1964) *Groups and Their Graphs*, Random House, New York.

1965

Busacker, R. G. and Saaty, T. L. (1965) *Finite Graphs and Networks: An Introduction With Applications*, McGraw-Hill, New York.
Harary, F., Norman, R. Z. and Cartwright, D. (1965) *Structural Models: An Introduction to the Theory of Directed Graphs*, John Wiley & Sons, New York.

1966

Tutte, W. T. (1966) *Connectivity in Graphs*, University of Toronto Press, London.

1967

Ore, O. (1967) *The Four-Color Problem*, Academic Press, New York.

1968
Moon, J. W. (1968) *Topics on Tournaments*, Holt, Rinehart and Winston, New York.

1969
Harary, F. (1969) *Graph Theory*, Addison-Wesley, Reading, MA.
Knödel, W. (1969) *Graphentheoretische Methoden und ihre Anwendungen*, Springer-Verlag, Berlin.
Nebeský, L. (1969) *Algebraic Properties of Trees*, Universita Karlova, Praha.
Roy, B. and Horps, M. (1969) *Algébre Moderne et Théorie des Graphes Orientées vers les Sciences Èconomique et Sociales*, Dunod, Paris.
Zykov, A. A. (1969) *Teoriia Konechnykh Grafov*, Nauka, Novosibirsk.

1970
Elmaghraby, S. E. (1970) *Some Network Models in Management Science*, Springer-Verlag, Berlin.
Harris, B. (ed.) (1970) *Graph Theory and Its Applications*, Academic Press, New York.
Moon, J. W. (1970) *Counting Labelled Trees, Canadian Mathematical Congress*, Montreal.
Wagner, K. (1970) *Graphentheorie*, Bibliographisches Inst., Mannheim.

1971
Behzad, M. and Chartrand, G. (1971) *Introduction to the Theory of Graphs*, Allyn and Bacon, Boston.
Marshall, C. W. (1971) *Applied Graph Theory*, Wiley-Interscience, New York.
Maxwell, L. M. and Reed, M. B. (1971) *The Theory of Graphs: A Basis for Network Theory*, Pergamon Press, New York.
Nakanishi, N. (1971) *Graph Theory and Feynman Integrals*, Gordon and Breach, New York.
Price, W. L. (1971) *Graphs and Networks: An Introduction*, Auerbach, New York.

1972
Johnson, D. E. and Johnson, J. R. (1972) *Graph Theory With Engineering Applications*, Ronald Press, New York.
Mayeda, W. (1972) *Graph Theory*, Wiley-Interscience, New York.
Read, R. C. (ed.) (1972) *Graph Theory and Computing*, Academic Press, New York.
Sachs, H. (1972) *Einführung in die Theorie der endlichen Graphen II*, Teubner, Leipzig.
Wilson, R. J. (1972) *Introduction to Graph Theory*, Academic Press, New York.

1973
Berge, C. (1973) *Graphs and Hypergraphs*, North-Holland, London.
Dörfler, W. and Mühlbacher, J. (1973) *Graphentheorie für Informatiker*, de Gruyter, Berlin.
Harary, F. and Palmer, E. M. (1973) *Graphical Enumeration*, Academic Press, New York.
White, A. T. (1973) *Graphs, Groups and Surfaces*, North-Holland, Amsterdam.

1974
Biggs, N. (1974) *Algebraic Graph Theory*, Cambridge University Press, Cambridge.
Malkevitch, J. and Meyer, N. (1974) *Graphs, Models, and Finite Mathematics*, Prentice-Hall, Englewood Cliffs, NJ.
Ringel, G. (1974) *Map Color Theorem*, Springer-Verlag, Berlin.
Sache, A. (1974) *La Théorie des Graphes*, Presses Universitaire de France, Paris.

1975

Christofides, N. (1975) *Graph Theory, An Algorithmic Approach*, Academic Press, London.

Cori, R. (1975) *Un Code pour les Graphes Planaires et ses Applications*, Société Mathématique de France, Paris.

Fulkerson, D. R. (ed.) (1975) *Studies in Graph Theory, Parts 1 and 2*, Mathematical Association of America, New York.

1976

Biggs, N. L., Lloyd, E. K. and Wilson, R. J. (1976) *Graph Theory 1736–1936*, Oxford University Press, London.

Bondy, J. A. and Murty, U. S. R. (1976) *Graph Theory with Applications*, North-Holland, New York.

Noltemeier, H. (1976) *Graphentheorie mit Algorithmen und Anwendungen*, de Gruyter, Berlin.

Teh, H. H. and Shee, S. C. (1976) *Algebraic Theory of Graphs*, Lee Kong Chien Institute of Mathematics and Computer Science, Singapore.

Trudeau, R. J. (1976) *Dots and Lines*, Kent State University Press, Kent, OH.

Weisfeiler, B. (1976) *On Construction and Identification of Graphs*, Springer-Verlag, Berlin.

1977

Andrasfai, B. (1977) *Introductory Graph Theory*, Akadémiai Kiado, Budapest.

Chartrand, G. (1977) *Graphs as Mathematical Models*, Prindle, Weber & Schmidt, Boston.

Fiorini, S. and Wilson, R. J. (1977) *Edge-Colourings of Graphs*, Pitman, London.

Giblin, P. J. (1977) *Graphs, Surfaces and Homology*, Chapman & Hall, London.

Graver, J. E. and Watkins, M. E. (1977) *Combinatorics With Emphasis on the Theory of Graphs*, Springer, New York.

Saaty, T. L. and Kainen, P. C. (1977) *The Four-Color Problem: Assaults and Conquests*, McGraw-Hill, New York.

1978

Beineke, L. W. and Wilson, R. J. (eds) (1978) *Selected Topics in Graph Theory*, Academic Press, London.

Bollobás, B. (1978) *Extremal Graph Theory*, Academic Press, New York.

Capobianco, M. and Molluzzo, J. C. (1978) *Examples and Counterexamples in Graph Theory*, North-Holland, New York.

Minieka, E. (1978) *Optimization Algorithms for Networks and Graphs*, Marcel Dekker, New York.

Roberts, F. S. (1978) *Graph Theory and Its Applications to Problems of Society*, SIAM, Philadelphia.

1979

Behzad, M., Chartrand, G. and Lesniak-Foster, L. (1979) *Graphs & Digraphs*, Prindle, Weber & Schmidt, Boston.

Bollobás, B. (1979) *Graph Theory: An Introductory Course*, Springer, New York.

Carrie, B. (1979) *Graphs and Networks*, Clarendon Press, Oxford.

Chachra, V., Ghare, P. M. and Moore, J. M. (1979) *Applications of Graph Theory Algorithms*, North-Holland, New York.

Even, S. (1979) *Graph Algorithms*, Computer Science Press, Potomac, MD.

Wilson, R. J. and Beineke, L. W. (eds) (1979) *Applications of Graph Theory*, Academic Press, London.

1980

Cameron, P. J. and van Lint, J. H. (1980) *Graphs, Codes, and Designs*, Cambridge University Press, Cambridge.

Golumbic, M. C. (1980) *Algorithmic Graph Theory and Perfect Graphs*, Academic Press, New York.

Graham, R. L., Rothschild, B. L. and Spencer, J. H. (1980) *Ramsey Theory*, Wiley Interscience, New York.

Haggard, G. (1980) *Excursions in Graph Theory*, University of Maine, Orono.

1981

Coxeter, H. S. M., Frucht, R., and Powers, D. L. (1981) *Zero-Symmetric Graphs*, Academic Press, New York.

Temperley, H. N. V. (1981) *Graph Theory and Applications*, Halstead Press, New York.

1982

Berge, C. (1982) *The Theory of Graphs and Its Applications*, Greenwood Publishing Group, Westport, CT.

Boffey, T. B. (1982) *Graph Theory in Operations Research*, MacMillan Press, London.

1983

Barnette, D. (1983) *Map Colorings, Polyhedra, and the Four-Color Problem*, Mathematical Association of America, New York.

Beineke, L. W. and Wilson, R. J. (eds) (1983) *Selected Topics in Graph Theory 2*, Academic Press, London.

1984

Aigner, M. (1984) *Graphentheorie: Eine Entwicklung aus der 4-Farben Problem*, Teubner, Stuttgart.

Gondran, M. and Minoux, M. (1984) *Graphs and Algorithms*, Wiley-Interscience, New York.

Mehlhorn, K. (1984) *Graph Algorithms and NP-Completeness*, Springer-Verlag, Berlin.

Mirkin, B. G. and Rodin, S. N. (trans. H. L. Beus) (1984) *Graphs and Genes*, Springer Verlag, Berlin.

Tutte, W. T. (1984) *Graph Theory*, Addison-Wesley, Reading, MA.

Walther, D. (1984) *Ten Applications of Graph Theory*, Kluwer, Hingham, MA.

White, A. T. (1984) *Graphs, Groups and Surfaces*, Revised edn, North-Holland, Amsterdam.

1985

Berge, C. (1985) *Graphs*, Second revised edn, North-Holland, New York.

Bollobás, B. (1985) *Random Graphs*, Academic Press, New York.

Chartrand, G. (1985) *Introductory Graph Theory*, Dover Publications, Mineola, NY.

Fishburn, P. C. (1985) *Interval Orders and Interval Graphs*, John Wiley & Sons, New York.

Gibbons, A. (1985) *Algorithmic Graph Theory*, Cambridge University Press, Cambridge.

Palmer, E. M. (1985) *Graphical Evolution*, John Wiley & Sons, New York.

1986

Chartrand, G. and Lesniak, L. (1986) *Graphs & Digraphs*, 2nd edn, Wadsworth & Brooks/Cole, Menlo Park, CA.

Lovász, L. and Plummer, M. D. (1986) *Matching Theory*, Akademiai Kiado, Budapest.

Yap, H. P. (1986) *Some Topics in Graph Theory*, London Mathematical Society Lecture Notes 108, London.

1987
Aigner, M. (L. Boron, C. Christenson, and B. Smith, translators) *Graph Theory: A Development from the 4-Color Problem*, BCS Associates, Moscow, ID.
Gross, J. L. and Tucker, T. W. (1987) *Topological Graph Theory*, Wiley-Interscience, New York.
Zykov, A. A. (1987) *Osnovy Teorii Grafov*, Nauka, Moscow.

1988
Beineke, L. W. and Wilson, R. J. (eds) (1988) *Selected Topics in Graph Theory 3*, Academic Press, London.
Gould, R. (1988) *Graph Theory*, Benjamin Cummings, Menlo Park, CA.

1989
Berge, C. (1989) *Hypergraphs: Combinatorics of Finite Sets*, Elsevier Science and Technology, San Diego.
Brouwer, A. E., Cohen, A. M. and Neumaier, A. (1989) *Distance Regular Graphs*, Springer-Verlag, Berlin.
Lau, H. T. (1989) *Algorithms on Graphs*, Tab Books, Blue Ridge Summit, PA.

1990
Bosak, J. (1990) *Decompositions of Graphs*, Kluwer Academic, Boston.
Buckley, F. and Harary, F. *Distance in Graphs*, Addison-Wesley, Redwood City, CA.
Fleischner, H. (1990) *Eulerian Graphs and Related Topics*, Part 1, Vol. 1, *Annals of Discrete Mathematics* **45**, North-Holland, Amsterdam.
Graham, R. L., Rothschild, B. L. and Spencer, J. H. (1990) *Ramsey Theory*, 2nd edn, Wiley-Interscience, New York.
Hartsfield, N. and Ringel, G. (1990) *Pearls in Graph Theory: A Comprehensive Introduction*, Academic Press, Boston.
König, D. (R. McCoart, translator) (1990) *Theory of Finite and Infinite Graphs*, Birkhäuser, Boston.
McHugh, J. A. (1990) *Algorithmic Graph Theory*, Prentice-Hall, Englewood Cliffs, NJ.
Ore, O. and Wilson, R. J. (1990) *Graphs and their Uses* (updated), Mathematical Association of America, New York.
Steinbach, P. (1990) *Field Guide to Simple Graphs*, Design Lab, Albuquerque.
Wilson, R. J. and Watkins, J. J. (1990) *Graphs: An Introductory Approach: A First Course in Discrete Mathematics*, John Wiley & Sons, New York.
Zykov, A. A. (1990) *Fundamentals of Graph Theory*, BCS Associates, Moscow, ID.

1991
Clark, J. and Holton, D. A. (1991) *A First Look at Graph Theory*, World Scientific, New Jersey.
Fleischner, H. (1991) Eulerian Graphs and Related Topics, Part 1, Vol. 2, *Annals of Discrete Mathematics* **50**, North-Holland, Amsterdam.
Voss, H.-J. (1991) *Cycles and Bridges in Graphs*, Kluwer Academic, Norweel, MA.

1992
Foulds, L. R. (1992) *Graph Theory Applications*, Springer, New York.

Thulasiraman, K. and Swarmy, M. N. S. (1992) *Graphs: Theory and Applications*, John Wiley & Sons, New York.

1993

Ahuja, R. K., Magnanti, T. L. and Orlin, J. (1993) *Network Flows*, Prentice-Hall, Englewood Cliffs, NJ.

Biggs, N. (1993) *Algebraic Graph Theory*, 2nd edn, Cambridge University Press, Cambridge.

Chartrand, G. and Oellermann, O. R. (1993) *Applied and Algorithmic Graph Theory*, McGraw-Hill, New York.

Holton, D. A. and Sheehan, J. (1993) *The Petersen Graph*, Cambridge University Press, Cambridge.

1994

Hartsfield, N. and Ringel, G. (1994) *Pearls in Graph Theory: A Comprehensive Introduction*, Revised and augmented, Academic Press, Boston.

Trudeau, R. J. (1994) *Introduction to Graph Theory*, Dover Publications, Mineola, NY.

1995

Bonnington, C. P. and Little, C. H. C. (1995) *The Foundations of Topological Graph Theory*, Springer, New York.

Jensen, T. R. and Toft, B. (1995) *Graph Coloring Problems*, John Wiley & Sons, New York.

Lovász, L., Graham, R. L. and Grötschel, M. (eds) (1995) *Handbook of Combinatorics*, Elsevier Science, Amsterdam.

Mahadev, N. V. R. and Peled, U. N. (1995) *Threshold Graphs and Related Topics*, Annals of Discrete Mathematics **56**, North-Holland, Amsterdam.

Prisner, E. (1995) *Graph Dynamics*, Longman, Essex.

1996

Chartrand, G., and Jacobson, M. S. (eds) (1996) *Surveys in Graph Theory*, Congressus Numerantium, Winnipeg.

Chartrand, G., and Lesniak, L. (1996) *Graphs & Digraphs*, 3rd edn, Chapman & Hall, London.

West, D. B. (1996) *Introduction to Graph Theory*, Prentice-Hall, Upper Saddle River, NJ.

Wilson, R. J. (1996) *Introduction to Graph Theory*, 4th edn, Addison-Wesley Longman, Edinburgh Gate, Harlow, England.

1997

Balakrishnan, V. K. (1997) *Schaum's Outline of Graph Theory*, McGraw-Hill, Boston, MA.

Beineke, L. W. and Wilson, R. J. (eds) (1997) *Graph Connections: Relationships Between Graph Theory and Other Areas of Mathematics*, Oxford University Press, Oxford.

Chung, F. R. K. (1997) *Spectral Graph Theory*, American Mathematical Society, Providence, RI.

Scheinerman, E. R. and Ullman, D. H. (1997) *Fractional Graph Theory: A Rational Approach to the Theory of Graphs*, John Wiley & Sons, New York.

Zhang, C.-Q. (1997) *Integer Flows and Cycle Covers of Graphs*, Marcel Dekker, New York.

1998

Asratian, A. S., Denley, Y. M., and Häggkvist, R. (1998) *Bipartite Graphs and Their Applications*, Cambridge University Press, Cambridge.

Chung, F. R. K., and Graham, R. L. (1998) *Erdös on Graphs*, A. K. Peters, Wellesley, MA.

Haynes, T. W., Hedetniemi, S. T. and Slater, P. J. (1998) *Fundamentals of Domination in Graphs*, Marcel Dekker, New York.

Haynes, T. W., Hedetniemi, S. T. and Slater, P. J. (1998) *Domination in Graphs Advanced Topics*, Marcel Dekker, New York.

Read, R. C., and Wilson, R. J. (1998) *An Atlas of Graphs*, Oxford University Press, New York.

Tutte, W. T. (1998) *Graph Theory As I Have Known It*, Clarendon Press, Oxford.

1999

Gross, J. and Yellen, J. (1999) *Graph Theory and Its Applications*, CRC Press, Boca Raton, FL.

Kolchin, V. F. (1999) *Random Graphs*, Cambridge University Press, Cambridge.

McKee, T. A. and McMorris, F. R. (1999) *Topics in Intersection Graph Theory*, SIAM, Philadelphia, PA.

Novak, L. and Gibbons, A. (1999) *Hybrid Graph Theory and Network Analysis*, Cambridge University Press, Cambridge.

2000

Balakrishnan, R. and Ranganathan, K. (2000) *A Textbook of Graph Theory*, Springer, New York.

Diestel, R. (2000) *Graph Theory*, 2nd edn, Springer, New York.

Imrich, W. and Klavzar, S. (2000) *Product Graphs*, John Wiley & Sons, New York.

Janson, S., Luczak, T. and Rucinski, A. (2000) *Random Graphs*, John Wiley & Sons, New York.

Merris, R. (2000) *Graph Theory*, John Wiley & Sons, New York.

Wallis, W. D. (2000) *A Beginner's Guide to Graph Theory*, Birkhäuser, Boston, MA.

2001

Berge, C. (2001) *The Theory of Graphs*, Dover Publications, Mineola, NY.

Bollobás, B. (2001) *Random Graphs*, 2nd edn, Cambridge University Press, Cambridge.

Mohar, B. and Thomassen, C. (2001) *Graphs on Surfaces*, Johns Hopkins University Press, Baltimore.

Ramirez-Alfonsin J. L. and Reed, B. A. (2001) *Perfect Graphs*, Wiley-Interscience, New York.

Tutte, W. T. (2001) *Graph Theory*, Cambridge University Press, Cambridge.

West, D. B. (2001) *Introduction to Graph Theory*, 2nd edn, Prentice-Hall, Upper Saddle River, NJ.

White, A. T. (2001) *Graphs of Groups on Surfaces: Intersections and Models*, North-Holland, Amsterdam.

2002

Arlinghaus, S., Arlinghaus, W. C., and Harary, F. (2002) *Graph Theory and Geography: An Interactive View E-Book*, John Wiley & Sons, New York.

Bang-Jensen, J. and Gutin, G. (2002) *Digraphs*, Springer, New York.

Bollobás, B. (2002) *Modern Graph Theory*, Springer, New York.

Borgelt, C. and Kruse, R. (2002) *Graphical Models: Methods for Data Analysis and Mining*, Wiley, New York.

Molloy, M. and Reed, B. (2002) *Graph Colouring and the Probabilistic Method*, Springer, New York.

Voloshin, V. (2002) *Coloring Mixed Hypergraphs: Theory, Algorithms and Applications*, American Mathematical Society, Providence, RI.

Wilson, R. A. (2002) *Graph Colourings and the Four-Colour Theorem*, Oxford University Press, Oxford.

2003

Bornholdt, S. and Schuster, H. G. (eds) (2003) *Handbook of Graphs and Networks: From the Genome to the Internet*, Wiley, New York.

Buckley, F. and Lewinter, M. (2003) *A Friendly Introduction to Graph Theory*, Prentice-Hall, Upper Saddle River, NJ.

Lauri, J. and Scapellato, R. (2003) *Topics in Graph Automorphisms and Reconstruction*, Cambridge University Press, Cambridge.

Wilson, R. (2003) *Four Colors Suffice: How the Map Problem Was Solved*, Princeton University Press, Princeton, NJ.

2004

Chartrand, G., and Lesniak, L. (2004) *Graphs & Digraphs*, 4th edn, CRC Press, Boca Raton, FL.

Cvetkovic, D., Rowlinson, P. and Simic, S. (2004) *Spectral Generalizations of Line Graphs*, Cambridge University Press, Cambridge.

Golumbic, M. and Trenk, A. (2004) *Tolerance Graphs*, Cambridge University Press, Cambridge.

Gross, J. and Yellen, J. (eds) (2004) *Handbook of Graph Theory*, CRC Press, Boca Raton, FL.

Lando, S. K. and Zvonkin, A. K. (2004) *Graphs on Surfaces and Their Applications*, Springer, New York.

Marchette, D. R. (2004) *Random Graphs for Statistical Pattern Recognition*, Wiley, New York.

Watkins, J. J. (2004) *Across the Board: The Mathematics of Chess Problems*, Princeton University Press, Princeton, NJ.

References

Each item in the bibliography is followed by one or more numbers in square brackets indicating the pages on which the item is referenced. The starred entries denote books.

Acharya, B. D.
        See [WAS1].

Alekseev, V. B.
[AG1]   (with V. S. Gonchakov) Thickness of arbitrary complete graphs. *Mat. Sbornik* **101** (1976) 212–30.   [158]

Anderson, I.
[A1]   Perfect matchings of a graph. *J. Combin. Theory* **10B** (1971) 183–6.   [237]

Appel, K.
[AHK1]   (with W. Haken and J. Koch) Every planar map is four-colorable. *Illinois J. Math.* **21** (1977) 429–567.   [221]
        See also [221].

Arumugam, S.
        See [JA1].

Battle, J.
[BHKY1]   (with F. Harary, Y. Kodama and J. W. T. Youngs) Additivity of the genus of a graph. *Bull. Amer. Math. Soc.* **68** (1962) 565–8.   [167]

Behzad, M.
[B1]   *Graphs and Their Chromatic Numbers*. Doctoral thesis, Michigan State University (1965).   [215]

Beineke, L. W.
[B2]   The decomposition of complete graphs into planar subgraphs, in *Graph Theory and Theoretical Physics*. Academic Press, New York (1967) 139–54.   [158]
[B3]   Derived graphs and digraphs, in *Beiträge zur Graphentheorie*. Teubner, Leipzig (1968) 17–33.   [108]
[BH1]   (with F. Harary) The genus of the n-cube. *Canad. J. Math.* **17** (1965) 494–6.   [168]

[BH2]   (with F. Harary) The thickness of the complete graph. *Canad. J. Math.* **17** (1965)
            850–9.   [158]
[BHM1]   (with F. Harary and J. W. Moon) On the thickness of the complete bipartite
            graph. *Proc. Cambridge Philos. Soc.* **60** (1964) 1–5.   [158]
[BR1]   (with R. D. Ringeisen) On the crossing number of products of cycles and graphs
            of order four. *J. Graph Theory* **4** (1980) 145–55.   [157]
[BW1]   (with R. J. Wilson) On the edge-chromatic number of a graph. *Discrete Math.*
            5 (1973) 15–20.   [211]
            See also [RB1].

Berge, C.
[B4]   Two theorems on graph theory. *Proc. Nat. Acad. Sci. USA* **43** (1957) 842–4.
                                                                                      [234, 239]
[B5]   Förbung von Graphen, deren sämtliche bzw. deren ungeraden Kreise starr sind.
            Wissenschaftliche Zeitung, Martin Luther Univ., Halle Wittenberg **114** (1961).
                                                                                           [203]
*[B6]   *The Theory of Graphs and its Applications.* Methuen, London (1962). [56, 274]
*[B7]   *Graphs and Hypergraphs.* North-Holland, London (1973).
                                                                        [99, 203, 278, 279, 284]

Bielak, H.
[BS1]   (with M. M. Syslo) Peripheral vertices in graphs. *Studia Sci. Math. Hungar.*
            18 (1983) 269–75.   [1, 23]

Biggs, N. L.
[BLW1]   (with E. K. Lloyd and R. J. Wilson) *Graph Theory 1736–1936.* Oxford University
            Press, London (1976).   [85, 92]

Blass, A.
[BH3]   (with F. Harary) Properties of almost all graphs and complexes. *J. Graph Theory*
            3 (1979) 225–40.   [341, 344]

Blažek, J.
[BK1]   (with M. Koman) A minimal problem concerning complete plane graphs, in
            Theory of Graphs and its Applications. Academic Press, New York (1964)
            113–7.   [150]

Bollobás, B.
[BC1]   (with E. J. Cockayne) Graph-theoretic parameters concerning domination,
            independence, and irredundance. *J. Graph Theory* **3** (1979) 241–9.   [285]
[BC2]   (with E. J. Cockayne) The irredundance number and maximum degree of a
            graph. *Discrete Math.* **49** (1984) 197–9.   [277]

Bondy, J. A.
[B8]   Pancyclic graphs. *J. Combin. Theory* **11B** (1977) 80–4.   [101]
[B9]   Small cycle double covers of graphs. *Cycles and Rays.* Kluwer Academic Publishers,
            Dordrecht (1990) 21–40.   [91]
[BC3]   (with V. Chvátal) A method in graph theory. *Discrete Math.* **15** (1976) 111–36.
                                                                                      [95, 96, 98]

Brandt, S.
[B10]   Subtrees and subforests of graphs. *J. Combin. Theory* **61B** (1994) 63–70. [307]

Brooks, R. L.
[B11]   On coloring the nodes of a network. *Proc. Cambridge Philos. Soc.* **37** (1941) 194–7.   [196]

Brualdi, R. A.
[BC4]   (with J. Csima) A note on vertex- and edge-connectivity. *Bull. Inst. Comb. Appl.* **2** (1991) 67–70.   [70]

Bruck, R. H.
[BR2]   (with H. Ryser) The nonexistence of certain finite projective planes. *Canad. J. Math.* **1** (1949) 88–93.   [254]

Buckley, F.
[BMS1]   (with Z. Miller and P. J. Slater) On graphs containing a given graph as center. *J. Graph Theory* **5** (1981) 427–34.   [22]

Burr, S. A.
[B12]   An inequality involving the vertex arboricity and edge arboricity of a graph. *J. Graph Theory* **10** (1986) 403–4.   [68]
[BR3]   (with J. A. Roberts) On Ramsey numbers for stars. *Utilitas Math.* **4** (1973) 217–20.   [325]

Camion, P.
[C1]   Chemins et circuits hamiltoniens des graphes complets. *C. R. Acad. Sci. Paris* **249** (1959) 2151–2.   [125]

Caro, Y.
            See [254].

Cayley, A.
[C2]   On the colouring of maps. *Proc. Royal Geog. Soc.* **1** (1879) 259–61.   [220]
[C3]   A theorem on trees. *Quart. J. Math.* **23** (1889) 376–8. Collected papers, Cambridge, **13** (1892) 26–8.   [60]
            See also [220].

Chartrand, G.
[CH1]   (with F. Harary) Graphs with prescribed connectivities, in *Theory of Graphs: Proceedings of the Colloquium Held at Tihany*, Hungary. Budapest (1968) 61–3.
                                                                                            [72]
[CHO1]   (with H. Hevia and O. R. Oellermann) The chromatic number of a factorization of a graph. *Bull. Inst. Combin. Appl.*, **20** (1997), 33–56.   [266]
[CHJKN1]   (with A. M. Hobbs, H. A. Jung, S. F. Kapoor and C. St. J. A. Nash-Williams) The square of a block is hamiltonian-connected. *J. Combin. Theory* **16B** (1974) 290–2.   [106]
[CKLS1]   (with S. F. Kapoor, L. Lesniak and S. Schuster) Near 1-factors in graphs. *Congress. Numer.* **41** (1984) 131–47.   [240]

[CKOR1]   (with S. F. Kapoor, O. R. Oellermann and S. Ruiz) On maximum matchings in
          cubic graphs with a bounded number of bridge-covering paths. *Bull. Austral.
          Math. Soc.* **36** (1986) 441–7.   [241]
[CK1]   (with H. V. Kronk) The point arboricity of planar graphs. *J. London Math. Soc.*
          44 (1969) 612–6.   [67]
[CPS1]   (with A. D. Polimeri and M. J. Stewart) The existence of 1-factors in line graphs,
          squares, and total graphs. *Indag. Math.* **35** (1973) 228–32.   [88, 253]

Chetwynd, A. G.
[CH2]   (with A. J. Hilton) Star multigraphs with three vertices of maximum degree.
          Math. Proc. Cambridge Philos. Soc. **100** (1986) 303–17.   [212]

Chudnovsky, M.
[CRST1]   (with N. Robertson, P. Seymour and R. Thomas) The strong perfect graph
          theorem. Preprint.   [203]

Chvátal, V.
[C4]   On Hamilton's ideals. *J. Combin. Theory* **12B** (1972) 163–8.   [96]
[C5]   Tough graphs and Hamiltonian circuits. *Discrete Math.* **5** (1973) 215–28.
                                                                                      [82]
[C6]   Tree-complete graph ramsey numbers. *J. Graph Theory* **1** (1977) 93.   [324]
[CE1]   (with P. Erdös) A note on hamiltonian circuits. *Discrete Math.* **2** (1972) 111–3.
                                                                                      [97]
          See also [BC3].

Cockayne, E. J.
[CH3]   (with S. T. Hedetniemi) Towards a theory of domination. *Networks* **7** (1977)
          247–61.   [274, 289]
[CFPT1]   (with O. Favaron, C. Payan and A. Thomason) Contributions to the theory of
          domination, independence and irredundance in graphs. *Discrete Math.* **33** (1981)
          249–58.   [290]
          See also [BC1], [BC2] and [280].

Csima, J.
          See [BC4].

de Jaenisch, C. F.
*[D1]   *Applications de l'Analyse Mathematique an Jenudes Echecs. Petrograd* (1862).
                                                                                      [273]

De Morgan, A.
          See [216], [217] and [220].

Descartes, B.
[D2]   A three colour problem. *Eureka* **9** (1947) 21.   [203]

Dirac, G. A.
[D3]   A property of 4-chromatic graphs and some remarks on critical graphs. *J. London
          Math. Soc.* **27** (1952) 85–92.   [194, 205]

[D4]   Some theorems on abstract graphs. *Proc. London Math. Soc.* **2** (1952) 69–81.
                                                                                    [97]
[D5]   In abstrakten Graphen vorhande vollstöndige 4-Graphen und ihre unterteilungen.
            Math. Nachr. **22** (1960) 61–85.   [78]
[D6]   On rigid circuit graphs. *Abh. Math. Sem. Univ. Hamburg* **25** (1961) 71–6.
                                                                                    [201]
[D7]   Homomorphism theorems for graphs. *Math. Ann.* **153** (1964) 69–80.
                                                                                    [298]
[D8]   Short proof of Menger's graph theorem. *Mathematika* **13** (1966) 42–4.   [75]
[DS1]   (with S. Schuster) A theorem of Kuratowski. *Nederl. Akad. Wetensch. Proc.* Ser. A
            57 (1954) 343–8.   [137]
            See also [299].

Duke, R. A.
[D9]   The genus, regional number, and Betti number of a graph. *Canad. J. Math.*
            18 (1966) 817–22.   [183]

Dyck, W.
[D10]   Beiträge zur Analysis Situs. *Math. Ann.* **32** (1888) 457–512.   [175]

Edmonds, J.
[E1]   A combinatorial representation for polyhedral surfaces. *Notices Amer. Math. Soc.*
            7 (1960) 646.   [175]

Eggleton, R. B.
[EG1]   (with R. K. Guy) The crossing number of the n-cube. *Notices Amer. Math. Soc.*
            17 (1970) 757.   [155]

Elias, P.
[EFS1]   (with A. Feinstein and C. E. Shannon) A note on the maximum flow through a
            network. *IRE Trans. Inform. Theory* **IT-2** (1956) 117–9.   [79]

Enomoto, H.
[EJKS1]   (with B. Jackson, P. Katernis and A. Saito) Toughness and the existance of
            k-factors. *J. Graph Theory* **9** (1985) 87–95.   [239]

Erdös, P.
[E2]   Some remarks on the theory of graphs. *Bull. Amer. Math. Soc.* **53** (1947) 292–4.
                                                                                    [321]
[E3]   Graph theory and probability II. *Canad. J. Math.* **13** (1961) 346–52.   [204]
[E4]   Extremal problems in graph theory, in *A Seminar in Graph Theory.* Holt, Rinehart
            & Winston, New York (1967) 54–9.   [299, 300]
[EG2]   (with T. Gallai) On maximal paths and circuits of graphs. *Acta Math. Acad. Sci.
            Hungar.* **10** (1959) 337–56.   [305]
[EG3]   (with T. Gallai) Graphs with prescribed degrees of vertices (Hungarian). *Mat.
            Lapok* **11** (1960) 264–74.   [14]
[EK1]   (with P. J. Kelly) The minimal regular graph containing a given graph, in *A
            Seminar on Graph Theory* (F. Harary, ed.). Holt, Rinehart & Winston, New York
            (1967) 65–9.   [8]

[ES1]   (with H. Sachs) Reguläre Graphen gegebener Taillenweite mit minimaler
           Knotenzahl. *Wiss. Z. Univ. Halle, Math-Nat.* **12** (1963) 251–8.   [309]
[ES2]   (with G. Szekeres) A combinatorial problem in geometry. *Compositio Math.*
           2 (1935) 463–70.   [318]
[EW1]   (with R. J. Wilson) On the chromatic index of almost all graphs. *J. Combin.*
           Theory **23B** (1977) 255–7.   [211]
           See also [CE1] and [305].

Eroh, L.
[E5]   Rainbow Ramsey Numbers. Doctoral thesis, Western Michigan University
           (2000).   [327]

Errera, A.
[E6]   Du colorage des cartes. *Mathesis* **36** (1922) 56–60.   [241]

Euler, L.
[E7]   Solutio problematis ad geometriam situs pertinentis. *Comment. Academiae Sci. I.*
           Petropolitanae **8** (1736) 128–40.   [85, 87]
[E8]   Demonstratio nonnullarum insignium proprietatum quibus solida hedris
           planis inclusa sunt praedita. *Novi Comm. Acad. Sci. Imp. Petropol.* **4** (1758)
           140–60.   [128]
           See also [127].

Fáry, I.
[F1]   On the straight line representations of planar graphs. *Acta Sci. Math.* **11** (1948)
           229–33.   [131]

Feinstein, A.
           See [EFS1].

Fink, J. F.
[F2]   Random Factors and Isofactors in Graphs and Digraphs. Doctoral thesis, Western
           Michigan University (1982).   [255]
[FR1]   (with S. Ruiz) Every graph is an induced isopart of a circulant. *Czech. Math. J.* **36**
           (1986) 172–6.   [256]

Fleischner, H.
[F3]   The square of every two-connected graph is hamiltonian. *J. Combin. Theory*
           16B (1974) 29–34.   [106]

Ford, L. R., Jr.
[FF1]   (with D. R. Fulkerson) Maximal flow through a network. *Canad. J. Math.*
           8 (1956) 399–404   [79]

Frucht, R.
[F4]   Herstellung von Graphen mit vorgegebener abstrakten Gruppe. *Compositio Math.*
           6 (1938) 239–50.   [45]

Fulkerson, D. R.
        See [FF1].

Gaddum, J. W.
        See [NG1].

Gallai, T.
[G1]   Über extreme Punkt- und Kantenmengen. *Ann. Univ. Sci. Budapest, Eötvös Sect. Math.* **2** (1959) 133–8.   [241]
[G2]   On directed paths and circuits, in *Theory of Graphs*. Academic Press, New York (1968) 115–8.   [199]
        See also [EG2] and [EG3].

Ghouila-Houri, A.
[G3]   Une condition suffisante d'existence d'un circuit Hamiltonien. *C. R. Acad. Sci. Paris* **156** (1960) 495–7.   [103]

Gilbert, E. N.
[G4]   Random graphs. *Ann. Math. Stat.* **30** (1959) 1141–4.   [341]

Goddard, W.
[GHS1]   (with M. A. Henning and H. C. Swart) Some Nordhaus-Gaddum-type results. *J. Graph Theory* **16** (1992) 221–31.   [281]

Golomb, S. W.
[G5]   How to number a graph, in *Graph Theory and Computing*. Academic Press, New York (1972) 23–37.   [266]

Gonchakov, V. S.
        See [AG1].

Graham, R. L.
[GS1]   (with N. J. A. Sloane) On additive bases and harmonious graphs. *SIAM J. Alg. Disc. Meth.* **1** (1980) 382–404.   [268, 269]

Grinberg, E. J.
[G6]   Plane homogeneous graphs of degree three without hamiltonian circuits. *Latvian Math. Yearbook* **4** (1968) 51–8.   [146]
        See also [149].

Guthrie, Francis
        See [217] and [219].

Guthrie, Frederick
[G7]   Note on the colouring of maps. *Proc. Royal Soc. Edinburgh* **10** (1880) 727–8.   [217]
        See also [217].

Guy, R. K.
[G8]   A combinatorial problem. *Bull. Malayan Math. Soc.* **7** (1960) 68–72.   [150]
[G9]   Crossing number of graphs, in *Graph Theory and Applications*. Springer-Verlag, New York (1972) 111–4.   [151, 152]
            See [EG1].

Hadwiger, H.
[H1]   Über eine Klassifikation der Streckenkomplexe. *Vierteljschr. Naturforsch Ges Zürich* **88** (1943) 133–42.   [205]

Hajnal, A.
[HS1]   (with J. Surányi) Über die Auflösung von Graphen in vollstöndige Teilgraphen. *Ann. Univ. Sci. Budapest, Eötvös Sect. Math.* **1** (1958) 113–21.   [201]

Hajós, G.
[H2]   Über einen Satz von K. Wagner zum Vierfarbenproblem. *Math. Ann.* **153** (1964) 47–62.   [205]

Haken, W.
            See [AHK1] and [221].

Hakimi, S. L.
[H3]   On the realizability of a set of integers as degrees of the vertices of a graph. *J. SIAM Appl. Math.* **10** (1962) 496–506.   [12]

Halin, R.
[H4]   Bemerkungen über ebene Graphen. *Math. Ann.* **53** (1964) 38–46.   [143]

Hall, P.
[H5]   On representation of subsets. *J. London Math. Soc.* **10** (1935) 26–30.   [235, 236]

Hamilton, W. R.
            See [92–93] and [217].

Harary, F.
*[H6]   *Graph Theory.* Addison-Wesley, Reading, MA (1969).   [56]
[HKS1]   (with P. C. Kainen and A. J. Schwenk) Toroidal graphs with arbitrarily high crossing numbers. *Nanta Math.* **6** (1973) 58–67.   [155]
[HM1]   (with L. Moser) The theory of round robin tournaments. *Amer. Math. Monthly* **73** (1966) 231–46.   [121, 125]
[HN1]   (with C. St. J. A. Nash-Williams) On eulerian and hamiltonian graphs and line graphs. *Canad. Math. Bull.* **8** (1965) 701–10.   [108]
[HN2]   (with R. Z. Norman) The dissimilarity characteristic of Husimi trees. *Ann. of Math.* **58** (1953) 134–41.   [37]
[HT1]   (with W. T. Tutte) A dual form of Kuratowski's Theorem. *Canad. Math. Bull.* **8** (1965) 17–20, 373.   [143]
            See also [BHKY1], [BH1], [BH2], [BHM1], [BH3], [CH1] and [48].

Havel, V.
[H7]   A remark on the existence of finite graphs (Czech.) *časopispěst. Mat.* **80** (1955)
            477–80.   [12]

Heawood, P. J.
[H8]   Map-colour Theorem. *Quart. J. Math.* **24** (1890) 332–9.   [220, 229]

Hedetniemi, S. T.
            See [CH3].

Heffter, L.
[H9]   Über das Problem der Nachbargebiete. *Math. Ann.* **38** (1891) 477–508.
                                                                                 [175, 230]

Hendry, G. R. T.
[H10]   On graphs with prescribed median I. *J. Graph Theory* **9** (1985) 477–81.   [22]

Henning, M. A.
            See [GHS1].

Hevia, H.
            See [CHO1].

Hierholzer, C.
[H11]   Über die Möglichkeit, einen Linienzug ohne Wiederholung und ohne
            Unterbrechnung zu umfahren. *Math. Ann.* **6** (1873) 30–2.   [87]

Hilton, A. J.
[H12]   Recent progress in edge-colouring graphs. *Discrete Math.* **64** (1987) 303–7.
                                                                                      [212]

            See also [CH2].

Hobbs, A. M.
            See [CHJKN1].

Hoffman, A. J.
[HS2]   (with R. R. Singleton) On Moore graphs with diameters two and three. *IBM J.
            Res. Develop.* **4** (1960) 497–504.   [312]

Holton, D. A.
*[HS3]   (with J. Sheehan) *The Petersen Graph*. Cambridge University Press (1993).
                                                                                 [308, 313]

Isaacs, R.
[I1]   Infinite families of nontrivial trivalent graphs which are not Tait colorable. *Amer.
            Math. Monthly* **82** (1975) 221–39.   [212]

Izbicki, H.

[I2]   Reguläre Graphen beliebigen Grades mit Vorgegebenen Eigenschaften. *Monatsh. Math.* **64** (1960) 15–21.   [46]

Jackson, B.

[J1]   Hamilton cycles in regular 2-connected graphs. *J. Combin. Theory* **29B** (1980) 27–46.   [97]
        See also [EJKS1].

Jaeger, F.

[J2]   On nowhere-zero flows in multigraphs. *Congress. Numer.* **15**, Winnipeg (1976) 373–8.   [229]

[JP1]   (with C. Payan) Relations du type Nordhaus-Gaddum pour le nombre d'absorption d'un graphe simple. *C. R. Acad. Sci. Ser. A.* **274** (1972) 728–30.   [280]

Jamison, R. E.

[JJL1] (with T. Jiang and A. C. H. Ling) Constrained Ramsey numbers on graphs. *J. Graph Theory* **42** (2003), 1–16.   [327]

Jones, D. M.

[JRS1] (with D. J. Roehm and M. Schultz) On matchings in graphs. *Ars Combin.* **50** (1998), 65–79   [245]

Joseph, J. P.

[JA1]   (with S. Arumugam) A note on domination in graphs. Preprint.   [282]

Jung, H. A.

        See [CHJKN1].

Jungerman, M.

[J3]   A characterization of upper embeddable graphs. *Trans. Amer. Math. Soc.* **241** (1978) 401–6.   [189]

Kainen, P. C.

        See [HKS1] and [153].

Kapoor, S. F.

        See [CHJKN1], [CKLS1] and [CKOR1].

Karaganis, J. J.

[K1]   On the cube of a graph. *Canad. Math. Bull.* **11** (1968) 295–6.   [105]

Katernis, P.

        See [EJKS1].

Kelly, J. B.

[KK1]   (with L. M. Kelly) Paths and circuits in critical graphs. *Amer. J. Math.* **76** (1954) 786–92.   [203]

Kelly, L. M.
    See [KK1].

Kelly, P. J.
    See [EK1] and [47–48].

Kempe, A. B.
[K2]   On the geographical problem of the four colors. *Amer. J. Math.* **2** (1879) 193–200.   [220]
[K3]   How to color a map with four colors. *Nature* **21** (1880) 399–400.   [220]
    See also [220–221].

Kirchhoff, G.
[K4]   Über die Auflösung der Gleichungen, auf welche man bei der Untersuchung der linearen Verteilung galvanischer Ströme geführt wird. *Ann. Phys. Chem.* **72** (1847) 497–508. Gesammelte Abhandlungen, Leipzig (1882) 22–3.   [63]

Kirkman, T. P.
[K5]   On a problem in combinations. *Cambridge and Dublin Math. J.* **2** (1847) 191–204.   [254]
    See also [93].

Kleinert, M.
[K6]   Die Dicke des n-dimensionalen Würfel-Graphen. *J. Combin. Theory* **3** (1967) 10–15.   [158]

Kleitman, D. J.
[K7]   The crossing number of $K_{5,n}$. *J. Combin. Theory* **9** (1970) 315–23.   [153]

Koch, J.
    See [AHK1] and [221].

Kodama, Y.
    See [BHKY1].

Koman, M.
    See [BK1].

König, D.
[K8]   Über Graphen und ihre Anwendung auf Determinantheorie und Mengenlehre. *Math. Ann.* **77** (1916) 453–65.   [7, 247]
[K9]   Graphen und Matrizen. *Math. Riz. Lapok.* **38** (1931) 116–19.   [235, 242]
*[K10]   *Theorie der endlichen und unendlichen Graphen*, Teubner, Leipzig. Reprinted Chelsea, New York (1950).   [7, 45]

Kotzig, D.
    See [259] and [267].

Kronk, H. V.
[KRW1]   (with R. D. Ringeisen and A. T. White) On 2-cell embeddings of complete
           n-partite graphs. *Colloq. Math.* **36** (1976) 295–304.   [190]
           See also [CK1] and [205].

Kuratowski, K.
[K11]   Sur le problème des courbes gauches en topologie. *Fund. Math.* **15** (1930)
           271–83.   [137]

Landau, H. G.
[L1]   On dominance relations and the structure of animal societies. III. The condition
           for a score structure. *Bull. Math. Biophys.* **15** (1953) 143–8.   [119, 122]

Lesniak, L.
[LS1]   (with H. J. Straight) The cochromatic number of a graph. *Ars Combin.* **3** (1977)
           39–46.   [206]
           See [CKLS1].

Lloyd, E. K.
           See [BLW1].

Lovász, L.
[L2]   On chromatic number of finite set-systems. *Acta Math. Acad. Sci. Hungar.*
           79 (1967) 59–67.   [204]
[L3]   Normal hypergraphs and the perfect graph conjecture. *Discrete Math.* **2** (1972)
           253–67.   [203]
[L4]   Three short proofs in graph theory. *J. Combin. Theory* **19B** (1975) 269–71.
                                                                                         [196]
*[L5]   *Combinatorial Problems and Exercises*. North-Holland, Amsterdam (1979).
                                                                                         [302]

Mader, W.
[M1]   3n-5 edges do force a subdivision of K$_5$. *Combinatorica* **18** (1998), 569–595.
                                                                                         [299]

Manvel, B.
[M2]   On Reconstruction of Graphs. Doctoral thesis, University of Michigan (1970).
                                                                                          [51]

Marczewski, E.
[M3]   Sur deux propriétés des classes d'ensembles. *Fund. Math.* 33 (1945) 303–7. [200]

Matthews, M. M.
[MS1]   (with D. P. Sumner) Hamiltonian results in $K_{1,3}$-free graphs. *J. Graph Theory*
           8 (1984) 139–46.   [83]

May, K. O.
[M4]   The origin of the four-color conjecture. *Isis* **56** (1965) 346–8.   [216]

McCuaig, W.
[MS2]   (with B. Shepherd) Domination in graphs with minimum degree two. *J. Graph Theory* **13** (1989) 749–62.   [277]

McDiarmid, C. J. H.
[MR1]   (with B. Reed) On total colourings of graphs. *J. Combin. Theory B.* To appear.
                                                                                        [216]

McKay, B. D.
[M5]   Computer reconstruction of small graphs. *J. Graph Theory* **1** (1977) 281–3.
                                                                                        [48]

McKee, T. A.
[M6]   Recharacterizing Eulerian: intimations of new duality. *Discrete Math.* **51** (1984)
            237–42.   [89]

Menger, K.
[M7]   Zur allgeminen Kurventheorie. *Fund. Math.* **10** (1927) 95–115.   [75]
            See also [55].

Meyniel, M.
[M8]   Une condition suffisante d'existence d'un circuit Hamiltonien dans un graph
            oriente. *J. Combin. Theory* **14B** (1973) 137–47.   [102]

Miller, Z.
            See [BMS1].

Moon, J. W.
[M9]   On subtournaments of a tournament. *Canad. Math. Bull.* **9** (1966) 297–301.
                                                                                        [125]
[M10]   On independent complete subgraphs in a graph. *Canad. J. Math.* **20** (1968)
            95–102.   [305]
            See also [BHM1].

Moser, L.
            See [HM1] and [121].

Mycielski, J.
[M11]   Sur le coloriage des graphes. *Colloq. Math.* **3** (1955) 161–2.   [203]

Nash-Williams, C. St. J. A.
[N1]   Decomposition of finite graphs into forests. *J. London Math. Soc.* **39** (1964) 12.
                                                                                        [69]

            See also [CHJKN1], [HN1] and [106].

Nijenhuis, A.
[N2]   Note on the unique determination of graphs by proper subgraphs. *Notices Amer. Math. Soc.* **24** (1977) A-290.   [48]

Nordhaus, E. A.
[NG1]   (with J. W. Gaddum) On complementary graphs. *Amer. Math. Monthly* **63** (1956) 175–7.   [205]
[NSW1]   (with B. M. Stewart and A. T. White) On the maximum genus of a graph. *J. Combin. Theory* **11B** (1971) 258–67.   [187, 190]

Norman, R. Z.
          See [HN2].

Oellermann, O. R.
          See [CHO1] and [CKOR1].

Ore, O.
[O1]   Note on Hamilton circuits. *Amer. Math. Monthly* **67** (1960) 55.   [94]
*[O2]   *Theory of Graphs.* Amer. Math. Soc. Colloq. Pub., Providence, RI **38** (1962).
                                                                            [274–276]
[O3]   Hamilton connected graphs. *J. Math. Pures Appl.* **42** (1963) 21–7.   [99]

Payan, C.
[P1]   Sur le nombre d'absorption d'un graphe simple. *Cahiers Centre Études Recherche Opér.* **17** (1975) 307–17.   [277, 336]
[PX1]   (with N. H. Xuong) Domination-balanced graphs. *J. Graph Theory* **6** (1982) 23–32.   [277]
          See also [CFPT1] and [JP1].

Petersen, J.
[P2]   Die Theorie der regulören Graphen. *Acta Math.* **15** (1891) 193–220.
                                                                       [239, 240, 249]
[P3]   Sur le théoréme de Tait. *L'Intermédiaire des Mathematiciens* **5** (1898) 225–7.
                                                                                 [246]

Plesník, J.
[P4]   Critical graphs of given diameter. *Acta Fac. Rerum Natur. Univ. Comenian. Math.* **30** (1975) 71–93.   [73]

Plummer, M. D.
          See [106].

Polimeni, A. D.
          See [CPS1].

Pósa, L.
          See [299] and [300].

Prüfer, H.
[P5]   Neuer Beweis eines Satzes über Permutationen. *Arch. Math. Phys.* **27** (1918) 142–4.   [60]

Ramsey, F.
[R1]   On a problem of formal logic. *Proc. London Math. Soc.* **30** (1930) 264–86.
                                                                                    [317]

Rédei, L.
[R2]   Ein kombinatorischer Satz. *Acta Litt. Szeged* **7** (1934) 39–43.   [124]

Reed, B.
[R3]   Paths, stars, and the number three. *Combin. Probab. Comput.* **5** (1996), 277–295.
                                                                                    [277]
            See also [MR1].

Ringeisen, R. D.
[R4]   Determining all compact orientable 2-manifolds upon which $K_{m,n}$ has 2-cell
            imbeddings. *J. Combin. Theory* **12B** (1972) 101–4.   [179, 190]
[RB1]   (with L. W. Beineke) The crossing number of $C_3 \times C_n$. *J. Combin. Theory* **24B**
            (1978) 134–6.   [155]
            See also [BR1] and [KRW1].

Ringel, G.
[R5]   Über drei kombinatorische Probleme am n-dimensionalen Würfel und
            Würfelgitter. *Abh. Math. Sem. Univ. Hamburg* **20** (1955) 10–19.   [168]
[R6]   Problem 25. Theory of graphs and its applications. *Proc. Int. Symp. Smolenice
            1963*, Prague (1964) 162.   [259]
[R7]   Das Geschlecht des vollstöndigen paaren Graphen. *Abh. Math. Sem. Univ.
            Hamburg* **28** (1965) 139–50.   [167]
*[R8]   *Map Color Theorem.* Springer-Verlag, Berlin (1974).   [167]
[RY1]   (with J. W. T. Youngs) Solution of the Heawood map-coloring problem. *Proc.
            Nat. Acad. Sci. USA* **60** (1968) 438–45.   [167, 230]
            See also [153] and [259].

Robbins, H. E.
[R9]   A theorem on graphs, with an application to a problem in traffic control. *Amer.
            Math. Monthly* **46** (1939) 281–3.   [112]

Roberts, J. A.
            See [BR3].

Robertson, N.
[RSST1]   (with D. Sanders, P. D. Seymour and R. Thomas) The four-colour theorem.
            J. Combin. Theory **870** (1997), 2–44.   [221]

[RST1]   (with P. D. Seymour and R. Thomas) Hadwiger's conjecture for K_6-free graphs.
            Combinatorica **13** (1993) 279–361.   [205]
            Also see [CRST1].

Roehm, D. J.
            See [JRS1].

Rosa, A.
[R10]   On certain valuations of the vertices of a graph, in *Theory of Graphs, Proc. Internat. Sympos. Rome 1966*. Gordon and Breach, New York (1967) 349–55.
[256, 257, 259, 263, 264]

Ruiz, S.
[R11]   Isomorphic decompositions of complete graphs into linear forests. *J. Graph Theory* **9** (1985) 189–91.   [259]
[R12]   Randomly decomposable graphs. *Discrete Math.* **57** (1985) 123–8.   [254]
            See also [CKOR1] and [FR1].

Ryser, H.
*[R13]   *Combinatorial Mathematics*. Wiley, New York (1963).   [254]
            See also [BR2].

Sachs, H.
            See [ES1].

Saito, A.
            See [EJKS1].

Sampathkumar, E.
            See [WAS1].

Sanders, D.
            See [RSST1].

Schultz, M.
            See [JRS1].

Schuster, S.
            See [CKLS1] and [DS1].

Schwenk, A. J.
            See [HKS1] and [297].

Sekanina, M.
[S1]   On an ordering of the set of vertices of a connected graph. *Publ. Fac. Sci. Univ. Brno.* **412** (1960) 137–42.   [105]

Seymour, P. D.
[S2]   On multicolourings of cubic graphs, and conjectures of Fulkerson and Tutte. *Proc. London Math. Soc.* **38** (1979) 423–60.   [215]
[S3]   Sums of circuits, in *Graph Theory and Related Topics*. Academic Press, New York (1979) 341–55.   [91]
[S4]   Nowhere-zero 6-flows. *J. Combin. Theory* **30B** (1981) 130–5.   [229]
            Also see [CRST1], [RSST1] and [RST1].

Shannon, C. E.
            See [EFS1].

Sheehan, J.
> See [HS3].

Shepherd, B.
> See [MS2].

Singleton, R. R.
> See [HS2].

Slater, P. J.
[S5] Medians of arbitrary graphs. *J. Graph Theory* **4** (1980) 389–92. [22]
> See also [BMS1].

Sloane, N. J. A.
> See [GS2].

Sós, V.
> See [305].

Spencer, J. H.
[S6] *Ten Lectures on the Probabilistic Method.* Capital Press, Montpelier (1994).
 [345]

Stewart, B. M.
> See [NSW1].

Stewart, M. J.
> See [CPS1].

Stockmeyer, P. J.
[S7] The falsity of the reconstruction conjecture for tournaments. *J. Graph Theory* **1** (1977) 19–25. [49]

Straight, H. J.
> See [LS1].

Sumner, D. P.
> See [MS1].

Surányi, J.
> See [HS1].

Swart, H. C.
> See [GHS1].

Syslo, M. M.
> See [BS1].

Szekeres, G.
[S8] Polyhedral decomposition of cubic graphs. *Bull. Austral. Math. Soc.* **8** (1973) 367–87. [91]

376 ■ Graphs & digraphs

[SW1]   (with H. S. Wilf) An inequality for the chromatic number of a graph. *J. Combin. Theory* **4** (1968) 1–3.   [198]
        See also [ES2].

Szele, T.
[S9]   Kombinatorikai vizsgálatok az irányitott teljes gráffel kapcsolatban. *Mat. Fiz. Lapok* **50** (1943) 223–56.   [336]
[S10]   Kombinatorische Untersuchungen über gerichtete vollstöndige Graphen. *Publ. Math. Debrecen.* **13** (1966) 145–68.   [125]

Tait, P. G.
[T1]   Remarks on the colouring of maps. *Proc. Royal Soc. London* **10** (1880) 729.
                                                                                       [222]
        See also [147].

Thomas, R.
        See [CRST1], [RSST1] and [RST1].

Thomassen, C.
[T2]   Landau's characterization of tournament score sequences, in *The Theory and Applications of Graphs*. Wiley, New York (1981) 589–91.   [119]
[T3]   A theorem on paths in planar graphs. *J. Graph Theory* **7** (1983) 169–76.
                                                                                       [147]
[T4]   Every planar graph is 5-choosable. *J. Combin. Theory* **62B** (1994), 180–181.
                                                                                       [225]
        See also [125].

Toida, S.
[T5]   Properties of an Euler graph. *J. Franklin Inst.* **295** (1973) 343–5.   [89]

Turán, P.
[T6]   Eine Extremalaufgabe aus der Graphentheorie. *Mat. Fiz. Lapok* **48** (1941) 436–52.   [294, 296]
[T7]   A note of welcome. *J. Graph Theory* **1** (1977) 7–9.   [153]

Tutte, W. T.
[T8]   On hamiltonian circuits. *J. London Math. Soc.* **2** (1946) 98–101.   [147]
[T9]   A contribution to the theory of chromatic polynomials. *Canad. J. Math.* **6** (1954) 80–91.   [228]
[T10]   A short proof of the factor theorem for finite graphs. *Canad. J. Math.* **6** (1954) 347–52.   [237]
[T11]   A theorem on planar graphs. *Trans. Amer. Math. Soc.* **82** (1956) 99–116.
                                                                                       [147]
[T12]   On the algebraic theory of graph colorings. *J. Combin. Theory* **1** (1966) 15–50.
                                                                                    [224, 229]
        See also [HT1].

Ulam, S. M.
        See [47–48].

Vasak, J.
[V1]   The thickness of the complete graph having $6m + 4$ points. Preprint.   [158]

Veblen, O.
[V2]   An application of modular equations in analysis situs. *Math. Ann.*
            14 (1912–1913) 86–94.   [89]

Vizing, V. G.
[V3]   On an estimate of the chromatic class of a *p*-graph. *Diskret. Analiz.* **3** (1964)
            25–30.   [209]
[V4]   Critical graphs with a given chromatic class. *Diskret. Analiz.* **5** (1965) 9–17.
                                                                                          [213, 214]
[V5]   A bound on the external stability number of a graph. *Doklady A. N.* **164** (1965)
            729–31.   [279]
            See also [282].

Wagner, K.
[W1]   Bemerkungen zum Vierfarbenproblem. *Jber. Deutsch. Math. Verein.* **46** (1936)
            21–2.   [131]
[W2]   Über eine Eigenschaft der ebene Komplexe. *Math. Ann.* **114** (1937) 570–90.
                                                                                          [143, 205, 221, 222]

Walikar, H. B.
[WAS1]   (with B. D. Acharya and E. Sampathkumar) Recent developments in the theory of
            domination in graphs. Mehta Research Institute Allahabad, MRI *Lecture Notes in
            Math.* **1** (1979).   [278]

Weinstein, J. M.
[W3]   On the number of disjoint edges in a graph. *Canad. J. Math.* **15** (1963) 106–11.
                                                                                          [243]

White, A. T.
*[W4]   *Graphs of Groups on Surfaces, Interactions and Models.* North-Holland, Amsterdam
            (2001).   [154]

Whitney, H.
[W5]   Congruent graphs and the connectivity of graphs. *Amer. J. Math.* **54** (1932)
            150–68.   [71, 77]

Wilf, H. S.
            See [SW1].

Williamson, J. E.
[W6]   Panconnected graphs II. *Period. Math. Hungar.* **8** (1977) 105–16.   [100]

Wilson, R. J.
            See [BW1], [BLW1] and [EW1].

Wilson, R. M.
[W7]   Decompositions of complete graphs into subgraphs isomorphic to a given graph. *Proceedings of the Fifth British Combinatorial Conference* (1975) 647–59.

[255]

Woodall, D. R.
[W8]   Sufficient conditions for circuits in graphs. *Proc. London Math. Soc.* **24** (1972) 739–55.   [103]

[W9]   Cyclic-order graphs and Zarankiewicz's crossing-number conjecture. *J. Graph Theory* **17** (1993), 657–671.   [153]

Xuong, N. H.
[X1]   How to determine the maximum genus of a graph. *J. Combin. Theory* **26B** (1979) 217–25.   [189]
           See also [PX1].

Youngs, J. W. T.
[Y1]   Minimal imbeddings and the genus of a graph. *J. Math. Mech.* **12** (1963) 303–15.   [175]
           See also [BHKY1] and [RY1].

Zaks, J.
[Z1]   The maximum genus of cartesian products of graphs. *Canad. J. Math.* **26** (1974) 1025–35.   [191]

Zarankiewicz, K.
[Z2]   On a problem of P. Turán concerning graphs. *Fund. Math.* **41** (1954) 137–45.

[153]

Zykov, A. A.
[Z3]   On some properties of linear complexes (Russian). *Mat. Sbornik* **24** (1949) 163–88. *Amer. Math. Soc. Translations* No. 79 (1952).   [203]

Index

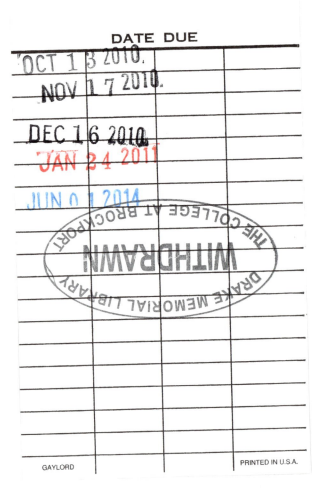